THE SCIENTIST'S ATOM AND THE PHILOSOPHER'S STONE

BOSTON STUDIES IN THE PHILOSOPHY OF SCIENCE

VOLUME 279

THE SCIENTIST'S ATOM AND THE PHILOSOPHER'S STONE

How Science Succeeded and Philosophy Failed to Gain Knowledge of Atoms

ALAN CHALMERS

Springer

Dr Alan Chalmers
University of Sydney
Unit for History and Philosophy of Science
N.S.W. 2006
Australia
achalmers@usyd.edu.au

Flinders University
Philosophy Department
PO Box 2100
Adelaide 5001
Australia
alan.chalmers@flinders.edu.au

ISBN 978-90-481-2361-2 e-ISBN 978-90-481-2362-9
DOI 10.1007/978-90-481-2362-9
Springer Dordrecht Heidelberg London New York

Library of Congress Control Number: 2009928422

Printed on acid-free paper

Springer is part of Springer Science+Business Media (www.springer.com)

Preface

In 1989 it became necessary for me to design a senior undergraduate course in the history of the physical sciences for a new programme in History and Philosophy of Science at the University of Sydney. I chose to survey the history of atomism from Democritus to the twentieth century, thereby giving students a taste of the varying contexts in which science has been practised. The work I needed to do to prepare for that course soon developed into a major research project. This book is the outcome.

One of the original sources subject to a critical reading by my students and myself was Robert Boyle's essay 'On the grounds and excellency of the mechanical hypothesis'. My appraisal of that essay appeared as a journal article in 1993. It represented a critical moment in the evolution of my thoughts on the history of atomism. In my article I distinguished between Boyle's account of the fundamental make-up of the physical world that he called the mechanical philosophy and the fruits of his experimentation, best represented by the pneumatics that he supported by experiments with his air pump. It was that distinction that gave me the basis for the thesis defended in this book. I came to see modern atomic theory as the rather recent legacy of experimental science as it emerged in the seventeenth century rather than a tradition of speculative philosophy dating back to Democritus and extending to seventeenth-century mechanical philosophy and beyond.

The title of my article of 1993, 'The lack of excellency of Boyle's mechanical philosophy' made some historians uncomfortable. Presumably they suspected that I was using contemporary standards to cast judgement on Boyle's case. There is certainly a fundamental problem involved here. How can one defend a philosophical thesis about the changing nature of science by invoking the history of science without projecting onto past science distinctions and norms that had yet to be constructed? The fact that my early research had not totally come to grips with that problem was brought home to me in 1998 at a time when I was a Senior Research Fellow at the Dibner Institute for the History of Science and Technology, an institution that, regrettably, has since been disbanded. I read a paper in which I presented as a problem the fact that, on the one hand, Boyle distinguished between his mechanical philosophy and his experimental science and yet, on the other, claimed that his mechanical philosophy had experimental support. The following day I was subject to the customary grilling by the other Fellows. My confidence in my position was shaken when one of the Postdoctoral Fellows asked me if I thought Boyle was confused. I had no satisfactory answer to that question. I still lacked an answer in 2007, when I taught a course on the history of atomism to a class of very discerning graduate students at the University of Pittsburgh. I presented the problem raised by my inquisitor at the Dibner to my students, inviting them to respond to it. No definitive answer emerged. I believe I can now dissolve the problem and I do so in Chapters 6 and 8 of this book. It draws on a distinction between two kinds of empirical support, a strong one involved in Boyle's experimental science and a weaker one involved in his defence of the mechanical philosophy. I do not have to impose that distinction on Boyle's work because it can be seen to have been made by Boyle himself, once one has learnt where to look. I hope my book displays how history and philosophy of science can be integrated in a way that combines rigorous philosophical argument with history of science that lives up to the highest standards of scholarly research.[1]

Boyle distinguished between philosophical theories about the ultimate structure of matter and less ambitious claims subject to experimental investigation and verification exemplified in the law that bears his name. But his appreciation of the distinction did not have, and could not possibly have had, the implications for him that the distinction between philosophy and science has for us. Modern debates about the ultimate structure of matter are conducted by philosophers as part of what they call 'metaphysics'. The distinction between their practice and that of scientists is recognised and, indeed, institutionalised insofar as the two practices are accommodated in different university faculties. The outcomes of the deliberations of modern metaphysicians have implications only for a handful of addicts and are of little concern to, and have minimal effect on, anyone else. Their ponderings tend to be a source of bemusement if not amusement for scientists. The situation was very different in the seventeenth-century situation in which Boyle worked. Natural philosophers were engaged in attempts to comprehend and give a theoretical basis for the new social formations and also in attempts to rewrite Christian theology following the discrediting of the Aristotelian philosophy with which it had become entwined in the Middle Ages. A mechanical philosopher like Thomas Hobbes put his version of the mechanical philosophy to work in the *Leviathan* just as Boyle put his mechanical philosophy to work in the theology that he constructed as part of his counter to the atheism that threatened him. Appeal to Boyle's law or the circulation of the blood would not have served such purposes at all. Seventeenth-century metaphysics, in the guise of the mechanical philosophy, was not the marginal and specialised activity that the metaphysics conducted in modern philosophy has come to be. Nevertheless, a distinction between philosophical metaphysics and experimental philosophy emerged, and was made explicit, in the seventeenth century. I believe we learn much about science by recognising the way in which, by the beginning of the twentieth century, a general atomic theory of matter that was experimentally supported had come about in a way that owed little to the philosophical versions of atomism that had origins in Ancient Greece. That is what my book is intended to demonstrate.

My research into the history of atomism has been carried out in a number of institutional contexts and I would like to acknowledge the support I have received. I have already referred to my period as a Senior Research Fellow at the Dibner

Institute for the History of Science and Technology. I held a similar position at the Center for the Philosophy of Science at the University of Pittsburgh in 2003/4 and was an Erskine Fellow at the University of Canterbury, New Zealand in 2000. I have also been hosted by the Centre for the Philosophy of Natural and Social Science at the London School of Economics, the Department of Philosophy at the University of Bristol and, for shorter periods, at the Department for History of Science at Harvard University and the Faculty of Life Sciences at the University of Manchester. For a decade after my retirement from the Unit for the History and Philosophy of Science at the University of Sydney late in 1999, my base was the Department of Philosophy at Flinders University, Australia, where, as an Honorary Senior Research Fellow, I was able to retain a foothold in academia and avail myself of that university's resources. At Flinders I was also able to hone my philosophical arguments by exploiting the expertise of the philosophers, especially Greg O'Hare, from whom I could get an insightful summary of the latest analytic philosophy in a way that spared me the need to search the literature, and George Couvalis, whose dogged refusal to be convinced by my arguments constituted an ever-present challenge that led to improvements in them. Also worthy of special mention is Rodney Allen whose support extended further than supplying the coffee.

There are many individuals from whom I have learnt, and I am sure to forget to mention some of them. Alan Musgrave read the entire manuscript of the first draft of this book and supplied detailed comments and criticisms that were invaluable. One of his marginal notes on my manuscript reads 'decide what you want to say and say it', which exemplifies the forthright and hence productive character of his critique. I am indebted to the recent scholarship of William Newman and Ursula Klein, which I draw on extensively. Unknown to them, I have had innumerable, almost daily, virtual conversations with them. Regrettably, our personal confrontations have been few, although they have both been generous in responding to my correspondence. Other individuals whose help and encouragement warrants mention are, in alphabetical order, Peter Anstey, Keith Bemer, Jed Buchwald, Hasok Chang, Karen Rue Hauck, Keith Hutchison, Deborah Mayo, Robert Nola, John Norton, Denis Pozega, Andrew Pyle, Nicholas Rasmussen, Jonah Schupbach and Neil Thomason.

I would like to recall the debt I owe to the late Heinz Post. He was Head of the Department of History and Philosophy of Science at Chelsea College, University of London and supervised my PhD thesis on Clerk Maxwell's electromagnetism. That department, which flourished under Post and did not survive long after his retirement, was designed to offer HPS courses to students with a good science degree. Those of us who obtained our doctorates under Post's influence like to think of ourselves as the 'Chelsea School', although the College has since been absorbed into Kings College, London. Included in our number are Noretta Koertge, John Pickstone, Harvey Brown, Simon Saunders, Steven French, Harmke Kaminga, Giora Hon, and Eric Scerri. Our work bears witness to the influence of an inspirational teacher. Post had an interest in the more recent history of atomism and I recall that he was encouraged by a publisher to write a book on the topic. It has been left to me, his grateful student, to produce that book and I like to think that it is one he would have endorsed.

Special thanks are due to Sandra Grimes. Her generous and constant support and encouragement were not always received with the appreciation and acknowledgement they deserved by an irascible and easily rattled author.

Note

1. I am a member of a group of seventeen international scholars brought together on the initiative of John Norton at the University of Pittsburgh and Don Howard at the University of Notre Dame under the banner &HPS. The aim is to foster the integration of history and philosophy of science in a way that lives up to the highest standards of both disciplines. To this end a conference is held every 2 years involving papers that can serve as exemplars of how this can be done.

Contents

Chapter 1
Atomism: Science or Philosophy?

Abstract Modern science includes a detailed theory of atoms and their structure. That theory, which goes well beyond what is directly observable, is nevertheless vindicated by experiment, living up to the stringent standards distinctive of science since its emergence in the seventeenth century. Speculations about an atomic structure of matter were prominent in the speculations of the Ancient Greek philosophers. However, it is very misleading to see the theories of the likes of Democritus as an anticipation of modern atomism. It is also a mistake to see modern atomism as emerging as a result of the development of its ancient precursor over the centuries. The methods of experimental science are quite distinct from the methods involved in the development of philosophical matter theories, from those of Leucippus and Democritus up to those of the seventeenth-century mechanical philosophers and beyond. A scientific version of atomism did not emerge until well into the nineteenth century and we learn much about the nature of science by appreciating this.

1.1 Introduction

There are about a million, million, million, million atoms in a typical coin. This has been established by modern science. What is more, much is known about the inner structure of atoms, knowledge that helps to account for the spectra of the radiation emanating from excited substances, for chemical combination, for how metals conduct electricity and so on. Given the minute sizes of atoms, which lie way, way beyond what could possibly be observed directly, how on earth could it be established that there are atoms? Whatever the difficulties standing in the way of the acquisition of this knowledge, they have been overcome to the extent that, not only can atoms be counted, but also their inner structure can be precisely specified. They are made up of a nucleus of protons and neutrons surrounded by electrons that are governed by quantum mechanical principles, principles quite different to those governing the world of our experience. This book tells the story of how knowledge of minute atoms became possible.

The protons, neutrons and electrons that make up atoms and the quantum mechanical principles that account for their behaviour are twentieth-century discoveries. Given this, and given the difficulties facing the project of unearthing

A. Chalmers, *The Scientist's Atom and the Philosopher's Stone*, Boston Studies in the Philosophy of Science 279, DOI 10.1007/978-90-481-2362-9_1,

knowledge of atoms that I have tried to dramatise in the previous paragraph, it seems startling to have to acknowledge that atomic theories were elaborated and defended in Ancient Greece in the fifth century BC, two and a half thousand years ago. Democritus, building on the ideas of his teacher, Leucippus, developed a view of the universe as consisting of nothing other than numerous invisible and unchanging atoms moving and colliding in the void and sometimes combining to form the macroscopic bodies of our experience. How was it possible for Democritus to anticipate the recognition that matter is composed of atoms? I believe the answer to this conundrum lies in the fact that Democritean atomism was far from an atomic theory that could do significant explanatory work and which could be empirically defended. Democritus's atoms are unchangeable and without inner structure and are akin to miniature inert stones. They bear little resemblance to the intricately structured quantum mechanical atoms of modern physics and are incapable of explaining much for that reason.

The atoms known to modern science are structured, potent and subject to change and they interact with and via fields. By contrast, the atoms of Democritus are inert and changeless, and reality consists of nothing other than the sum total of such atoms in the void. Democritean atoms interact only by coming in contact and there is no room for anything like the fields of modern science. However, these marked differences in the content of modern and ancient atomism is not the most important feature that distinguishes them. One additional feature is the extent to which the atoms of the Ancient Greeks were intended to represent the ultimate and only constituents of the world. They were invoked to explain the possibility of change in general whilst being themselves changeless. The credentials of modern atomic theory do not include the capacity to give ultimate accounts of the only constituents of the world. Who knows what inner structure of electrons will be revealed using the next generation of particle accelerators? Also, modern matter theory involves fields as well as particles. A second feature that involves a qualitative distinction between Ancient Greek atomic theories of the ultimate structure of matter and contemporary atomic theory is the nature of the case made for them. The case for contemporary atomism appeals, for example, to J. J. Thompson's experiments involving the deflection of cathode rays by electric and magnetic fields, that enabled him to estimate the ratio of the charge to mass of the particles constituting the rays, and Jean Perrin's experiments on Brownian motion, that established that gases are composed of a specifiable number of molecules in random motion. By contrast, Democritus's case for his atoms as the ultimate and changeless constituents of the world appealed to some very general intuitions about the nature of reality and change. Leucippus and Democritus, together with other Ancient atomists such as Epicurus, and also mechanical philosophers such as Pierre Gassendi and Robert Boyle who revived a version of Ancient atomism in the seventeenth century, offered a philosophical account and defence of atomism that went far beyond what could be adequately defended empirically. This contrasts with the experimental case made by scientists in support of modern atomic theory.

A key theme of this study is the difference between accounts of the structure of matter sought by philosophers and those substantiated in experimental science. Such a distinction is hardly something that needs stressing in a contemporary context. Science and philosophy are practiced within different Faculties in most universities. The former involves practical work requiring laboratories and elaborate equipment. The latter requires access to libraries and the facility to interact with other philosophers. The scientist can mock the armchair philosophers who think they can further knowledge simply by thinking and arguing and can take delight in the story of Thales, the first philosopher, walking into a pit whilst contemplating the stars. On the other hand, the philosopher can be scornful of the senior undergraduate scientist in his or her class who does not even know where the University Library is! The difference between distinct practices of science and philosophy that is now institutionalised began to emerge at the time of the increased use of experiment in the seventeenth century as a key tool for probing fundamental questions about the nature of the world. The capabilities of experimental science were to expand beyond anything that could possibly have been anticipated in the seventeenth century to the extent that many of the questions about the fundamental structure of reality that had been considered to be the province of philosophy were answered by science. The philosophical atomist's miniature stones were replaced by the scientist's quantum mechanical atoms.

Versions of Ancient atomism were revived in the seventeenth century by so-called mechanical philosophers such as Pierre Gassendi and Robert Boyle. Many of those philosophers were also at the forefront of the emerging emphasis on experiment as a key tool in the production of knowledge of the material world. I maintain that a gulf separated these two enterprises to an extent that was not adequately appreciated or acknowledged by the mechanical philosophers and continues to be inadequately appreciated today. The atoms of the mechanical philosophers resembled miniature inert stones just as those of the Ancient atomists did. The atoms in Boyle's philosophy, for instance, had an unchanging shape and size, had some degree of motion or rest, and were all made of universal matter characterised in terms of its impenetrability. The only source of activity and change latent in the natural world was the motion of the atoms. It is perhaps not surprising, from a modern point of view, that there was scant experimental evidence for these atoms and that explanations of phenomena that appealed to them were ineffective or inadequately defended. The state of affairs contrasts markedly with the status, for example, of the knowledge of air pressure defended by Boyle's experiments on air, especially those employing his air pump. Boyle's experiments clinched the claims that air has a pressure and that it is the cause of the height of the mercury in a barometer. The status of that experimental knowledge and the way in which that status was established by Boyle contrasts markedly with the corresponding status and mode of defence of his claims about atoms. This distinction, that I will elaborate and defend in detail later in this book, provides me with a key motif for my epistemological history of atomism. I raise the question of when knowledge of atoms was clinched in the same kind of way and to the same extent as knowledge of air pressure was and I answer, 'late in the nineteenth and early in the twentieth centuries'.

1.2 Science and Philosophy Transcend the Evidence for Them

At face value it would appear that science differs from philosophy insofar as the latter kind of knowledge is borne out by observation and experiment in a way that philosophical knowledge is not. The controlled functioning of modern technologies involving lasers, microchips and hydrogen bombs provides ample evidence that scientific knowledge has a validity that has no analogue in philosophy. The reality of lasers leaves no room for scientists to sensibly doubt the quantum mechanical nature of the stimulated emissions that are responsible for their functioning whereas philosophers endlessly debate the question of the nature of the mind and its relation to the brain, the relationship between facts and values, and whether, in observing a table, we are presented with a sighting of a table, a mental image of a table or a belief in the presence of a table or whatever. Whilst philosophers are wise to ensure that their claims are compatible with science, they do not expect to settle their disputes by appeal to observation and experiment in a way that scientists typically do. All this makes common sense.

A problem that needs to be faced stems from the fact that scientific knowledge is general knowledge no less than philosophical knowledge is. The Ancient Greeks knew how to make mercury by grinding cinnabar in a copper dish. They knew that heavy objects fall to the ground and also how to correlate the seasons with the positions of the sun in the ecliptic. But such knowledge is not sufficiently general to meet the demands of the philosopher or the scientist. Aristotle sought to comprehend why stones fall to the ground in terms of his theory of how the four elements constitute an earth-centred terrestrial domain and Newton did so by appeal to his universal law of gravitation. Both these claims transcend the evidence for them. If science differs from philosophy by being empirically confirmed then we need an account of how its generalities can be justified by appeal to empirical evidence in a way that those in philosophy cannot.

Scientific and philosophical claims about the world go far beyond the evidence on which they are based in two ways. They go deeper, as it were, to claim the existence of unobservable things, and they generalise beyond the circumstances in which evidence is identified. The evidence-transcending nature of philosophy such as involved in Aristotle's attempt to explain all terrestrial phenomena in terms of the interaction of four elements or that of the mechanical philosophers to reduce all phenomena to the motions of universal, inert matter is blatant. But it is characteristic of scientific claims too. In the late eighteenth and early nineteenth centuries chemists identified a range of gases such as oxygen, nitrogen and hydrogen that are not directly observable. Further, basing their claims on a few well-designed experiments, they included gases in their general accounts of the formation of compounds from elements, including an account of combustion that involved combination with oxygen. If we are to insist that evidence-transcendence is warranted in science in a way that it is not in philosophy then we need an account of how scientific claims are confirmed that will enable the distinction to be maintained.

1.3 How the Claims of Science are Confirmed

The demands that a theory has correct empirical consequences or that it be merely compatible with empirical evidence are much too weak to capture what is distinctive about science. One problem is that false theories can have true consequences. The hypothesis that the sun orbits a stationary earth was borne out by a range of evidence, and Aristotle's theory of the four elements entailed that stones will fall to the ground and flames rise in the air, but those theories are false nevertheless. A related problem concerns the possibility of there being alternative theories compatible with the same data. The stationary-earth theory correctly predicts that a stone dropped from a tower will land at its foot. But once Galileo had shown that this would also be the case for a steadily spinning earth, the experiment could not count as evidence for either a stationary or steadily spinning earth. A third point is that theories, if they have sufficient leeway, can be made compatible with the evidence by means of suitable adjustments. If we are free to pick the circular orbits corresponding to the cycles and epicycles in Ptolemy's astronomy so that they fit observations of planetary positions then that fit, in itself, is not genuine evidence for the theory.

An account of theory confirmation that meets the worries raised in the previous paragraph needs to capture some suitably demanding relationship between theory and evidence. Some inter-related ideas that go some of the way are as follows.[1] Evidence counts in favour of a theory only if that evidence is acquired in a way that constitutes a genuine test of that theory. A genuine test of a theory will be such insofar as the theory is unlikely to pass it if it is false. A theory will not be tested against data if the theory is contrived to fit it rather than following naturally from it, and, even if the data does follow naturally from it, it will not be tested against the data if there is an alternative theory that fits the data equally naturally and equally well. These thoughts seem to capture intuitions about the tower argument and Ptolemaic astronomy that I mentioned in the previous paragraph. But they are not adequate as they stand.

More needs to be said about the demand that evidence follow in a natural, rather than contrived, way from the theory it is meant to test. Theories alone rarely imply any evidence that might serve as a test of them. They need to be supplemented by a range of supplementary laws and data before they can do so. Consider, for example, what it takes to test Newtonian astronomy against some observed positions of the planet Mars. The fundamental assumptions of the theory are Newton's three laws of motion plus the universal law of gravitation. Before an orbit for Mars can be derived from the theory a range of observations of past positions of Mars relative to the sun and, once development of the theory is sufficiently advanced, relative to the other planets too, need to be fed in. Observations need to be adjusted to allow for refraction in the earth's atmosphere and to take account of the fact that the position from which the readings are taken varies from moment to moment because of the motion of the earth. Newton's astronomy can be tested against some predicted position of Mars only by adding a host of assumptions and observations such as these. What is the difference between adding these assumptions to Newton's theory to make a test possible, and adding epicycles to Ptolemy's theory in my previous example? The difference seems to be that the assumptions added in the Newtonian case have support independent of the information gathered in the test situation. In my example involving Ptolemy's theory, which is to some extent a caricature of the historical situation,[2] epicycles are added to the theory solely to bring about a match between theory and data. There is no support for the addition independent of the data supposedly predicted or explained.

Single tests are rarely, if ever, conclusive. Theories, in conjunction with appropriate hypotheses and observations, can yield correct predictions even though they are false and some alternative true and they can also make incorrect predictions even though they are true. The assumption that light travels as waves in an aether made many correct predictions in the nineteenth century in spite of the fact that there is no aether, and Newton's astronomy, combined with the necessary additional information, clashed with observations of the orbit of Uranus, not because of failings of the central theory but because the influence of the yet to be discovered planet Neptune had not been taken into account. False theories can have true consequences and failed predictions can be due to shortcomings in auxiliary assumptions or observations added to the theory rather than in the theory itself.

The uncertainties involved in theory testing can be ameliorated only by further testing. One strategy is to test a theory in a variety of circumstances involving differing sets of auxiliary assumptions. The basic laws involved in Newtonian astronomy can be tested by observing that the period of a pendulum varies with height above the earth's surface in just the way predicted by that theory. Here the auxiliary assumptions, such as an estimate of the radius of the earth, are quite different from those needed in the astronomy example. The fact that Newton's theory gets it right about the existence of Neptune, the return of Halley's comet, the lack of sphericity of the earth and the variation of the earth's gravity with height is a sure sign that it has passed severe tests, so much so that when Cavendish added further support to the theory by measuring the attraction between laboratory-sized objects the positive result was pretty much a forgone conclusion. The logical gap between theory and evidence notwithstanding, it is rarely the case that a theory that has survived a few crucial tests that differ in kind turns out to be totally on the wrong track. If it were on the wrong track then the existence of a wide variety of evidence in its favour would involve a remarkable and unexplained coincidence.

So we have a rough characterisation of a significant test of a theory. Such a test involves confirming predictions deduced from a theory in conjunction with independently testable and successfully tested auxiliary assumptions. The so-called tacking paradox is an indication that more needs to be said. Given the characterisation of a severe test that I have proposed, the tests of Newton's theory that I have mentioned above, and others like them, are also tests of the theory consisting of Newton's three laws of motion and the law of gravitation plus the claim that there is a devil with four legs spreading evil about the world. The suggestion is of course silly. But why, exactly, is it silly? The answer is surely that the addition of the devil hypothesis adds nothing to the successful content of the theory. If we ask of the augmented theory, 'could it pass these tests if it were false?' then the answer is, of course it could. It could pass the tests if the devil had only two legs or if there were no devil at all.

It is no coincidence that the augmented theory yields correct predictions because they follow from the unaugmented part. We need to separate the portion of a theory responsible for its testable predictions from the redundant part. In my contrived example it is trivially obvious how to do this. But this is not always the case and there is a serious issue involved.

Deborah Mayo, whose work convinced me of the importance of the idea that theories can be partitioned into those parts that have been and those that have not been tested, has usefully illustrated the point by reference to testing of the General Theory of Relativity.[3] Subsequent to Einstein's formulation of his General Theory of Relativity, investigation of that theory's structure made possible the separation of the assumption that space-time is curved from assumptions about the cause and degree of the curvature. Some testable consequences of General Relativity follow from the former assumption alone. They stand whether Einstein's own more specific theory about the curvature is right or not. Consequently, successful tests of those predictions constitute tests of curved space-time but not of Einstein's theory as a whole. Further tests are necessary to test Einstein's General Theory of Relativity against alternatives. Current theoretical and experimental work on General Relativity is construed with this kind of problem in mind, for example by Clifford Will (1993 and 1996). Theorists explicitly seek to partition the theory into the parts that have been tested and those that have not. A similar situation arose when Einstein formulated electromagnetic theory in a way that dispensed with the aether and challenged physicists to produce experimental evidence for the prediction of which the addition of the aether made a difference. Their failure to do so constituted a case for dropping the aether. Partitioning of a theory into separate parts is not obvious, is not easy, and cannot always be done. But in those instances where it can be done it is possible to identify which parts of a theory are tested and which not by specified tests. A simpler example is the removal of absolute space from Newtonian mechanics. Once it was realised that all the tests of Newtonian mechanics including absolute space could be passed by the theory minus that assumption, then absolute space was dropped as part of that science.

So far I have argued that a scientific theory is confirmed if (i) a range of kinds of prediction that follow from it in conjunction with independently testable and successfully tested auxiliary assumptions are vindicated by experiment and (ii) the successful tests cannot be accounted for by some specified sub-set of the theory. It is no part of this position that the confirmation of a theory in this sense shows it to be true in an unqualified sense. The fact that a theory has survived tests so far is no guarantee that it will not fail new kinds of test in the future. By the turn of the twentieth century even Newton's theory proved to have its shortcomings. It failed to account, for example, for the motions of fast moving electrons in discharge tubes, where the variations in mass with velocity, un-anticipated in Newton's theory, become consequential. It would be absurd to deny Newton's theory the status of 'scientific knowledge' for this reason. Scientific knowledge typically gets corrected or absorbed as a limiting case of a more adequate theory. However, I claim that items of scientific knowledge that have been significantly confirmed in something like the way I have indicated continues to have a range of applicability that is absorbed into and explained by the more adequate knowledge that transcends and replaces it. So the fact that scientific knowledge is fallible, improvable and replaceable does not undermine the distinction I am invoking between science and philosophy.

On my account, philosophical, as opposed to scientific, claims about the structure of mater are not confirmed by, but at best only accommodated to, the phenomena. Aristotle's account of terrestrial matter as composed of the four elements, air, earth, fire and water, was of this kind. That account has not survived as a limiting case of contemporary science. But then, it was never significantly confirmed in the way I have argued is typical of science.

When contemporary philosophers identify some claim as 'an empirical matter' (perhaps some claim about the functioning of sight in the context of the philosophy of perception) they mean that it is a matter for science to decide and so outside of the domain of their philosophy. In my attempt to outline a sense of experimental confirmation involved in science I seek to make explicit a distinction that is taken for granted in the contemporary academic scene. But this was not the case historically. The emergence of scientific knowledge that was both general and experimentally confirmed as distinct from what was referred to as natural philosophy is very much tied up with the story of the emergence of atomism as a scientific theory. The account of confirmation that I have sketched and that has become distinctive of science will be used in the following chapters to inform my investigation and evaluation of atomic theories of the past.

1.4 Inference to the Best Explanation

Theories can be assessed in terms of their explanatory power. On this view, theories are adequate to the extent that they explain a wide range of phenomena, the wider the range the better the theory. In our quest for knowledge we should opt for the theory with the greatest explanatory power.

This account is in need of some sharpening up if it is to be up to the task of distinguishing between science and philosophy. It may well be the case that in the middle of the fifth century BC ancient atomism was the best explanation of change. A contemporary philosopher may well argue that his or her philosophy of perception gives the best explanation of the relevant facts. But in neither of these cases is it appropriate to regard the explanatory power that we are conceding for the sake of argument as sufficient to confer on the theories involved the status we have learnt to demand of science. The explanatory power exhibited, for example, by modern quantum mechanics and its ability to explain chemical bonding, line spectra, lasers, the spectrum of black body radiation and the tunnel effect exhibited by alpha-particle radiation is of a qualitatively different kind. Whilst philosophers may well have to rest content with inference to the best explanation, scientists aspire to do better and infer the right explanation. What is needed to make sense of these intuitions is some demanding standards for what is to count as an adequate explanation which are met by science but not by philosophy.

The account of confirmation sketched in the previous section helps to formulate two demands that can appropriately be made of an explanation in science. Firstly, a phenomenon or event is explained by a theory only if it follows from that theory, in conjunction with auxiliary assumptions, in a natural way. Any auxiliary assumptions used in the derivation need to have independent support. The powders that result from burning metals are heavier than the samples of metal from which they originate. Chemists who understood combustion as the driving off of phlogiston explained the increase in weight by assuming the phlogiston to be replaced by air denser than it. This is not an adequate scientific explanation so long as there is no evidence, independent of the combustion experiments, for the low density of phlogiston and its replacement by denser air. Secondly, and along similar lines, it can be insisted that a theory only adequately explains a phenomenon that naturally follows from it if there is independent support for the theory. The particle theory of light explains why light travels in straight lines. But is it the right explanation? An affirmative answer can be given to the extent that there is independent support for the particle theory. It would be an enormous co-incidence if a theory that can naturally explain a wide range of phenomena is giving explanations that are on the wrong track. Once we have an adequate account of confirmation in science then it can be exploited, in the way I have tried to do here, to make a distinction between explanations that are merely the best available and explanations that have strong claims to be the right ones.

A scientific theory explains a phenomenon if that phenomenon is a natural consequence of it, and it can be argued to be the right explanation to the extent that it can explain other, independent, phenomena in a similar way. Philosophical accounts of the way of the natural world fall short of this insofar as they are merely accommodated to the phenomena. Modern philosophies of perception, for instance, take, or should take, heed of the latest scientific findings about perception and should be constructed in a way that does not clash with that science. A philosophy of perception that cannot be accommodated to scientific findings is inadequate for that reason. But the rival accounts that can be accommodated to those findings go beyond what is sanctioned by science just because they are merely accommodations to science. In making this distinction I do not aim to discredit philosophy. Perception is in many respects a puzzling phenomenon. There is the question of exactly what it is that we are presented with in an act of perception and how that presentation relates to the object perceived. There is also the issue of whether perceptions in the 'mind' commit us to a mental as distinct from a physical world. These are questions that should not be dismissed simply on the grounds that science cannot answer them.

The claim that science aspires to the right explanations of phenomena needs to be qualified in the same kind of way that claims that scientific theories can be confirmed needed to be qualified. Theories turn out to have their limits and need to be modified or transcended. However, significantly confirmed theories need to live on as limiting cases of their successors. Newton's theory provides an explanation of the precession of the equinoxes that is approximately correct because that theory does follow as a limiting case of relativity theory. If something like this were not the case then the fact that Newton's theory was capable of yielding explanations of a wide range of phenomena meeting the stringent demands I have outlined above would be a mystery. Explanations in science have claims to be the right ones to the extent that the theories appealed to in those explanations have been confirmed in the demanding way that has come to be characteristic of scientific as opposed to other kinds of knowledge.

1.5 Science Involves Experimental Activity and Conceptual Innovation

Science, as it has evolved as a discipline distinct from philosophy, is not a passive, armchair activity designed to apprehend the world as revealed by the unaided senses. Testing the adequacy of scientific claims involves active experimental intervention. What is more, the construction of the conceptual apparatus needed to frame scientific claims requires intellectual innovation. The story of the path from philosophical to scientific atomism will involve identification of the emergence of the appropriate kind of experimentation and the appropriate modes of conceptualisation.

Evidence bearing on scientific laws and theories typically involves intervening in and interrogating nature in a deliberate way. Common-sense knowledge that objects fall to the ground is borne out by acquaintance with everyday happenings, but scientific versions of the law of fall, that freely falling bodies move with a uniform acceleration, need to be tested against experimental as opposed to mere observational evidence. Times and distances of fall need to be measured and interference from non-gravitational forces such as friction or air-resistance needs to be eliminated, controlled or allowed for. Galileo's experiments involving the rolling of balls down inclined planes were early attempts to provide what is necessary. Experimental evidence for scientific knowledge claims is not any old kind of observational evidence, but a special kind of evidence generated in demanding circumstances. To seek and vindicate scientific knowledge we need to 'twist the lions tale', as Francis Bacon put it at the dawn of the scientific revolution.

The law of fall also serves to illustrate my second point. Its formulation requires precise notions of uniform velocity and uniform acceleration. At the time he was conducting the inclined-plane experiments invoked in the previous paragraph, Galileo was still struggling to fashion adequate notions of these concepts and the mathematics able to cope with them. Boyle supported a version of the law that bears his name by varying the pressure on a volume of air trapped by mercury in a U-tube. But the notion of pressure involved was not simply given. It is quite a tricky one[4] that gradually evolved as Boyle and his contemporaries struggled conceptually as well as experimentally with phenomena involving air pressure.

A third point involves the recognition that learning from experiment typically involves prior knowledge of the experimental situation. Newton provided powerful arguments for the inverse square law of gravitation by appealing to detailed observations of the motions of the planets, but his arguments involved assuming the three laws of motion. Further, correction of observed planetary positions needed to allow for the earth's motion and for refraction in the atmosphere. Sources of error can only be eliminated or allowed for insofar as they are known about. Possession of knowledge is a precondition for its acquisition and improvement. There is a sense in which science pulls itself up by its bootstraps.[5] An understanding of how preconditions necessary to make possible an atomic theory that is experimentally testable came to be satisfied is a key focus of this book.

1.6 Reading the Past in the Light of the Present

The rough characterisation of science I have sketched in the preceding sections, and my distinction between scientific and philosophical knowledge of the natural world, is a contemporary perspective. I need to be careful not to impose this perspective on the past in a way that is illegitimate and misleading from a historical, and, indeed, from a philosophical point of view. Writing a history of science that simply picks out those claims and practices that come close to living up to a contemporary conception of science would not in itself be particularly instructive. It would not serve to explain how those claims and practices came to be set in place, nor would it establish how their status was viewed at the time.

There is a sense in which my study of the history of atomism is informed by a contemporary perspective. I aim to throw light on the nature of science, and aim to do so by studying how a scientific knowledge of atoms became possible. I claim that atomism prior to the nineteenth-century amounted to something less than scientific knowledge and I intend to show that it acquired the status of a scientific theory late in the nineteenth and early in the twentieth centuries. In order to diagnose the situation in this way I first need to establish just what claims and modes of argument past versions of atomism actually involved and identify the historical process that led to them. I intend my characterisations of past versions of atomism to meet the highest standards a historian would aspire to.

In a way, it is the view that I contest, namely, that Ancient Greek atomism and the corpuscular theories of the seventeenth century were important anticipations of modern atomism and set in train a historical process that led to it, that is guilty of illegitimately projecting present knowledge onto the past. It is as if those early speculations about atoms must have been meritorious and productive because there are atoms! Lancelot Whyte (1961, p. 3), in an extended essay on atomism, exemplifies aspects of the view I oppose when he writes that the 'conception of atom has been the spearhead of the advance of science' so that 'the fertility of the Greek atomic philosophy proves the power of speculative reason'. A. G. van Melsen (1960, p. 83) asserts that the seventeenth century 'owes its outstanding importance to the fact that *scientific* atomic theory came into existence'. Even William Newman (2006), a leading contemporary historian of chemistry whose work I draw on extensively in later chapters, sees Boyle's atomic chemistry as putting that area of investigation on a path that led to Lavoisier. An implication of the case defended in this book is that these views are historically mistaken and are so because they fail to appreciate the qualitative difference between scientific and philosophical claims about the structure of matter and portray the former as emerging from the latter.

I will be contesting the view, expressed by van Melsen, that the seventeenth-century witnessed the emergence of a scientific version of atomism. However, I do locate in that century the serious beginnings of a split between scientific and philosophical modes of understanding. Not only do I find the distinction implicit in seventeenth-century practice but also I find some of the practitioners making the distinction explicit. I will be arguing that most of the points I have made in this introduction about the nature of science were made at some place or other by Robert Boyle when he distinguished between ‘matters of fact’ and philosophical claims. He even made a distinction that maps onto the one I make between accommodation and confirmation.

The appreciation and formulation of a distinction is one thing. What is made of it is another. Seventeenth-century intellects were intent on articulating world-views that would underpin and help comprehend the new social order and also to recast Christian theology in light of the undermining of the Aristotelian philosophy with which it had become entwined. Empirically confirmed knowledge that conformed to my characterisation of science such as the circulation of the blood or Boyle’s law was not up to such tasks. It was to be several centuries before it was apparent that science was capable of yielding a matter theory of the generality sought by seventeenth-century natural philosophers that was also confirmable by experiment. By that time, science and philosophy were separated institutionally and many of the questions concerning the structure of matter that had been seen as in the province of philosophy became answered by science.

1.7 Writing History of Science Backwards

One important piece of evidence for modern atomism came in the form of J. J. Thomson’s experiments with cathode rays completed in 1897. Those experiments established that cathode rays are beams of minute charged particles and yielded a measure of the ratio of charge to mass of those particles. It is pertinent to ask what conditions needed to be fulfilled for Thomson’s experiments to be possible. As we will see in Chapter 13, some of those preconditions concerned the availability of the appropriate technology, such as that involving the production of suitably low pressures in the vacuum tubes employed in the production of cathode rays. On the theoretical side, one crucial piece of knowledge that Thompson needed and presupposed was what is now known as the Lorentz force law, the law that specifies the force experienced by a charged body moving in electric and magnetic fields of specified strength. It is only by employing instances of that law, combined with Newton’s second law of motion that he also needed to presuppose, that Thomson was able to deduce information about the charge and mass of the cathode particles from observed deflections of the cathode rays. The Lorentz force law was in fact relatively novel in 1897. Arguments for it emerged in the work of Oliver Heaviside,

H. A. Lorentz and Thomson himself as they struggled with what proved to be a difficult problem of the interaction between charged bodies and the electromagnetic field.

Having identified preconditions for the possibility of Thomson's experiments the process can now be taken a stage further. One can ask, for instance, what the preconditions were for making sense of the arguments for the Lorentz force law produced by Lorentz, Heaviside and Thomson. One of those preconditions was the notion of the electromagnetic field itself, which had emerged in the work of James Clerk Maxwell. Maxwell's work itself built on Faraday's conception of lines of force. Our request for the preconditions for the possibility of establishing various claims in science lead us on paths back through the history of science.

Thomson's experiments were just some of many that contributed, in the late nineteenth and early twentieth centuries, to experimental knowledge of atoms that is now taken for granted. Those other experiments had their own preconditions. Atomism applied to chemistry took for granted chemical elements and formulae and theories of ionisation built on knowledge of electrolysis dating back to Humphry Davy and Michael Faraday whilst the notion of energy levels in atoms and molecules relied on measurements of spectra in the light emitted from gases in discharge tubes. For each component of twentieth-century knowledge of atoms, paths can be traced backwards in history that result from the repeated request for the theoretical and experimental preconditions for the various moves made. I claim that were a history of atomism to be written in this way, then the story that resulted would be vastly different from one which traces a path from the speculations about atoms found in Democritus forwards through the mechanical philosophers, and beyond. I doubt if the atomism of the Ancients would figure in the backwards-written history at all!

I am not going to follow my own advice in this book and write the history of atomism backwards by tracing preconditions in the way I have suggested above. I have already indicated that, were I to do so then it is doubtful whether much of the history of philosophical atoms from the Ancients on would figure in the story at all. Because of that, a backwards-written history would not enable me to fruitfully draw a contrast between philosophical and scientific atomism in the way that I aspire to do in order to illustrate some instructive differences between the two modes of knowledge.

1.8 The Structure of the Book

Our investigation of atomism begins in Ancient Greece. Chapters 2 and 3 describe and assess the atomic theories of Democritus and Epicurus. Those theories attempted to characterise a reality behind the appearances in an ingenious way that confronted philosophical problems. Whatever their merits, these were philosophical rather than scientific theories. They were not confirmed by observational evidence in any significant sense. In the Ancient Greek context atomism was just one of a range of attempts to give a general characterisation of the ultimate nature of reality. Chapter 4 situates Greek atomism in its context with special attention given to Aristotle's philosophy. It was, of course, Aristotle's philosophy that, in the main, became generally adopted in Western Europe prior to the Scientific Revolution. What is less appreciated is the extent to which the more empirically-orientated works of Aristotle contain germs of an atomic theory that were influential in medieval philosophy and fed into a version of atomism that was very different from that of Democritus and Epicurus and had stronger claims than the latter to be empirically based.

The beginnings of a kind of atomism in Aristotle alluded to above were to be built on in medieval Europe in ways that have only been adequately appreciated in recent decades. William Newman is a major contributor to the new historical picture and I draw heavily on his work. Chapter 5 deals with two areas in which the relevant Aristotelian texts were deployed in novel ways, alchemy and the medieval theory of natural minima. The chemistry of the scientific revolution owes more to the development of alchemy than is typically appreciated. If Newman is right, then there was an important tradition in medieval alchemy that incorporated an atomic theory of matter. In the early seventeenth century, Daniel Sennert, a German philosopher and professor of medicine, constructed an atomic theory of chemistry which drew on this tradition and combined it with a second tradition having its roots in Aristotle, the theory of natural minima. Natural minima were atoms insofar as they were least parts of homogeneous substances, but they differed markedly from Democritean atoms, as we shall see. Sennert's atomism is described and assessed in the closing sections of Chapter 5.

Atomism in the second half of the seventeenth century, championed by philosophers such as Pierre Gassendi and Robert Boyle, is typically seen as part of the Scientific Revolution and as involving a revival of Ancient Greek atomism. Insofar as its proponents construed their atomism as embedded in the new anti-Aristotelian natural philosophy called the 'mechanical philosophy' they themselves construed things in this way. My account, in Chapter 6, of the mechanical philosophy and the atomism embedded in it challenges this picture. Focussing on the work of Boyle, I distinguish between the mechanical philosophy and the new experimental science and argue that the latter owed little to the former. I argue that the mechanical philosophy was supported and fruitful to a much lesser extent than is typically supposed. Boyle's experimental science was progressive, sure enough, but it was able to be so by drawing on the work of artisans, alchemists and a range of philosophers such as Van Helmont and Sennert, usually presumed to be, and presented by Boyle as, the opposition. If my distinction between science and philosophy is taken seriously then atomism as characterised by the mechanical philosophers was no more part of what has become known as science than that of Democritus, and it needs to be distinguished from the experimental advances of the seventeenth-century that did constitute major beginnings of a practice that resembles and marks the beginnings of modern science.

I stick with the theme of atomism and the mechanical philosophy in Chapter 7 to describe Newton's elaboration and transformation of it. Once again, I urge that it is important to distinguish between natural philosophy, of which Newton's atomism was a version, and the new science, of which the mechanics of Newton's *Principia* was and remains an exemplary instance. I argue that Newton's atomism was not supported by evidence in a way that his mechanics and parts of his optics undoubtedly were. I also concur with the view of some recent historians that Newton's atomism, though influential in the eighteenth century, was unproductive,

Chapter 8 is devoted to the origins of modern chemistry. I revisit the work of Boyle, who has been referred to as the 'father of chemistry' to argue that his mechanical philosophy was in fact relatively unproductive in chemistry. In a sense, the limitations of Boyle's chemistry can be attributed to the extent to which he integrated it into his mechanical atomism. There is much to be said for the position recently defended by Ursula Klein, whose work I freely draw on. According to her, the notion of chemical combination that was to prove central for chemistry emerged out of craft practices of metallurgy and pharmacy insofar as those practices involved breaking compounds into their components and reconstituting them from their components. Klein portrays a table schematising such reactions, published in 1718 by Etienne Geoffroy, as capturing the essentials of those developments and setting the scene for further developments that were to lead to the chemistry of Lavoisier later in the century in a way that owed no debt to atomism.

John Dalton introduced atoms into chemistry early in the nineteenth century, with one kind of atom for each element. There was a sense in which this atomism made contact with experiment insofar as it predicted and explained why substances combine in constant proportions by weight. My main objective in Chapter 9 is to argue that the considerable advances in nineteenth century chemistry did not owe much to Dalton's atomism. They came about through a use of chemical formulae not dependent on atomism of the kind Dalton had proposed. My case exploits the historical work of Alan Rocke and Ursula Klein.

In Chapter 10 I contrast my view about the emergence of definitive chemical formulae and the relative atomic weights of elements that followed from them with the traditional one. The common story is that, in 1858, Stanislao Cannizzaro put atomic chemistry in good shape by showing how Avogadro's hypothesis (equal volumes of gases at the same temperature and pressure contain equal numbers of molecules) could be used to calculate relative molecular weights and how these, combined with measurable equivalent weights, could be used to determine atomic weights and formulae. I challenge this story on the grounds that the problem of atomic determination was not the key one that it is presumed to have been, that, in any case, organic chemists solved the problem chemically with no need for Avogadro's hypothesis, and that, what is more, Cannizzaro's method did not give chemists the structured chemical formulae they needed. My intent in this chapter is not merely to challenge the received view on historical grounds but to help elucidate the kind of theory that nineteenth century chemistry was. I claim that it was not dependent on knowledge of atoms. The development of nineteenth-century chemistry was a precondition for, not the result of, the introduction of atoms into chemistry.

In Chapter 11 we turn our attention away from chemistry to physics, more specifically to the rise of thermodynamics and the kinetic theory. Thermodynamics was a phenomenological theory based on two fundamental laws, the conservation of energy and the increase in entropy. It made significant progress in the last few decades of the nineteenth century. In particular it was able to solve two problems in chemistry that had confounded atomists, namely, the measurement of chemical affinities and the anomalous vapour densities of some gases that could be explained thermodynamically by appeal to thermal dissociation. The kinetic theory of gases also met with successes in the 1860s but it did have problems. Its predictions were not completely borne out by measurements of the specific heats of gases and it had difficulty coping with irreversibility. (Heat flows only from a hot to a cold body. But the reverse process, conceived of as arising from the reversal of the molecular motions, is a perfectly valid Newtonian mechanical one.) According to the kinetic theory the second law of thermodynamics is only statistically true, but its proponents had no evidence, apart from irreversibility itself, that this was the case.

Any reasonable doubts about the existence of the molecules of the kinetic theory were dispelled by Jean Perrin's classic experiments on Brownian motion. In Chapter 12 I discuss how these experiments gave strong evidence for the existence of molecules and established the (qualified) truth of the kinetic theory in a way that lived up to the most stringent demands on what it takes to confirm a theory. There is no doubt that Perrin gained experimental access to molecules. A decade earlier, Pieter Zeeman and J. J. Thomson had gained experimental contact with electrons via experiments on spectra and cathode rays. These experiments, discussed in Chapter 13, gave experimental access not only to atoms but to components of atoms. What made these experiments possible were developments in physics and technology in the nineteenth century.

There is a sense in which the existence of atoms was firmly established early in the twentieth century, but they are very different from the kinds of entities envisaged in the philosophical tradition. Atoms are not fundamental insofar as they have an inner structure which was already being explored experimentally in the years immediately following those that mark the ending of this book. The task of exploiting the electronic structure of atoms to explain their stability, chemical bonding and spectra was still one for the future at the time our story ends. The tasks that lay ahead were scientific not philosophical tasks and the accomplishment of many of them were to pose more headaches for mechanistically-inclined philosophers. Many of the intuitions that had been transformed into fundamental principles by the Ancient Greeks and taken over by the mechanical philosopher, such as the idea that there is only one kind of matter and that it is impenetrable, turn out to be false. There is a range of fundamental particles, some of them charged and some of them not, not to mention fields. Wave-functions representing electrons superimpose and add up and alpha particles emitted in radioactive decay tunnel through the potential barrier holding them in an atomic nucleus. What is more, many of the properties possessed by atomic or sub-atomic particles, such as the half-integral spin of the electron, differ markedly from the shape and size attributed to atoms by Democritus, Epicurus and the mechanical philosophers. The old philosophical concepts were undermined, not by an improvement in philosophical argument but by way of a clash with the findings of science. These are the issues summarised in the concluding chapter.

1.9 A Note on Terminology

The word 'atom' stems from the Greek word for something that cannot be broken down. On that definition the modern atom is not worthy of the name. My book is a history of atomism which stresses the difference between the atoms of science and the atoms of philosophy. To be able to describe it in this way I need to use the terms 'atomism' and 'atom' in a suitably vague way so that, for instance, the atoms of Democritus, natural minima and the modern atom all qualify as atoms. Atoms, in the general sense in which I use it, are discrete parts of macroscopic objects or substances whose properties serve to account for the wholes they are parts of. Where I need to be more precise I use terms such as 'Ancient Greek atoms', 'natural minima', 'Daltonian atoms', 'the modern atom' or 'electron'.

Notes

1. The construction of an adequate account of confirmation in science is a major issue on which there is a vast literature. Contemporary accounts which I largely endorse and draw on here are due to John Worrall and Deborah Mayo. See, for instance, Worrall (2002), Mayo, (1996) and Mayo (2002). I have outlined some quibbles with Mayo's position in Chalmers (2002a) but the quibbles should not disguise the broad agreement.
2. The status of Ptolemy's theory was superior to that suggested by my caricature. Some of its claims did meet the requirement that there be independent support. For instance, the epicycles added to account for retrograde motion had the additional consequence that the planets be brightest, because nearest to the earth, when retrogressing, a 'natural' prediction that was confirmed.
3. See Mayo (2002).
4. If downwards pressure is exerted on a perforated tennis ball filled with water, the water is ejected in all directions, not just in the direction of the applied force.
5. Clark Glymour (1980) introduced the term 'bootstrapping' to characterize the way scientific progress is made via the testing of hypotheses.

Chapter 2
Democritean Atomism

Abstract An atomic structure of matter was proposed by Leucippus and Democritus in Ancient Greece in the fifth century BC as a response to an argument of Parmenides to the effect that change is impossible. For the ancient atomists permanent reality consists of portions of being (atoms) each characterised by an unchanging shape and size. Change involves the motion and rearrangement of atoms. In effect, ancient atoms were like idealised stones moving and colliding in the void and sometimes becoming entangled. As well as the physical, stone–like, atoms proposed as a response to Parmenides, some early Greek philosophers proposed individual magnitudes as components of continuous magnitudes as a response to Zeno's paradoxes. These 'atoms' were distinct from physical atoms and I doubt if Democritus included them in his atomic theory. The Ancient Greeks, including the atomists, defended their accounts of the ultimate structure of reality by extracting principles from common sense that could plausibly be construed as self-evident and drawing their consequences by logical reasoning. The connection between the early philosophical atomic theories and experience was tenuous.

2.1 Philosophy as the Refinement of Common Sense by Reason

What is the nature of being, not the being of a metal as opposed to a stone, not the being of an inanimate object as opposed to a living being, but of being in general? What is the world really like at rock bottom as opposed to what it appears to be like unreflectively? How is change possible, not just particular examples of change, such as the change of water into steam or of an olive seed into an olive tree, but change in general? What is the nature of the good, not a good wine or a good act, but the good in general? These questions are so general and abstract as to make a grasp of their meaning difficult. It was by raising questions such as this that some individuals in Greece and Italy in the sixth century BC with a peculiar turn of mind and sufficient time on their hands gave birth to what has become known as philosophy. They began to talk of what everyone understands ('leaves exist and unicorns do not, and the former change from green to brown') in language hardly anyone understands ('it is or it is not', 'change is the functioning of potential as potential'). The first

A. Chalmers, *The Scientist's Atom and the Philosopher's Stone*, Boston Studies in the Philosophy of Science 279, DOI 10.1007/978-90-481-2362-9_2,

atomic theories arose, in the writings of Leucippus and Democritus, as a response to problems posed by some questions that were very abstract indeed.

The main characteristic, and the chief novelty, of the views aired by the ancient philosophers was the extent to which they were developed and defended by appeal to reason or logic rather than being rendered plausible by appeal to mythical stories or personified gods. Certain assertions, taken as a starting point, perhaps because of their apparent self-evidence ('you cannot travel from A to B without passing all the points in between A and B'), or because they are borne out by experience, ('the world is subject to change') are subject to critical scrutiny by tracing their consequences. An assertion can be rendered problematic by showing it to have absurd consequences or by showing it to clash with some other 'self-evident' assertion or fact of experience. A philosophy is a set of general assertions about the world that is designed to withstand critical scrutiny of this kind.

One of the most striking consequences of this intellectual activity is the extent to which seemingly obvious assertions turn out to be problematic. 'Everything that happens has a cause' is problematic because it leads to an infinite series of causes, x_2 being invoked as the cause of x_1, x_3 invoked as the cause of x_2 and so on. Other puzzles emerge from reasoning that is more mathematical in character. It is a consequence of Pythagoras's theorem that it is impossible to choose a small unit of measure such that the side of a square is made up of m of those units and its diagonal is made up of n of them, where n and m are whole numbers. A perfect square implies irrational numbers. Other problems emerge in the context of infinite series. Every number has its square. 4 is the square of 2, 9 is the square of 3 and so on. What is more, every perfect square has a square root. So each perfect square can be paired with the number that is its square root, and, conversely, each number can be paired with its square. It would seem to follow from this that there are the same number of square roots as there are numbers. But this clashes with the seemingly obvious fact that the perfect squares are a subset of the sequence of natural numbers. In the sequence 1, 2, 3 etc, only 1, 4, 9 etc are perfect squares. Two seemingly obvious notions clash with each other. The removal of problems of the kind illustrated by my examples was the driving force behind the philosophies of the Ancients.

A natural assumption implicit in much everyday life is the assumption that the world is as it appears to be to the senses. Fire burns, heavy objects fall to the ground, the sun traverses the sky once a day and so on. But a little reflection reveals that there must be more to the world than this. A magnetised needle looks or feels no different to an unmagnetised needle. There must be something different about the magnetised one that makes it distinct from the other, but, whatever it is, it is not evident to the senses. The typical behaviour of a leaf is to turn from green to brown in the autumn. There must be something about leaves that is responsible for such behaviour and which renders a leaf different from a stone or an egg. Again, whatever it is, it is not something evident to the senses. It would appear that there is more to the world than meets the eye, and the philosophers of antiquity aimed to use reason to fathom what this 'more' amounts to.

In these opening paragraphs I have attempted to give the flavour of the kind of thinking initiated by the so-called 'Presocratics', those pioneer philosophers who preceded Socrates. Socrates died at the beginning of the fourth century BC, the century that was to witness the construction of the systematic philosophies of Plato and Aristotle. It is time to turn our attention to how the deliberations of the Presocratics prepared the way for two of their number, Leucippus and his pupil Democritus, to postulate atomic theory.

2.2 Parmenides and the Denial of Change

According to the position urged by Parmenides early in the fifth century BC, the universe is an eternal, unchanging, homogeneous sphere. All evidence to the contrary is an illusion. As we shall see, the reasoning that led Parmenides to this conclusion involved the critical and rational interrogation of some common sense ideas of the kind I described in the previous section. A characterisation of Parmenides' conclusion as startling and counter-intuitive would be an understatement. Nevertheless, most of the presuppositions assumed and conclusions reached by Parmenides were shared by Leucippus and Democritus. It took only a minor modification in the Parmenidean position for them to arrive at their conclusion that the universe consists of atoms in the void. The precise historical facts about, and most appropriate logical structure to attribute to, Parmenides' position are still matters of scholarly dispute.[1] In the following I attempt to outline the main line of argument in a way that is not too contentious and which provides an appropriate background for comprehending the origins of atomism.

According to Parmenides, 'it is or it is not' (Kirk et al. 1999, p. 250). This is a quotation from the first part of a two-part treatise by Parmenides written in verse, outlining 'the way of truth', as opposed to the false way of the mistaken opinions of mortals described in the second part. From the context it is clear that Parmenides intends his assertion to express the fact that existence and non-existence, or perhaps being and non-being, are mutually exclusive alternatives. There is surely a range of interpretations of the claim that render it uncontentious. The distinctive feature of Parmenides' position stems from his denial of the possibility of the 'is not', that is, his denial of the possibility of non-being. For Parmenides, the assertion that there is such a thing as non-being has no intelligible content. It is nonsense. One cannot have an intelligible thought of, nor make intelligible assertions about the non-existent for there is nothing for the thought or assertion to pick out as its referent. As Parmenides put it, 'you could not know what is not – that cannot be done – nor indicate it' (Kirk et al. 1999, p. 245) and again, 'What is there to be said and thought must needs be: for it is there for being, but nothing is not' (Kirk et al.1999, p. 247). I am reminded of a cartoon that amused me in my youth. An ornithologist, gesticulating towards the higher reaches of a tree, declares to his understandably bemused audience, 'and this bird has no characteristic features whatsoever'. The onlookers are bemused because they have scant idea what it is they are meant to be looking for. From a Parmenidian perspective, asserting the existence of non-being is even more lacking in intelligible content than the ornithologist's declaration, for at least the audience knows the ornithologist is referring to a bird. Claims about the existence of nothing, or non-being, do not even have that degree of specificity.

Having ruled out non-being from our alternatives we are left only with being. This, for Parmenides, is enough to establish that change of any kind is an impossibility. For any change involves the coming to be of something that was not and the ceasing to be of something that was. The change from A to B involves the coming to be of B and the ceasing to be of A. But the coming to be of B implies that B did not exist prior to its becoming B. This has been ruled out by the rejection of the 'is not'. The non-existence of B cannot be intelligibly asserted. In a similar fashion, the ceasing to be of A is ruled out by the non-intelligibility of asserting the non-existence of A. That that is, is and that that is not, is not, and that's it. Change is impossible.

Further thoughts on the impossibility of coming to be from what is not can be found in Parmenides' tract. If there is to be an intelligible notion of change, then it would seem to require that the preconditions for the result of a change be present in the situation that changes. The change of a seed into a tree implies that there is something present in the seed responsible or setting the preconditions for the change into a tree. If this were not the case, then anything might change promiscuously into anything else and there would indeed be no distinction between a hawk and a handsaw. But there can be no specific preconditions for change residing in non-being. Nor can there be any reason stemming from non-being why a change to any kind of being should happen at some time rather than another. The emergence of being from non-being is unintelligible. 'For what birth will you seek for it?' wrote Parmenides of change from non-being to being. 'How and whence did it grow? I shall not allow you to say nor to think from not being; for it is not to be said nor thought that it is not; and what need will have driven it later rather than earlier, beginning from the nothing, to grow?' (Kirk et al. 1999, pp. 249–250). The coming to be of something from nothing is impossible, and, insofar as any change involves some such coming to be, change in general is impossible.

The rejection of all change and of non-being leads Parmenides fairly straightforwardly to his view of the universe as an eternal, unchanging, homogeneous slab of undifferentiated being. Parmenides' view that the Universe is spherical is less well-supported, and the view of his follower, Melissus, that the universe is an infinite expanse of being seems to follow more naturally from Parmenides' assumptions. It was Melissus, also, who drew connections between the homogeneity and oneness of the universe and the rejection of void, understood as non-being. For if there is no void, then there is nothing to separate one portion of being from another.

The soundness of the Parmenidean argument can be, and has been, challenged on a variety of grounds. Most of them are tied up with the vagueness and ambiguities associated with the notion of 'being' as employed by Parmenides. 'To be' can be interpreted in an existential sense ('to be or not to be') or in a predicative sense (to be green). A green leaf becoming brown does not invite the conclusion that something comes from nothing in the way that Parmenides' talk of becoming in general and of generation from nothing does. So one could accept the force of a Parmenidean–style argument against the intelligibility of something coming from nothing without having to accept his rejection of change in general. Another shortcoming of Parmenides' case lies in his tendency to treat 'to be' and 'to exist' as synonymous. We shall see how Leucippus and Democritus open the door to their atomism by challenging that identification.

Parmenides' case has more force than it otherwise does if we concede to him an assumption that seems implicit in it, namely, that there is only one kind of being and one way of being real. Many philosophers from the Presocratic philosophers onwards have found a version of that assumption attractive. The first Presocratic philosopher, Thales, proposed that the world is the manifestation of one underlying reality that he identified with water. The early atomists assumed all atoms to be made of the one universal matter, an assumption that persisted in the views of the seventeenth-century mechanical philosophers, through to the views of Newton and beyond. There are tensions between the assumption that there is just one way of being real and the existence of change. Imagine a piece of copper wire of some definite extension. There is no doubting its material existence. However, because it is a piece of copper wire it is capable of being stretched, and might well be stretched in the future. Part of what the wire is here and now is its capability of being stretched just as part of what it is here and now is to posses a specified extension. It would seem that having the capacity to be stretched (which the wire may never exhibit if it is never subject to an appropriate force) is a different kind of property than having a definite extension. Aristotle distinguished between actual and potential being in this kind of context, whilst modern philosophers distinguish between dispositional and categorical properties. Taken at face value, these distinctions seem to imply that there is more than one way of being real. Philosophers through the ages and including modern ones have been suspicious of including potential being and dispositional properties in a characterisation of being and have tried to explain them away. But dispensing with them lands one with the problem of making intelligible sense of change. How are we to make sense of the stretching of a wire if its degree of elasticity is not a real property of it? Is it not the case that the constant of proportionality in Hooke's law refers to, and gives a precise measure of, precisely that property? There is a tension between the view that there is only one sense of being real and the phenomenon of change. Parmenides, as we have seen, picked up on this. He attempted to remove it by denying change.

It may appear to some readers that my attempt to extract some value and contemporary relevance from the work of Parmenides is laboured and unduly charitable. After all, such a reader might observe, to think that one can rule out the possibility of a vacuum or void, which is one of the important consequences of Parmenides' argument, simply by investigating what is implied by the use of concepts such as being and existence in ordinary language is to misconstrue the significance of such analysis. If Parmenides' concepts of being and existence cannot accommodate change then he should modify them. There is some merit in this negative assessment of the force of Pamenides' case. However, anyone that holds it is going to find it difficult to avoid a similar judgement concerning the value of the reasoning of the Presocratic atomists. For their position involved an acceptance and deployment of most of Parmenides' arguments and conclusions. Only on one point did they disagree with him, and that was over the question of the intelligibility of the void. If one judges that Parmenides' philosophy involved an excessive and unhelpful degree of abstractness or generalisation or that it amounted to inappropriate emphasis on linguistic analysis then it will be difficult to resist judging ancient atomism in a similar light.

2.3 The Atomism of Leucippus and Democritus: The Basics

As we have seen, Parmenides concluded that change is impossible and that the evidence provided by the senses to the contrary must be illusory. Leucippus and Democritus fully accepted the terms of the debate as set by Parmenides, but showed how change could be rescued by accepting the existence of the void as well as of being. Once the void is admitted portions of being can be separated from other portions by it and change can be understood in terms of the motions and re-arrangements of the portions. The portions of being are themselves miniature Parmenidean worlds that are one and changeless for all the reasons that Parmenides' one, the universe as a whole, was argued to be changeless. Leucippus and Democritus called them atoms. When something changes, the change involves a re-arrangement of atoms, whilst the identity of the something that changes is due to the persistence of the atoms that are re-arranged. Change, and the veracity of the senses insofar as they indicate change, are saved by replacing Parmenides' undifferentiated expanse of being by innumerable unchanging atoms in the void.

This general atomistic solution to the problem posed by Parmenides relies on making sense of the void. Leucipuss and Democritus attempted to do this by making a distinction between being and existence. 'Since the void exists no less than body, it follows that what is not exists no less than what is' (Kirk et al. 1999, p. 414). Atoms are the only kind of being. They are full of being and nothing but being. Non-being, the void, is the absence of being. It exists where atoms are absent. According to Aristotle, as reported by Simplicius some thousand years later, Democritus 'calls space by these names – "the void", "nothing", and "the infinite", while each individual substance he calls "thing", the "compact" and "being"' (Kirk et al. 1999, p. 414). Once the void is identified with the absence of atoms then the move to an infinite void is a natural one.

As Parmenides made clear, talk of the void, the nothing, is a tricky business that can readily lead to the utterance of absurdities. It is doubtful whether the early atomists completely avoided the problem. However, whether or not there are regions empty of all matter seems to be a substantive issue and not one that can be settled by manipulating words. Democritean atomism, with its affirmation of the void, should be judged on its ability to give a general account of the way the world is. The fact that it could avoid the Parmenidean rejection of change as an illusion was a good start.

Atoms are changeless in themselves. They have no physical parts and cannot be split or penetrated because they contain no void that will serve to separate one region of an atom from another. This is the same kind of reasoning that Melissus employed to argue for the permanence of the Parmenidean one, the only difference being that, for Parmenides and Melissus, the presence of void as a separator is an impossibility, whereas the absence of void in an atom is, for Leucippus and Democritus, a contingent fact. This difference notwithstanding, atoms are full and changeless, and provide the permanence that lies behind all change. As a finite and located portion of being, an atom has a definitive shape and size. Perhaps because there is no reason for them to have one shape and size rather than another, atoms come in all kinds of shapes and sizes, indeed, in an infinity of shapes and sizes. Atoms are capable of movement in the void that separates them.

This much of the Democritean position follows from the most general assumptions involved in the response to Parmenides.[2] There is just one kind of being and there is also non-being, so change becomes a conceptual possibility in the way we have seen. Further basic features of atomism are arrived at by imagining atoms to be something akin to small stones or other solid bodies of everyday experience, with the important difference that atoms are absolutely solid and unbreakable whereas a stone is only relatively so. Atoms can move just as stones can, and when they collide they typically rebound. Atoms 'move by mutual collisions and blows' (Kirk et al. 1999, p. 424), but they also have shapes, such as protuberances and concavities, hooks and eyes, that make it possible for them to become 'entangled' or 'intertwined' to form relatively stable complexes.

> Atoms move in the infinite void, separate one from the other and differing in shapes, sizes, position and arrangement; overtaking each other they collide, and some are shaken away in any chance direction, whilst others, becoming intertwined one with another according to the congruity of their shapes, sizes, positions and arrangements, stay together and so effect the coming into being of compound bodies. (Kirk et al. 1999, p. 426)

It is in such a way that the macroscopic objects of our everyday experience are accounted for. Such objects are subject to change and destruction in a way that atoms are not because the complexes of atoms can be shaken or broken apart. The complexes of atoms are never totally permanent and one because always some void separates the atoms.

'On the nature of necessity: Democritus means by it the resistance and movement and blows of matter' (Kirk et al. 1999, p. 419) Here we have another basic feature of Democritean atomism that is natural once atoms are likened to stones and that is the notion of necessity it involves. The motions of two atoms after a collision is a result of, that is, is necessitated by, the sizes, shapes, speeds and direction of their motion immediately prior to the collision just as can be presumed to be the case with colliding stones. As for the causes of the motions prior to the collision, in the case of atoms these were in turn caused by previous collisions and so on back in infinite time. The Democritean world is a deterministic world that is infinite both spatially and temporally.

An interesting issue, and one on which contemporary as well as Ancient commentators on Democritus are divided, concerns the issue of whether weight is to be ascribed to Democritean atoms.[3] Amongst the Ancient commentators, we have, for instance, Aristotle bluntly declaring 'Demicritus says that each of the indivisible bodies is heavier in proportion to its excess [that is, in proportion to its size]' whilst, no less bluntly, Aetius assures us that 'Democritus says that the primary bodies (that is, the solid atoms) do not possess weight' (Kirk et al. 1999, p. 421). Insofar as atoms are packets of Parmenidean being, then it seems inappropriate and gratuitous to attribute weight to them in addition to some determinate shape and size. Further, insofar as weight is associated with a tendency for heavy objects to move downwards, it cannot sensibly be attributed to atoms in the void because in the void there is no downwards. So Aetius's denial that Democritean atoms have weight has plausibility. On the other hand, once we interpret Parmenidean being as the universal matter composing idealised stones, as is implicit in the notion of colliding and rebounding atoms, then Aristotle's claim that atoms have weight, a weight that is directly proportional to the size of an atom, makes perfect sense. The assumption that an atom has weight would seem to be necessary to capture the idea that a large atom will have more effect on the motion of a small one with which it collides than the small one will have on it. The weights of atoms form an essential component of what it is that determines the outcomes of their collisions.

The divergent views on whether Democritean atoms have weight or not can be reconciled by distinguishing two distinct notions of 'weight' which I will refer to as gravitational weight and unwieldiness. Gravitational weight is the tendency of heavy objects to move downwards. Atoms as such do not have gravitational weight. The weight of an earthly stone is something that, in Democritean terms, must be the result of collisions of atoms that constitute the whirl that is the earth. Unwieldiness is that property of an object that renders it difficult to start or stop. A window is more likely to be smashed by a large stone than a small one moving at the same speed because the large one has more weight (in the sense of unwieldiness). A motion of a large boulder in an avalanche will be scarcely affected by small stones that lie in its path, unlike the motion of the small stones, which will be drastically affected by the impact of the boulder. This difference is due to the differing weights (degrees of unwieldiness) of the stones. Unwieldiness is not the same as gravitational weight. My example of a stone being hit by a boulder could be replaced by one of a ship striking a dinghy at sea, where the gravitational weight of each of the latter is countered by the up-thrust of the sea.

The distinction between unwieldiness and gravitational weight is formalised and made precise in Newtonian physics, where it becomes the distinction between mass and weight. But the distinction makes common sense, as the examples I have chosen are designed to show, and does not require the acuity of a Newton for its appreciation. The distinction is not readily appreciated because, in common discourse, the word weight is used indiscriminately to refer to both notions. 'The large stone pushes the small one aside because it weighs more' makes perfect common sense, even if it makes a Newtonian physicist uncomfortable. Further, I suggest that my examples would have made as much sense to an Ancient Greek as they do to us. The Greeks, like us, had just one common word for weight. Their word was βαροζ. But when that word or a derivative was used in the Hippocratic works to maintain that 'his chest seems to sigh and to contain a heaviness that prevents it from moving', and in the Aristotelian work *Mechanical Problems* to assert that 'the smaller and the lighter is more easily moved than the larger and the heavier', its translation as unwieldiness rather than gravitational weight makes most sense of the text.[4] When Aetius denied that Democritean atoms had weight he meant gravitational weight. When Aristotle claimed that Democritean atoms had weight proportional to their size he meant weight in the sense of unwieldiness. The problem of interpretation is solved.

As I have indicated, attributing weight (in the sense of unwieldiness) to atoms goes beyond what is implied by interpreting them as packets of Parmenidian being. It involves interpreting them as something akin to idealised stones. If two packets of Parmenidean being collide it can certainly be concluded that they cannot penetrate each other, but it cannot be concluded that they rebound. Democritean atomism did augment the Parmeidean picture by adding the stone-like quality of atoms. For a Democritean atomist, being was matter-like, where matter is interpreted in something like its common sense.

2.4 Atomic Explanations of Properties

In the previous section I outlined the most basic features of Democritean atomism. They relied on acceptance of Parmenides' notion of being, which was such that there can only be one kind of it, added the notion of non-being as an existent, and further drew on analogies between portions of being and solid stones. The basic features of atomism arrived at in this way were able to accommodate change, conceived of in a totally general way. Parmenides' theory has the consequence that change is impossible whereas Democritean atomism allows change as a possibility. Experience shows that there is change, so the latter theory is preferable to the former unless all experience is dismissed as illusory. So far so good for Democritus. But his task hardly ended there. There are all kinds of properties and change exhibited by the world of experience and Democritus needed to show, at least in a general kind of way, how this is compatible with the view that the world is made up of nothing other than atoms, as Democritus conceived them, moving in the void, colliding and sometimes combining. Democritus did respond to the challenge in a variety of ways.

Democritean atomism has it that the universe is an aggregate of stone-like entities. This is hardly how things appear, on the face of it! Items in the world are hot and cold, have colours and tastes and smells, all of which are subject to change. Vegetables have life in a way that stones do not, whilst animals, and especially humans, have a form of life qualitatively more enhanced than vegetables and so on. For Democritean atomism to be viable, the vast variety of properties, substances and modes of change in evidence in the world need to be explicable in terms of, or reduced to, or somehow shown to arise from, arrangements and motions of atoms in the void, the atoms themselves being far too small to be detectable by the senses.

Properties of aggregates of individuals that are not properties of those individuals can arise as the result of aggregation in a way that poses no deep problem. Crowds can be dense and can surge in a way that the individual people that make up the crowd cannot, in spite of the fact that crowds consist of nothing other than moving people. The density of a material substance could be understood by Democritus to arise as a result of the ratio of space occupied by atoms to that occupied by void, whilst stability of macroscopic objects and substances could conceivably be understood as the result of the mode of interlocking of atoms. Democritus attributed the mobility of fire to the spherical shape of its atoms (Kirk et al. 1999, p. 427), presumably because the regular surfaces of spheres eliminate friction and the capability of getting caught up. I am not suggesting that arriving at the correct account of stability, density and mobility, or any other property of a macroscopic substance, was straightforwardly open to Democritus. It clearly was not. What I am saying is that some properties of aggregates can result from collections of individuals in a way that poses no special philosophical or other kind of problem, and that Democritus was free to avail himself of this in his attempts to account for properties of macroscopic bodies as resulting from properties of the atoms that compose them.

Not all properties can be construed as arising in a straightforward way as a result of aggregation, as Democritus was aware. There is a sense in which properties such as tastes, colours and smells do not straightforwardly exist in the bodies to which they are attributed at all, but rather exist in us as a result of our sensing them. Democritus is presumed to have been making a distinction of this kind when he declared 'by convention sweet, by convention bitter, by convention hot, by convention cold, by convention colour: but in reality atoms and void' (Kirk et al. 1999, p. 410). Sensible properties are not correctly understood as existing as properties of the objects sensed but are rather sensations in us caused by our interactions with those objects. Individuals tend to react in characteristic ways to the taste of sweet and bitter things and conventionally label those things sweet and bitter. This leaves room for individual differences, so what is sweet to some may be bitter to others. To be consistently atomistic, such an analysis of sensations required, of course, that the latter themselves be interpreted as states of mind composed of atoms. For a Democritean atomist the claim that reality consists of nothing other than atoms and the void needs to be taken quite literally and so applies to minds and souls as much as to stones and trees.

There were two general ways, then, in which Democritus could reconcile the variety of properties manifest in the world with the claim that there exists nothing other than atoms possessing shape and size moving in the void. He could explain how properties result from aggregates of atoms and he could explain how other sensible properties arise as a result of interactions with our senses.

2.5 Atomic Explanations of Specific Phenomena

It is one thing to outline in a schematic way how change is possible and what general character atomic explanations of properties manifest in the world might take. It is another thing to actually provide and defend definite atomic explanations of specified phenomena. Democritus did propose such atomic explanations. Perhaps not surprisingly, they were far from satisfactory.

Democritus did on occasions move beyond the general idea that phenomena are to be explained by reference to the shapes, sizes and motions of unchanging atoms.

He attempted to explain specific phenomena by attributing shapes and sizes to the atoms responsible for them. We have already mentioned his suggestion that the mobility of fire is due to the spherical shape of its atoms. He also gestured towards the explanation of some sensory experiences. 'Bitter taste is caused by small, smooth, rounded atoms, whose circumference is actually sinuous; therefore it is both sticky and viscous. Salt taste is caused by large, not rounded atoms, but in some cases jagged ones' (Kirk et al. 1999, p. 429). There are two kinds of problem with such speculations. Firstly, there is the question of whether the proposed shapes and sizes do serve to explain the phenomena they are designed to explain. Once we have settled on spherical shapes for fire atoms to account for its mobility, what scope is left to explain other properties of fire? Why should small, smooth atoms give rise to a bitter sensation? Even if the proposed shapes and sizes did explain phenomena there is the question of whether the proposed explanations are the right ones. What is wrong with the suggestion that the mobility of fire is due to the small size of its atoms, enabling it to move through crevices between the atoms of any obstacle it encounters? Conceivably, specifications of atomic shapes and sizes could be defended by showing that they were borne out by phenomena in addition to and independent of the one they were designed to explain. Democritus did not even come close to anything like this. The shapes and sizes he occasionally attributed to atoms were unsubstantiated guesses.

As well as speculating about specific shapes and sizes for atoms, Democritus, more ambitiously, offered detailed atomic mechanisms as an explanation of the origin of our cosmos and the functioning of the senses. Our world, as well as many other worlds more or less like it, are caused by the chance collisions of atoms in the void. Such collisions somehow form a whirl, with heavier atoms congregating at the centre of the whirl and becoming entangled to form the Earth. Visual sensations are due to 'images of the same shape as the object' that are 'continually streaming off from the objects of sight and impinging on the eye' (Kirk et al. 1999, p. 428). Once again, it is doubtful whether the proposed mechanisms are capable of explaining what they were designed to explain. A whirl would yield a cylindrically rather than spherically symmetric earth, and if, consistent with this, Democritus believed we lived on a flat earth, the whirl would seem to imply that heavy objects would be inclined to slide across the flat surface of the earth towards its axis rather than perpendicularly. It is difficult to see how the proposed mechanism for sight could accommodate the visualisation of shapes, sizes and colours. Apart from the explanatory adequacy of the models there is, once again, the question of the lack of empirical evidence in their favour.

2.6 Atomism as a Response to Zeno's Paradoxes

We have discussed atomism as it arose in response to the denial of the possibility of change by Parmenides. There is another, rather different, kind of atomism that arises as a response to various paradoxes posed by Zeno, a follower of Parmenides. Whether or not Democritus subscribed to this second version of atomism is an issue on which scholars are divided. In this section I discuss this second kind of atomism, which involves an attempt to block Zeno's paradoxes by postulating 'indivisible magnitudes', leaving aside whether Democritus proposed and championed it or not. I leave the historical question as a topic for 2.8.

Very little of Zeno's own writing has survived. Our main source for the content of Zeno's work is the same as that for the details of Democritus's position, namely, Aristotle.[5] The latter attributes four paradoxes of motion to Zeno. Briefly summarised, they go as follows.

The argument from dichotomy shows how we can never leave a room once in it, because before we can reach the door we must reach the half way point, before we can get to the half way point me must get half way to the half way point and so on for ever.

In the second paradox Achilles races the tortoise and makes the mistake of giving him a start. To catch the tortoise Achilles must first reach the point where the tortoise was, by which time the tortoise has moved on a distance. Achilles must again reach the point where the tortoise was and so on. Achilles must perform an infinite number of transits to where the tortoise was, and so cannot catch him.

The third paradox arises when we imagine an arrow at some point in its flight, occupying a region of space exactly equal to its length. Since it occupies only a single space it cannot be said to move. But at any point in its flight the arrow occupies a space exactly equal to its length. Again, at this instant the arrow is at rest. But if the arrow is at rest at any single instant in its flight then it is at rest throughout its flight.

In the fourth paradox we imagine three parallel rows of carriages, one stationary, one moving to the left and the other moving at the same speed to the right. Imagine each moving carriage travelling one carriage-length past the stationary ones in unit time. Relative to the stationary carriages the moving carriages have passed one carriage in unit time, but relative to each other they have passed two carriages in unit time. This might appear to imply that the moving carriages are moving with two different speeds at the same time. This is a confusion that arises if we fail to appreciate the relative character of motion implicit in common sense, and well appreciated, for example, by Aristotle. Relative to one set of moving carriages the other set of moving carriages is moving twice as fast as they are relative to the stationary ones. No paradox arises. The paradox is less easily dismissed if it is taken as posing problems for the view that there are indivisible magnitudes (atoms) of space and time. The carriages are assumed to be one atom of space long with the moving ones covering one atom of space in one atom of time. After one atom of time one carriage moves past one atom of the stationary ones to the left and the other past one atom of the stationary ones to the right. But that means that, relative to each other, in one atom of time an atom in each of the moving carriages has passed two atoms in the other moving carriage. So it must have passed just one of those carriages in half an atom of time, contrary to the assumption that time is atomic.[6]

There is a fifth paradox, perhaps more fundamental than the four paradoxes of motion, attributed to Zeno by Simplicius. The paradox is discussed by Aristotle (*Physics*, 1, 2, 316a, 23–316b, 9). I will refer to it as the paradox of division. It involves the division of a finite entity assumed to be continuous and hence infinitely divisible. Suppose the continuous entity is infinitely divided, with an infinity of parts resulting. Do these parts have a finite size or don't they? If they do not, then they will result in zero size when they are combined. If they do, then an infinite size will result from the combination of the infinite number of finite sized parts. In neither case is the original finite-sized entity recovered from the parts into which it has been presumed to have been infinitely divided.

These paradoxes can all be construed as arising as a consequence of a natural deployment of common-sense ways of talking. It makes perfect common sense to talk of covering half the distance to the door or dividing an object in half, although the notion of a precise half is an idealisation not readily achieved in practice. It also makes perfect common sense to talk of repeating one of these processes each time you have done it. The paradoxes arise to the extent that an infinite repetition makes sense. Since unrefined common sense leads to the paradoxes, then some refining of the concepts and reasoning involved becomes necessary.

Since the paradoxes arise as a result of infinite division they can be resolved if the path to infinite division can be blocked. If some, if not all, of space, time, motion and matter are not continuous, that is, if they are not infinitely divisible in the way they are presumed to be in the construction of the paradoxes, then perhaps the paradoxes can be avoided. Those who considered such a response frequently used the word atom (ατομοζ) to refer to indivisible magnitudes setting a limit to division. Aristotle, our main source as far as the transmission of the views of the Presocratic philosophers are concerned, does so in some of his discussion (and ultimate rejection) of indivisible magnitudes.

With one qualification to be specified at the end of this paragraph, it is uncontroversial to observe that physical atoms do not serve as a response to Zeno. To see this, suppose we accept the atomism of Leucippus and Democritus. Imagine that I am attempting to leave the room by successively halving the remaining distance to the door and suppose that I have nearly made it. Suppose I have only one atom to go. Then I still have to traverse half that atom, then half of that half and so on. I still have an infinite number of halvings to accomplish and Zeno's paradoxical conclusion that I cannot leave the room results. The division of atoms that is involved here is not a physical division insofar as it does not involve physically separating one part of an atom from another. Exactly what kind of division is involved is a tricky business that needs careful consideration. But first the qualification: Some ancients on occasions interpreted the paradox of division as involving physical division, worrying that infinitely divisible matter might be worn away to nothing.[7] If the paradox is interpreted in that way then physically indivisible atoms do serve to resolve it.

We have observed that it is not appropriate to treat the division involved in Zeno's paradoxes of motion as physical division. Apart from the qualification mentioned above, Zeno's paradoxes arise in the context of division that is other than physical division. Contemporary scholars who stress the need for the distinction typically characterise the former kind of division as conceptual or theoretical division, division that is made in thought or by the mind. On these readings, Democritus's atoms do not have physical parts insofar as they cannot be physically broken down, but they can be broken into parts by the mind. It is possible to conceive of their parts.[8] I do not deny that it is possible for the mind to conceive of parts of physically indivisible atoms. What is more, the idea that the parts so conceived can themselves be divided into parts ad infinitum puts demands on our conceptual resources. The problem of making precise conceptual sense of the continuum, highlighted by Zeno, was to pose a challenge for logicians and mathematicians for two and a half millennia.

I am dissatisfied with the opposition between physical division and division in thought as a means of grasping the distinction between the sense in which physical atoms are divisible, and the kind of divisibility involved in the generation of Zeno's paradoxes. It is not enough to acknowledge that physical atoms can be conceived of as having parts. *They have parts*. If Democritus was anything like correct, then long before there were humans present to conceive of atoms, physical atoms moved in the void, came together by chance and became entangled in a way that led to the formation of the world. Their combination requires that they have hooks and eyes or something of the kind that become interlocked. That is, atoms have parts. A large atom might shield a smaller one from bombardment from a stream of atoms by standing in their path, whereas a smaller one will not shield a larger one in the same way because *parts* of the larger one will protrude beyond the smaller. One atom might cover half of another, leaving the other half exposed. In another position it might further halve the exposed part and so on. If atoms are continuous then there are an infinite number of parts of an atom that may be left exposed by a second atom resting alongside it. If atoms are continuous then, by virtue of being continuous, they have an infinite number of parts.

If conceptual or theoretical division is misleading terminology to refer to division of physically indivisible atoms into parts, then how should we refer to that division and the resulting parts? Geometrical division and geometrical parts are terms that have some appeal but will not do. One problem is that the division of times and motions into parts are as much involved as the division of spatial dimensions in the generation of Zeno's paradoxes. But there is another problem, connected with the ambiguity with the term 'geometrical'. That term can be taken in a mathematical sense to refer to the geometry formalised by Euclid, in which lines are defined as continuous magnitudes that can be divided indefinitely. But it can also be interpreted as descriptive of physical space. The continuity or otherwise of apparently continuous magnitudes occurring in nature, or of space and time themselves, are substantive issues that cannot be decided by appeal to definitions.

The problems posed by Zeno arise from the continuity, understood as indefinite divisibility, of apparently continuous magnitudes. One atom partially overlapping another divides the latter into parts. When one atom passes another, whether there is a mind to contemplate it doing so or not, the overtaking atom successively passes parts of the one it overtakes. The question arises whether or not continuous motion marks off an infinite number of such parts, opening the way for Zeno to construct his paradoxes. Some were tempted to evade the paradoxes by denying continuity and positing 'indivisible magnitudes'. I will distinguish the kind of division involved here, which is not adequately understood as necessarily involving division by the mind, from physical division by referring to it as metrical division, and I will refer to the resulting parts as metrical parts. Those who attempted to block the way to Zeno's paradoxes by proposing indivisible magnitudes proposed metrical parts that cannot be metrically divided. We can best get a grip on the kind of atomism involved here by considering Aristotle's objections to it.

It is clear that the indivisible magnitudes invoked as a response to Zeno are quite different from the physical atoms invoked as a response to Parmenides. The latter have a variety of shapes and sizes and need to have in order to play their role of explaining change and continuity through change. Indivisible magnitudes have no shape, and, insofar as they have size, all have the same indivisible size. Atomic theories of space and time arising as a response to Zeno are distinct from physical atomic theories arising as a response to Parmenides.

2.7 Aristotle's Critique of Indivisible Magnitudes

Aristotle appreciated the problems that Zeno posed for the notion of a continuum understood as 'that which is divisible into divisibles that are infinitely divisible' (*Physics*, 6, 2, 232b, 25). He argued that it is not possible to construct a continuous magnitude out of indivisibles on the grounds that to do so requires that the indivisibles be laid part to part or edge to edge, an impossibility given that indivisibles lack parts or edges (6, 1, 231a, 21–231b, 5). Nor can indivisible points be laid in succession since, from the definition of a continuum, two indivisible points not coincident will have an infinity of indivisible points in between them (6, 1, 231b, 6–14). He argued that moving indivisibles entail indivisible units of time, since if one indivisible passes another in time, t, and time is continuous, then it will pass half that indivisible in half the time, thus dividing the indivisible. Given the consequence that indivisibles of time follow from indivisible spacial magnitudes, Aristotle concluded that this kind of atomism entails that motion must occur in jerks, which he clearly regarded as absurd (6, 1, 231b, 19–232a, 12). He argued that the present, separating the past and future, cannot have any duration, so that time is not made up of 'nows' and he used this to combat Zeno's arrow paradox (6, 3, 233b, 33–234b, 9). He also sought to resolve Zeno's other paradoxes of motion. He argued that, although a distance to be traversed can be divided into an infinity of parts by successive bisections he insisted that the available time can also be so divided, with each distance to be traversed matched by an interval of time, the infinite number of distances and times both having a finite sum (6, 2, 233a, 22–32 and 6, 9, 239b, 10–29)).

There remains the puzzle of how the passing of an infinite number of spaces or times can be accomplished, whether to reach the door or to catch the tortoise. Aristotle responded to this with a distinction between actual and potential division. Distances and durations, and any other continuous magnitudes, are said by Aristotle to be divisible potentially but not actually (8, 8, 263a, 15–263b, 9). The distance separating me from the door is potentially divisible. The halfway point potentially divides it in half. The division can be actualised, by severing the floor, by my passing half-way and then stopping, or by my merely making the division in thought. But, having so halved the distance, the remaining distance remains potentially divisible, and so on, however many halvings are actualised. The (potential) infinite divisibility of continuous magnitudes is not, for Aristotle, relative to some process of division. Further actual division of continuous magnitudes is always possible whatever the process of division because continuous magnitudes *are* infinitely divisible.[9]

Aristotle responds to Zeno's paradoxes of motion by insisting that continuous magnitudes such as lines, although potentially divisible at an infinity of points, are not made up of points and cannot be actually divided at an infinity of points. He further assumes that a line is not divided by continuous motion along it, since the motion, like the space and time it involves, is infinitely divisible. Segments of motion, however small, can be paired off with the times taken for that motion and the distances covered by it. Division at a point will require marking off the point in some way, whether by stopping at, or marking or contemplating it. An infinity of such actual divisions of a continuum is impossible, not because of limitations in our physical or mental resources, but, more fundamentally, because the continuum is infinitely divisible, with points not standing next to point, nor indivisible next to indivisible.

2.8 Did Democritus Propose Indivisible Magnitudes as a Response to Zeno?

We have discussed the brand of atomism that proposes indivisible magnitudes as a response to Zeno and we have described Aristotle's objections to it. There remains the historical question of whether Democritus himself actually proposed an atomism of this kind in addition to the physical atomism that he undoubtedly proposed as a response to Parmenides. This is a matter on which contemporary scholars disagree. In this section I attempt to sort the matter out. The reader not concerned with the historical and scholarly niceties may wish to skip this section.

Our main source of information on the content of Democritus's theories is Aristotle. The disputes about what kind or kinds of atomism to attribute to Democritus are in the main disputes about how to interpret what Aristotle has to say on the matter. Democritus's own writings are not available to us. There is also a second order difficulty. Aristotle wrote a work 'On Democritus' which has also been lost. However, there are commentaries on that lost work by Ancient Greek philosophers such as Simplicius that have come down to us.

David Furley (1967, pp. 79–101) reads Aristotle as attributing to Democritus a recognition of the distinction between the two kinds of atomism and defence of both positions. Kirk et al. (1999, p. 415) think that Furley has read more into the text than is justified. Jonathan Barnes (1996, pp. 352–360) thinks likewise. He reads Aristotle as attributing only physical atomism to Democritus. Richard Sorabji (1983, pp. 354–357) accounts for the fact that support for both Furley and Barnes can be fashioned from a reading of Aristotle's text by suggesting that Aristotle in fact failed to distinguish between physical and what he (Sorabji) calls conceptual divisibility.

My own position is that Aristotle certainly distinguished between the two types of atomism but there is no convincing evidence that he attributed to Democritus the second type, the type that invokes indivisibles as a response to Zeno.

The point that physical atoms, with various shapes and sizes, have parts and are divisible in some sense other than physically divisible seems so straight-forward that I find it difficult to accept that a philosopher so sensitive to fine distinctions such as Aristotle could have failed to appreciate the distinction. In any case, I can point to one passage where Aristotle makes the distinction explicit. The passage occurs in *On the Heavens* (306a, 30–306b, 2) in a context where he is discussing the Platonic view that attributes the properties of the elements to the characteristic shapes of their atoms.

> For any one who gives each element a shape of its own, and makes this the ground of distinction between the substances, has to attribute to them indivisibility; since division of a pyramid or a sphere must leave somewhere at least a residue which is not a sphere or a pyramid. Either, then, a part of fire is not fire, so that there is a body prior to the element – for every body is either an element or composed of elements – or not every body is divisible.

We need not get sidetracked into Plato's theory of the elements. What Aristotle is saying, quite explicitly, is that, to serve their role as atoms the latter must retain their characteristic shape. That is, they must be physically indivisible. He also points out that, should such an atom be divided, one of the resulting parts at least will lack the shape of the original atom and so be unable to play the role of that original atom. If the properties of fire stem from the sphericity of its atoms, then they must be physically indivisible, since division of a sphere cannot result solely in spheres. The spherical atoms are indivisible in one sense, corresponding to what I have called physical indivisibility, but divisible in another sense, which allows us to say that not both parts of a divided sphere can be a sphere.

Recognising that Aristotle distinguishes between types of division in one place is not sufficient to establish that he did not confuse the distinction in others. However, I believe it is possible to read Aristotle on the response to Parmenides and Zeno in a way that does not involve confusion.

Aristotle's discussion of and response to Zeno's paradoxes of motion in the *Physics* makes no reference to Democritus. In that place Aristotle does not spell out precisely what kind of division is at issue when he contemplates infinite division and indivisible magnitudes as a possible barrier to it. But, on my reading, this is not a failing or a sign that he has not adequately grasped the distinction between kinds of division. Aristotle does not need to distinguish between types of division because, as I have argued above, *the argument is independent of the type of division*. Aristotle's central discussion of Zeno's paradoxes of motion give grounds neither for the claim that he has failed to grasp the distinction between physical and other types of division nor that he is attributing indivisible magnitudes to Democritus.

In *Generation and Corruption* Aristotle does mention Democritus in the context of a discussion of indivisibles and infinite divisibility and here the task of interpreting Aristotle's meaning is more demanding. There is more scope for attributing confusion to Aristotle and some scope for reading him as attributing assumptions concerning indivisible magnitudes to Democritus. I shall resist both lines of argument.

Generation and Corruption is about coming-to-be and passing-away. Having introduced the problem and mentioned a range of solutions offered by his predecessors in Chapter 1 of Book 1 he proceeds in Chapter 2 to some details, concentrating especially on the question of divisibility and the possibility of coming-to-be and passing-away coming about through the association and dissociation of indivisible magnitudes. He rejects the latter idea. But he is full of praise for Democritus's theory which he clearly regards as the best available amongst those of his predecessors. I suggest that these two aspects of Aristotle's position, appearing alongside of each other in the chapter in question, are best reconciled by interpreting Aristotle as seeing the merits of Democritus's physical atomism as distinct from the commitment to indivisible magnitudes that he is opposing.

Aristotle complains that the theories of his predecessors fail to get to grips with the range of kinds of change in evidence in the world 'with the single exception of Democritus' (1, 2, 315a, 35). He acknowledges how the atomism of Leucippus and Democritus can explain coming-to-be and passing-away by the association and dissociation of particles with shape and how alteration can come about through changing arrangements of these particles and he hints also at how perception can be explained by reference to their motion. He identifies a distinctive feature of the theory of Leucippus and Democritus to be their identification of the primary 'reals' as *bodies* (1, 2, 315b, 30). Aristotle (1, 2, 316a, 5–14) praises Democritus for arriving at his theory on the basis of an intimate acquaintance with nature rather than on abstract reasoning.

> Lack of experience diminishes our power of taking a comprehensive view of the admitted facts. Hence those who dwell in intimate association with nature and its phenomena grow more and more able to formulate, as the foundations of their theories, principles such as to admit of a wide and coherent development: while those whom devotion to abstract discussion has rendered unobservant of the facts are too ready to dogmatise on the basis of a few observations. The rival treatments of the subject now before us will serve to illustrate how great is the difference between a 'scientific' and a 'dialectical' method of inquiry. For, whereas the Platonists argue that there must be atomic magnitudes 'because otherwise "The Triangle" will be more than one', Democritus would appear to have been convinced by arguments appropriate to the subject. i.e. drawn from the science of nature. Our meaning will become clear as we proceed.

All this makes perfect sense assuming Aristotle understood Democritus's theory to be the physical atomism I have summarised earlier in this chapter. The high standing of Democritus's method is meant to be made clear 'as we proceed'. What follows in Arsitotle's discussion is a critique of ideas of continua made up of indivisible magnitudes and of such continua being divisible through and through. Democritus's physical atoms are not indivisible in the sense presupposed in the discussion, so there is reason to suppose that Aristotle's critique is not directed at Democritus. Further, a key problem Aristotle raises in connection with the possibility of infinite division everywhere is that bodies will be worn away to nothing as a result of it

(1, 2, 316b, 25). This difficulty does not arise in Democritus's physical atomism because atoms cannot be physically divided. The way in which Aristotle winds up the discussion of infinite divisibility and indivisible magnitudes can be read as affirming Democritus's position

> Hence there are both 'association' and 'dissociation', though neither (a) into, and out of, atomic magnitudes (for that involves many impossibilities), nor (b) so that division takes place through and through – for this would have resulted only if point were 'immediately-next' to point: but 'dissociation' takes place into small (i.e. relatively small) parts and 'association' takes place out of relatively small parts.

I believe the context invites an interpretation of 'atomic magnitudes' as 'indivisible magnitudes' and of 'small parts' as 'physical atoms'.

Whilst Aristotle does see merit in Democritus's physical atomism as a response to Parmenides, he does not accept it. Later in *Generation and Corruption* (1, 8, 325b, 34–326b, 6) he offers some criticisms of atomism, mentioning Leucippus and Democritus by name. Aristotle indicates that he will not go into a detailed discussion of the assumption of 'indivisible solids' but will give a short digression on some of the difficulties. But, as we have seen, he has already offered a detailed discussion of atoms interpreted as indivisible magnitudes, suggesting that the latter are not to be identified with the 'indivisible solids' of Democritus's atomism. The details of Aristotle's objections are not my concern here. Suffice it to say that the discussion makes it crystal clear that the 'atoms' being subject to criticism are finite sized bodies of various shapes and sizes which are physically indivisible.

At two locations in *On the Heavens* (3, 4, 303a, 3–24 and 3, 7. 306a, 27–30) Aristotle describes atomism as being in conflict with the mathematical sciences, the first of the passages referred to making it clear that it is Democritean atomism that is in question. Furley (1967, p. 87) takes this as evidence that in these locations at least Aristotle must be taking atomism to involve indivisible magnitudes. 'When Aristotle says that the Democritean theory of indivisible magnitudes is in conflict with mathematics, can he mean anything other than theoretically indivisible magnitudes? How could physical atoms be in conflict with mathematics?' However, both passages make it quite clear that Aristotle cannot be talking about indivisible magnitudes, because he attributes various shapes and sizes to the atoms, and in the second passage, as I have discussed above, he even discusses the consequences of dividing them.

Aristotle does claim, in both passages, that physical atomism is in conflict with the 'mathematical sciences'. Those sciences included geometry, those parts of optics concerned with the passage of rays of light in straight lines and astronomy. Aristotle's view on the mathematical aspects of those theories was that the mathematics is an abstraction from the physical. Only physical bodies have shape, but their shape can be abstracted, in thought, from their physical being. Geometry is the theory of shapes so abstracted. On this view, geometry will clash with atomism insofar as the edges and sides of physical objects will be irregular and different from the continuous straight lines and planes assumed in geometry. Democritus's physical atomism clashes with geometry on this view. Talk of Democritus's theory clashing with geometry need not imply that that theory involved indivisible magnitudes of the kind invoked by some to counter Zeno, as Furley suggests.

There is one further passage from Aristotle that I need to consider. It is the one that fits least well into my view that Aristotle attributed only physical atomism to Democritus. The passage occurs early in the *Physics* (1, 3, 187a, 1).

> Are we then to say that the All is composed of indivisible substances? Some thinkers did, in point of fact, give way to both arguments. To the argument that all things are one if being means one thing, they conceded that not being is; to that from bisection, they yielded by positing atomic magnitudes.

I am unable to contest the view of contemporary scholars that Aristotle is here referring to the atomists. It certainly appears as though Aristotle is representing atomism as a response both to Parmenides and to Zeno, with 'atomic magnitudes' emerging as a response to the latter. I can maintain my thesis against such an implication by insisting that the atomic magnitudes referred to here are physical atoms and that the bisection involved is physical bisection. Physical indivisibility stands in the way of the infinite division that threatens to reduce matter to nothing. We have already seen that Aristotle viewed the paradox of division in that way in *Generation and Corruption*, fearing that, after such division, 'nothing would remain and the body will have passed into what is incorporeal' (1, 2, 316b, 25–27).

2.9 Democritean Atomism: An Appraisal

Leucippus and Democritus shared with their fellow Presocratics the recognition that there is more to the universe than is directly apparent to the senses. A leaf retains its identity as such through a wide variety of visible changes. There is something about the being of an olive seed that accounts for its capacity to grow into an olive tree and something about the being of a loadstone responsible for its magnetic properties, although neither mode of being is revealed directly to the senses. To understand the basic and enduring nature of the world we need to probe behind the evidence of the senses. The idea that the world of experience is to be explained by, or reduced to, some more fundamental reality was one shared by the Presocratic philosophers and their successors. How is such a fundamental account of reality to be arrived at and defended? Part of the answer involves an appeal to reason, the kind of reasoning that the Presocratic philosophers are justly famous for pioneering. But reasoning alone is clearly not enough. Truths about the nature of reality are not logical truths. Some principles were needed to constitute premises to which reasoning could be applied. The source of these principles was, in effect, common sense. Certain ideas having their origin there were taken as self-evident or highly plausible and were elevated to the status of principles. The adequacy of these principles could then be put to the test by drawing their consequences, investigating them for consistency, and comparing them with other claims regarded as self-evident.

Democritean atomism can readily be construed as falling into such a pattern. It assumes as self-evident that fundamentally there is only one kind of being and adds to this the existence of non-being as a consistent notion, and as something that can exist alongside being. These fundamental assumptions were fleshed out by further adding to them certain notions that have their origins in common sense experience of such things as colliding stones. It is presumed that one sample of being cannot co-exist with another and that when two such samples collide they rebound. The aim is to follow through the consequences of these assumptions to account, in general terms, for the various kinds of phenomena and change in evidence in the world.

A key problem for Democritus, and for the Presocratics generally, was posed by the status of the fundamental principles. On what grounds can it be asserted that there is just one kind of being? Other of the Presocratics did not regard this as self-evident. Empedocles, for instance, proposed four elements, air, earth fire and water, as the ultimate constituents of all things. A century later Aristotle was to argue that the being of a subject differs from the being of its predicates and insisted that form, as well as matter, needed to be included in the kinds of things existing in the world. Even if it is accepted that there is just one kind of being, it is not clear that it should take the form that it does in Democritus's theory. For Anaximines, for instance, observable reality comes about as the result of the compressions and rarefactions of a fundamental being that was air-like.

Competing proposals for the ultimate nature of the being underlying the appearances can be assessed by the extent to which they are compatible with, or serve to explain, common phenomena in evidence in the world. The mere existence of variety and change in the world posed a problem for Parmenides, and Democritean atomism was an improvement on it insofar as it could straightforwardly accommodate them, at least in general terms. Nevertheless, there were classes of phenomena that posed problems for Democritean atomism. Gravity and elasticity provide examples. Over two decades of teaching I invited my students to devise a system of Democritean atoms that can mimic the fact that an elastic substance returns to its original size both when stretched and compressed. I never received a satisfactory answer. Biological phenomena pose more serious problems, as does free will. Given the necessity associated with atomic collisions, the motions after collision being determined by those prior to them, there seems to be no room for Democritus to distinguish between an involuntary twitch of my arm and an instance of my purposefully raising it. There was certainly a wide range of phenomena that posed a serious challenge for Democritus. I have already pointed out, in 2.5, that his attempts to rise to the challenge with respect to the functioning of the senses and the origin of the world were highly problematic, and quite unsatisfactory if taken at all seriously. Democritean atomism was not tested against or confirmed by the evidence of the senses in any significant sense and there is even reason to qualify the claim that it could be accommodated to the evidence.

I think it is useful to counter any inclination to marvel at Democtitus's anticipation of the truth that there are indeed atoms in the following way. Why on earth would anyone wish to assent to the idea, then or now, that the world consists of nothing but showers and accumulations of miniature idealised stones? I stress this point to highlight the extent to which our task of tracing the origins of contemporary atomic theory has barely started with Democritus. In so doing, I do not wish to be taken as denigrating the efforts of Democritus and the other Presocratics. They were astute to recognise that there must be more to reality than meets the eye and extraordinarily ingenious in fashioning reasoning as a tool to help them get to the bottom of things. Using principles latent in common sense as a source of principles on which to bring such reasoning to bear was the only option they had.

Democritus exhibits an ambivalent attitude towards the role of the evidence of the senses in his scheme of things, but it is one that can be explained. He compares 'legitimate' reason to 'bastard' sensory evidence (Kirk et al. 1999, p. 412) and 'sometimes does away with what appears to the senses, and says that none of these appear according to the truth but only according to opinion' (Kirk et al. 1999, p. 410). On the other hand, Democritus has the senses warn, 'wretched mind, do you take your assurances from us and then overthrow us? Our overthrow is your downfall' (Kirk et al. 1999, p. 412). A negative assessment of the senses is warranted by the recognition that they at best yield only a superficial knowledge of the surface of things rather than knowledge of an underlying reality. Also, there are plenty of examples of their unreliability. What seems hot to one observer is only warm to another, what is sweet to one is bitter to another and so on. On the other hand, these very facts about the limits of evidence of the senses are based on other evidence provided by them. The fact that a loadstone has capabilities not revealed by its superficial appearance is itself evidenced by observations of its adherence to metals and its propensity to align north-south. The variability of sensory evidence from person to person is itself something that our sensory judgements give evidence for, whilst the faulty nature of our judgement of the shape of a distant tower is established by more reliable close-up observations. The details of our sensory reports on various instances of change may be limited or mistaken, but we can hardly be mistaken about the fact that there is change. Understood in this way, a critical attitude to the evidence of the senses can be maintained whilst recognising that they nevertheless yield knowledge of general features of the world that provides some testing ground for the consequences of atomic (and other) theories.

Given the resources available to them, coming up with significant truths about the structure of material reality was beyond the Presocratics. The most impressive progress made by a reasoned critique and extension of common sense came in areas connected with mathematics. Those developments had implications for theories about reality even though they did not constitute such theories. The Pythagorians established that, if there are material squares, their diagonals are incommensurable with their sides. Zeno showed that any theory that construes a phenomenon as continuous faces the challenge of developing an account of that continuum that avoids his paradoxes. As we have see, some saw it necessary to introduce atomic space and time to avoid his conclusions, although I have argued that Democritus should not be included among their number. Some of the conundrums raised by the Presocratics have ramifications to this day. It was well into the nineteenth century before a theory of the continuum up to the task of helping to dispose of Zeno's paradoxes was devised. The understandable failure of the Presocratics to come up with defendable claims about the deep structure of material reality need not stand in the way of a sense of awe at the magnitude of their achievements.

Notes

1. Kirk et al. (1999) contains translations of source material bearing on the views of Parmenides along with comments on their main doctrines and arguments as well as references to some of the disputes amongst contemporary scholars concerning how the texts are to be interpreted. The modification of Parmenides' views by Melisus are similarly treated. All my quotations from the Presocratics are from this source. A meticulous analysis of the logic of Parmenides' arguments is to be found in Barnes (1996, pp. 155–230).
2. It is difficult to separate the contributions of Leucippus and Democritus given the scarcity of available sources. It seems that the basic atomistic response to Parmenides was initiated by Leucippus and that Democritus added detail. For convenience I do an injustice to Leucippus and frequently refer to this first version of Ancient Greek atomism as Democritean Atomism.
3. This issue is discussed in more detail in Chalmers (1997).
4. The quotations come from P. Potter (1988, p. 17) and W. S. Hett (1936, p. 357) respectively.
5. Aristotle's formulation and discussion of Zeno's four paradoxes of motion are in Book 6 of the *Physics*, especially Chapters 2 and 8–10. There is a detailed discussion of atomism as a response to Zeno and Parmenides in the context of a general discussion of change in *Generation and Corruption*, 1, 2, 315a, 26–317a 32. Later in that work, 324b, 25–325b, 37, there is an outline of Democritean atomism as a response to Parmenides and also, 326a, 1–326b, 6, a critique of the notion of an atom. *On the Heavens* (1, 4, 271b, 9–11, 3, 3, 303a, 4–303b, 7 and 3, 7, 306a, 19–306b, 2), contains a few remarks about the conflict (or alleged conflict) between indivisible atoms and mathematics, presumably the infinite divisibility attributed to geometrical extension. Quotations from Aristotle throughout this book are taken from McKean (1968) unless otherwise specified.
6. I have a favourite interpretation of these paradoxes, based on remarks drawn form Brumbaugh (1964, pp. 64–67) which explains why there are four of them. I cannot defend a claim that this interpretation is what Zeno intended because we lack relevant access to what he actually said, and I am at a loss to answer the objection that no ancient commentator on Zeno ever said anything that supports my contention. The interpretation construes each of the paradoxes as ruling out one of the four possible combinations of assuming space and time to be either infinitely divisible or atomic. The argument from dichotomy rules out continuous space and atomic time. It assumes that the remaining distance to the door can always be further divided, and yields an infinite time to reach the door because each of the infinite number of times taken to cover the infinite number of distances must be each at least an atom of time long. Achilles and the tortoise rules out continuous time and atomic distance. It always takes Achilles a finite time, no matter how small, to reach where the tortoise was, and in that time the tortoise must have moved at least one atom of space. I have already interpreted the paradox involving moving carriages as ruling out atomic space in conjunction with atomic time. The moving arrow rules out continuous space and continuous time insofar as it rules out the possibility of the arrow moving to the *next* point in the *next* moment of time, because for a perfect continuum there is no next point, as Aristotle made clear. If there were a next point (spatially or temporarily) then in conjunction with the original point it would define an interval (of space or time) that could be infinitely divided.
7. Aristotle seems to interpret the paradox in that way in *Generation and Corruption*, 1, 2, 316b, 23–28.
8. Furley (1967, p. 4) distinguishes between physical division and 'theoretical division', where an object is theoretically divisible 'if parts can be distinguished within it by the mind'. Sorabji (1983, p. 352) makes the distinction in terms of two kinds of barrier to division, physical barriers and conceptual barriers. Pyle (1995, p. 20) distinguishes between physical division and division 'in thought' and Barnes (1996, p. 353) distinguishes between 'division by the axe' and 'division by the mind'.
9. My interpretation of Aristotle's position is in line with that of Lear (1979–80 and 1982).

Chapter 3
How does Epicurus's Garden Grow?

Abstract Epicurus proposed a physical atomism that was a modification of that of Democritus in the light of criticisms of the latter that Epicurus read into Aristotle. One of those modifications involved an emphasis on the priority of evidence provided by the senses. Epicurus rejected scepticism with respect to that evidence that he saw the Democriteans as encouraging. However, it is not the case that the atomism constructed by Epicurus was defended empirically in a way that surpassed what Democritus had accomplished. Epicurus was intent on developing an atomic theory that responded to Zeno as well as Parmenides. He attempted to avoid Zeno's paradoxes by assuming his atoms to be continua composed of indivisible magnitudes. The degree to which he was intent on countering conceptual puzzles connected with the problems posed by continuous magnitudes led to abstractions that were remote from anything empirically testable. Epicurus was led to a picture of atoms all falling in the infinite void at the same speed, one indivisible magnitude of space in one indivisible magnitude of time. He needed to modify this picture in a contrived way to accommodate the bulk of observable phenomena such as gravity. It is arguable that even billiard-ball collisions became a conceptual problem for Epicurus.

3.1 Epicureanism

In the late fourth and early third centuries BC the atomic theory of Democritus was modified by Epicurus in an attempt to protect it from criticism emanating largely from Aristotle. For Epicurus, a defence of atomism was a means to an end. He hoped to promote individual happiness by advocating a simple life amongst friends free of anxieties stemming from religion and fear of death and also from the promotion of unrealisable ideals embodied in traditional conceptions of such things as honour, fame romantic love, high culture (and Epicureanism in its modern sense). A universe formed by the chance coming-together of atoms is a world without divine purpose and without life after death. Coming to know it for what it is helps individuals achieve happiness free of anxiety and aspire to aims that are achievable. The Epicureans aimed to practice what they preached in the garden on the outskirts of Athens that was the locus of their living as well as their philosophising.

A. Chalmers, *The Scientist's Atom and the Philosopher's Stone*, Boston Studies in the Philosophy of Science 279, DOI 10.1007/978-90-481-2362-9_3,

Here my focus is not on Epicureanism generally, but on the content of and case made for atomism. Although much of Epicurus's own writing has been lost, what has survived is considerably more plentiful than is the case with Democritus. In addition to what has survived of the writing of Epicurus and his followers there is a detailed outline of the Epicurean philosophy in the form of a poem, *On the nature of things*, by the Roman, Lucretius, written about half a century BC.[1]

Like Democritus, and indeed, like the Ancient Greek philosophers generally, Epicurus takes certain ideas, for example, that there is only one kind of being and that void can exist, elevates them to the status of principles, and explores their consequences with logical rigour. In some respects, this resulted in Epicurean atomism being more removed from the world of common sense than that of Democritus. For example, as we shall see, it led Epicureans to the view that space and time are atomic, with all physical atoms moving at the same speed, covering one atom of space in one atom of time. In another respect, Epicurean atomism was closer to the world of common sense than that of Democritus, at least if we heed what the Epicureans preached. For it challenged those, like Democritus, who cast doubt on the status of the testimony of the senses. For the Epicureans the evidence of the senses has to be taken at its face value. If the senses reveal that worldly objects have colours and degrees of hotness then they really have such properties. There is a tension between the degree of abstraction involved in the basics of Epicurean atomism and the privileged status it recommends for the evidence of the senses, a tension that I will attempt to resolve.

3.2 Physical Atoms in the Void

The central tenet of atomism, that there exist only physical atoms in the void, which emerged in Democritian atomism as a response to Parmenides, re-appears in Epicurean atomism and for the same reason. Atoms are individual and indestructible packets of being with an unchanging shape and size. They, along with void, are the sole constituents of the universe. Their motion in the void accounts for change whilst their physical indivisibility accounts for an enduring and persisting reality through change.

One important development in Epicurean atomism involved a clarification of the notion of vacuum or void made in response to difficulties inherent in the Democritean account of it. Democritus held that void is the absence of atoms. Is this sufficient to make sense of atoms having a place and a motion? When an atom moves from one place to another, it would seem that void is created in the place vacated by the atom and annihilated in its new place, at odds with the basic Parmenidean idea, adopted by Democritus, that something cannot emerge from or dissolve into nothing. Another possibility, that the void flows around a moving atom like water around a fish, attributes to the void a reality it is meant to lack. A third possibility is that both the space vacated and that newly occupied by a moving atom exist permanently, and irrespective of the presence of the atom. This has the implication that two entities, body and space, can co-exist in the same place at the same time. Aristotle raised difficulties of this kind in his conceptual critique of the notion of void.[2] As Andrew Pyle points out,[3] Aristotle's critique has most force if the void is conceived of as having some kind of tenuous corporeal existence. The Epicurean response was to articulate the idea of void as an infinite expanse of the 'non-tangible' as opposed to the tangible atoms that occupy various locations in it. The void, as non-tangible, has no material-like or other causal properties and is indifferent to whether it is occupied by atoms or not. The position was succinctly expressed by Sextus Empiricus.

> Therefore, one must grasp that, according to Epicurus, of 'intangible substance', as he calls it, one kind is named 'void', another 'place', and another 'room', the names varying according to the different ways of looking at it, since the same substance when empty of all body is called 'void', when occupied by a body is named 'place', and when bodies roam through it becomes 'room'. But generally it is called 'intangible substance' in Epicurus' school, since it lacks resistant touch. (Long and Sedley, 1999, p. 28)

Physical atoms, then, are physically indivisible, have distinctive shapes and sizes and move in an all-pervasive void, both bringing about change and constituting a substrata that persists through change. In this respect, Epicurus responded to Parmenides in much the same way as Democritus had. However, this is not the end of the story. Epicurus pushed his atomism in novel directions, as we proceed to discuss.

3.3 Atoms and Indivisible Magnitudes

In the previous chapter I indicated how an atomism involving indivisible magnitudes and differing from physical atomism could be seen as arising as a response to Zeno's paradoxes. I was at pains to argue that Democritus did not propose such a version of atomism. Epicurus undoubtedly did, however, and did so in response to Aristotle's commentary on the paradoxes of motion and the paradox of division.

In his discussion of Zeno's paradoxes, Aristotle, quite brilliantly, developed a theoretical account of the continuum, the kind of continuum presupposed in geometry. He concluded that a line cannot be made up of points. A point is distinct from a line precisely because a line is divisible whereas a point is not. A point cannot have a magnitude, for if it did it would be divisible. And since a point cannot have magnitude it cannot have parts or edges. Given all this, it can be argued that points cannot be combined to form a line. Two points cannot be placed edge to edge or part to part to form the beginnings of a line because they do not have parts or edges. Nor can they be placed 'in succession', because two non-identical points define a line, a line which is, as such, infinitely divisible. Two points in succession will have an infinity of points in between them. For Aristotle (*Physics*, 6, 1, 231a, 30ff.), a line can be divided at a point into two segments. That point will be the end point of both of the resulting segments but is a part of each segment only in the sense that it constitutes their boundaries. Lines are not made up of points.

The argument presented here for the impossibility of constructing lines out of indivisible points is taken by Aristotle (*Physics*, 6, 1, 231b, 15–18) to apply to the construction of any continuous magnitude out of indivisible, partless, magnitudes. (Indivisible magnitudes are like points insofar as they have no parts but unlike them insofar as they have magnitude. As we shall see, Epicurus was able to exploit the difference in his response to Aristotle.) Aristotle drew connections between atomic space, time and motion. If space consists of indivisible magnitudes, then time must, also. If time varied continuously and space was atomic, then halving the time taken to traverse an atom of space would result in the passing of half an atom of space, contradicting the assumption of atomic space. Conversely, if space is continuous and time atomic, then whatever distance is covered in an atom of time, half that distance will be covered in less time, contradicting the assumption of atomic time. Aristotle concluded that if space is atomic, motion must occur in jerks, passing whole numbers of units of space at a time, because to pass part of a unit of space is counter to the assumption that space is atomic. If space and time are atomic (and Aristotle has argued that if one is, both are), it can never be said of something that it moves, but only that it has moved. At a particular moment, an object will occupy a definite place and can never be caught in the act of moving to the next place, one unit of space further along. Aristotle regarded the absurdity of these conclusions as arguments against the atomic character of space, time and motion.

One further argument of Aristotle (*Physics*, 6, 2, 233a, 6–7) should be mentioned in this place. The mere existence of any two uniform motions that differ from each other is sufficient to rule out atomic time and distance. Let A be an object moving with the swifter of the two motions and B an object moving with the slower. Suppose that A passes one atom of space in time t. Then B will travel a distance less than an atom of space, contrary to the assumption of atomic space. Suppose that B travel a distance d in an atomic unit of time. Then A will travel distance, d, in less than an atomic unit of time, counter to the assumption of atomic time. Whatever times and distances are taken as atomic, 'the quicker will divide the time and the slower will divide the length'.

As we shall see, Epicurus apparently accepted the force of all of Aristotle's arguments here but adopted, rather than denied, some of what Aristotle considered to be their absurd conclusions.

Epicurus aimed to avoid paradoxes of the kind formulated by Zeno by introducing indivisible magnitudes to block the path to infinite division that was the source of paradox. Epicurean atoms, like their Democritean forerunners, have no physical parts and cannot be physically divided. However, atoms were presumed to have theoretical or conceptual or mathematical parts, and the size of these parts could not fall below a certain minimum size. Atoms are made up of a finite number of metrical minima that are themselves indivisible. It is not difficult to see how minimum parts for atoms leads to minimum parts for space and time in the Epicurean scheme of things. It is difficult to see how an indivisible magnitude situated in a divisible space could fail to have parts corresponding to the parts of the space it occupies. Further, suppose two atoms were to pass each other moving continuously through the void. Then, immediately after the leading minimum part of each atom drew level, their continuous motion would result in part of those leading minima being traversed. But that contradicts the assumption that the minima have no parts. Just as Aristotle had argued, if space is made up of indivisible parts, so is time, with the consequence that motion takes place in jerks. The Epicureans simply accepted that motion takes place in jumps, covering whole numbers of units of space in whole numbers of units of time.[4] At any one time an atom will occupy exactly as many units of space as it contains. If it is moving, then at a later time it will occupy some other set of units of space equal to the number that it contains. An atom can never be caught in the acting of moving between units of space. At the atomic level motion is discontinuous and made up of jumps.

The Epicurean response to Aristotle so far involves simple acceptance of conclusions that Aristotle had found absurd. A second aspect of the Epicurean defence of indivisible parts of atoms found an ingenious way around one of Aristotle's arguments. As we have seen, Aristotle had argued that partless entities cannot be laid side by side to make up a whole because they lack parts or sides. Against this, Epicurus argued that partless entities can be placed in succession, with no other entities of the same kind in between them, to make up a whole, even though they will not be edge to edge or part to part. He appealed to an analogy to make the point that this notion is intelligible. It makes sense to conceive of the least size, the smallest patch on a painting, say, that can be perceptually discriminated by a human. It makes further sense to conceive of the whole painting as composed of a large number of these perceptual minima lying next to each other. But, in the situation so understood, the perceptual minima do not have perceivable edges or parts just because they are perceptual minima. The minima of an atom lie next to each other and make up the whole atom in a way analogous to the way in which the perceptual minima lie next to each other and compose the whole painting. Contrary to Aristotle, this is possible in spite of the fact that there is a sense in which neither the conceptual minima of atoms nor the perceptual minima of paintings have parts or edges.[5] What is more, it is no longer the case that two minima next to each other constitute an interval divisible into an infinite number of further minima, as Aristotle stressed was the case with geometrical points.

The minimum parts of atoms exist only as parts of atoms. They cannot exist independently of the atoms of which they are parts. Atoms owe their physical character to their characteristic shape and size that determine how they can become entangled and how they respond to collisions. Minima have no shape and do not vary in size. They don't even have edges.[6] Lucretius spells out the distinction between the physicality of physically indivisible atoms and the dependent, less than physical, nature of the minima of which they are composed.

> Then again, seeing that there is always a final extremity of that body which is below the threshold of our senses, it is presumably partless and of a minimal nature, and never was or could be separated by itself, since its very existence is as a part of something else: it is one part, the first, and is followed by similar parts in sequence, one after the other, filling out the nature of the body in dense formation. Since these cannot exist by themselves, they must stick together inextricably. Therefore the primary particles [atoms] are solid and uncompounded, being tightly packed conglomerations of minimal parts, not composed by assembling these but rather gaining their strength through being everlastingly uncompounded. Nature is still preventing anything from being prised away or subtracted from them, but preserves them as seeds of things.[7]

David Konstan (1979, pp. 395–407) has further elaborated on the difference between physical atoms and indivisible minima, finding some support for the validity of his suggestions in the Epicurean texts. If atoms are to collide or become entangled, as they must in Epicururus's theory, then they must come into contact. Two atoms in contact will have no void between them. This raises a conceptual problem. What is the difference, for example, between two cubic atoms in contact face to face and a single prismatic atom equal in size to the two cubes combined. The idea that atoms are one and physically indivisible because they contain no void, already found in Democritus, will not do. Konstan suggests that physical atoms have a unity by virtue of having edges. The edges of an atom are composed of the layers of minima comprising its outer surfaces. Two adjacent cubic atoms will involve two adjacent edges whereas the prismatic atom of twice their volume will have no edges in its interior. Whereas physical atoms have edges, indivisible minima do not. Consequently, those minima cannot exist in their own right as physical entities that collide or combine by becoming entangled. This makes sense of Lucretius' insistence, in the passage quoted above, that atoms are not compounded of, or composed by assembling, minima. Rather, they are 'everlastingly uncompounded' in spite of being made up of minima.

The atoms of Democritus, a version of which reappears in Empiricus, were far from what could be observed and lacked many of the properties, such as degree of elasticity, fragility and colour, possessed by objects of experience. But at least they had shape and size and could be imagined as idealised stones. Epicurus went even further beyond the observable. In his resolve to counter Zeno's paradoxes he postulated minima of atoms that did not even have shape or edges and did not vary in size. These indivisible minima, having the size of a minimum of space, totally lacked the physicality of idealised stones.

3.4 Atomic Speeds and Observable Speeds

As Sorabji (1983, p. 349) has observed, the arguments that led Epicurus to his minimal parts were of a conceptual kind, the kind that a philosopher can perform from an armchair. Some further details of Epicurus's position came about in a similar fashion, although, eventually, some selective recourse to experience became necessary.

We have seen in the previous section how Epicurus, in response to Aristotle's discussion of the consequences of Zeno's paradoxes for atomism, came to accept a cinematographic motion for atoms. Atoms jump from one unit of space to the next without ever being part way through the transition. A natural consequence of this, which the Epicureans adopted, is that atoms are forever in motion always at the same speed, covering one unit of space in one unit of time. To assume otherwise involves unpalatable assumptions. One way in which two atoms could differ in speed would be to have the slower one resting for several units of time in one place before moving on, falling behind an atom moving one unit of space each successive unit of time. This raises the question of why an atom should resume its motion after any one number of time units rather than another. Another possible way for atoms to differ in speed would involve, for instance, the faster one jumping more units of space per unit of time than the slower one. This has the unpalatable consequence that an atom can move from one place to another without traversing the places in between. The Epicurean assumption, that atoms invariably move a minimum unit of space each minimum unit of time, avoided these conceptual difficulties.

There is a problem here that even a philosopher in his armchair can appreciate the need to address. The waiter typically approaches his armchair at a speed in excess of that of the philosopher's aging colleague. It is clearly not the case that everything moves at an identical speed. What is more, it is typically the case that motion meets with resistance, as evidenced, for instance, by the failure of a projected ball of paper to complete the journey from armchair to waste paper basket. Epicurus had a response to these problems. For Epicurus, the relationship between the motion of individual atoms and the motion of a macroscopic object composed of atoms was like the relationship between the motion of individual bees and that of the swarm. All atoms composing an object constantly move at one speed, but, as in the case with the bees constituting the swarm, not all in the same direction. The atoms constituting a macroscopic object always move with their one speed but constantly change direction as the result of collisions. The resultant motion of the object is the net sum of the individual atomic motions. There is plenty of scope in this picture for macroscopic objects to move at different speeds.

The way in which Epicurus deals with resistance is quite clearly a response to Aristotle.[8] Aristotle had argued that all motion encounters resistance. According to him, when an object falls through a medium its speed of fall is determined by the magnitude of the weight and the resistance of the medium. If an object falls through two media offering different resistance, the ratio of the speeds of fall will be inversely as the ratio of the two resistances. But, since a void offers zero resistance, the speed of fall of the object in a vacuum will bear no ratio to its speed in a medium. That is, its speed will be infinite. Likewise, if two differing weights were to fall in a void they would experience no resistance and so would fall at the same speed, which, says Aristotle, is impossible. Epicurus agrees with Aristotle that all atoms moving freely in the void, falling or otherwise, will move at the same speed, the one speed with which atoms ceaselessly move. According to Epicurus, this speed will not be infinite, but will be immeasurably fast (Epicurus says unimaginably fast) compared to the motion of the swarm of atoms constituting a macroscopic object. However, the motion of a macroscopic object through a medium will encounter resistance as it collides with the atoms constituting the medium. Epicurus has no reason to challenge Aristotle's claim that the ratio of the speeds with which a given object falls through media of differing resistance varies as the ratio of those resistances.

3.5 Gravity

The Parmenideans aside, all of the Ancient Greek philosophers heeded the observable phenomena insofar as they accepted the need to accommodate the reality of motion and change into their characterisations of reality. Aristotle made the further demand that each characterisation needed to be able to distinguish between natural and forced changes and motions. The distinction made common sense. Some changes and motions occur by nature, that is, they occur by virtue of the nature of the thing that changes. Olive trees naturally grow into olive trees and stones naturally fall to the ground. It is plausible to regard the cause of the motion or change to reside in the nature of the thing that moves or changes. By contrast, casting a stone in the air or crushing an olive seed results in a forced motion, the cause of which is external to, and imposed on, the thing that changes or moves.[9] Aristotle claimed that Democritus could not adequately accommodate the distinction since, for him, all atomic motions are forced, coming about as the result of collisions with other atoms. Aristotle pressed this criticism in the context of gravity, developing a detailed account of his own to which Epicurus felt obliged to respond.

Aristotle was convinced that gravitational phenomena indicate that there is a natural centre, a point towards which heavy objects naturally fall and away from which light objects rise. Since an infinite space cannot have a centre, or provide any other natural place, Aristotle rejected such a notion and developed his theory involving a finite, spherical, earth-centred universe. Of the four elements that Aristotle presumed to be the components of all terrestrial bodies, earth and water were heavy, having a natural motion towards the centre, whilst air and fire were light, moving naturally away from the centre.[10]

Epicurus acknowledged the need to accommodate the gravitational phenomena highlighted by Aristotle. He accepted that there be a preferred direction in which heavy objects fall, but rejected the claim that this implied a centre. Epicurus maintained the infinite void basic to atomism but proposed that this void be not isotropic, but that there be a preferred direction in it. Atoms are presumed to move naturally (fall) in this one direction. What is more they do so at the incredibly fast speed with which all atoms ceaselessly move, one minimum of space in one minimum of time. This move has given Epicurus a direction of fall, but it has not yet yielded anything like the observable phenomena. This basic assumption of Epicurus involves a constant downward rain of atoms all moving parallel and at the same speed. This does not explain why stones fall towards the earth (catching the earth up, as it were, as the atoms composing the stone and the earth plummet downwards in the void) nor does it explain why some bodies fall naturally to earth and others rise above it.

Before he could explain why stones fall to the earth Epicurus needed to explain how there came to be an earth at all. How can atoms moving parallel at the same speed encounter each other to form any kind of macroscopic body? The notorious swerves introduced by Epicurus provided the answer. The direction of the downwards motion of an atom is presumed to be subject to small random deviations to one side. These deviations lead to collisions, and once a cascade of collisions ensues, atoms can become entangled to form macroscopic bodies and eventually, bodies as large as the earth. Details of how one gets from here to a theory of gravity that captures the main observable phenomena are not to be found in the writings of the Epicureans. Perhaps a large body like the Earth encounters much resistance from atoms with which it collides on its rapid descent through the void, whereas stones experience less resistance because they are shielded from these blows by the earth. Rare bodies rise because they are squeezed upwards as a result of denser ones pressing downwards.[11]

Epicurean swerves did not only serve the purpose of engendering atomic collisions. They also had an important bearing on what would now be described as the problem of free will. A universe composed entirely of atoms in the void in which the motion of an atom is a necessary consequence of collisions with other atoms is a completely determined world. How can human choice and responsibility find a place in such a world? Random swerves open up a space for such things by putting an end to total determinism. The swerves do not of themselves explain free will. They are more able to explain the involuntary twitching of my arm than purposeful lifting of it. However, they leave room for free will if, by means of that faculty, we are able to influence some swerves, thereby rendering some of them other than random. This issue is of central importance for Epicurean philosophy generally, which, after all, was designed as a guide to life. But it is not of central relevance to the concerns of this book and I do not discuss it further.

3.6 Explaining the Phenomena by Appeal only to Atoms and Void

Like Democritus, Epicurus is committed to the idea that the universe is made up of nothing other than atoms in the void, and again, like Democritus, he is obliged to show how the variety of phenomena evident in the observable world can be reconciled with his strong claim. There is a sense in which Epicurus made the job harder than it had been for Democritus. In his determination to develop a version of atomism that could dissolve Zeno's paradoxes, Empicurus was led by abstract reasoning to notions such as minima of atoms and of space and time, and a unique, cinematographic motion for atoms that far exceeds the speed of any observable body. Again like Democritus, Epicurus needed to show, at least in a notional way, how common phenomena such as those associated with gravity and properties of bodies such as their colour and degree of hotness could be traced back to atomic mechanisms. But for Epicurus, atomic collisions were themselves phenomena that needed explaining!

I suggested in the previous chapter that Democritean atomism incorporated the idea that atoms collide and rebound in a way akin to idealised stones. I do not suggest that Democritus was in possession of laws of motion that would allow him to deal with collisions in an exact way. He clearly was not. But the notion of necessity employed by Democritus, that implied that the motion of an atom is a necessary result of collisions with other atoms, can reasonably be assumed to incorporate the idea that the motion of atoms after collision is a necessary consequence of their shapes and sizes together with their speeds and directions of motion prior to collision. Epicurus may well have presumed that this is the case for colliding stones, but he could not assume it for colliding atoms. A crucial difference is that atoms move always at the same speed, before and after collision. Even more drastically, there is a sense in which atoms cannot collide in the common sense at all. When the edges of two atoms on a collision course reach adjacent minima of space then at least one of them must change direction. Atoms cannot collide in a sense that requires their edges to be in the same place.[12] Not only the collisions of stones, but also the collisions of atoms that an atomist presumes to underlie them, need some explaining in Epicurus's theory.

There is a strong sense in which, for an atomist like Epicurus, all properties of observable bodies need to be reduced to, or explained by reference to, the properties of atoms in the void. The properties attributed to atoms by Epicurus overlap with, but are not identical to, those attributed to atoms by Democritus. Epicurus assumed each atom to have an unchanging shape and size. It is also safe to assume that he attributed weight to atoms in proportion to their size, where weight is interpreted in the sense of the unwieldiness that I attributed to Democritean atoms in the previous chapter.[13] In addition, we have the one, unique and ceaseless, motion attributed to atoms by Epicurus, one unit of space each unit of time. Finally, we have gravitational weight, that property of an atom that urges it downwards.

The main argument for Epicurean atomism as opposed to the Democritean version is that the former offers a solution to Zeno's paradoxes whereas the latter does not. Once we move beyond the generality and degree of abstractness of the responses to Parmenides and Zeno, and look for a case for Epicurean atomism in terms of its explanatory power, then there is not much to be found. The Epicurean explanation of observable gravitational effects is totally unclear as far as the details are concerned, and makes no clear advance on Democritus, and anyone inclined to make a case that Epicurus seriously grappled with the problem must face up to the fact that Epicurus still worked with a flat-earth theory at a time when the case for a spherical one was strong enough to convince most of his contemporary philosophers. The Epicurean accounts of the formation of the world through the chance meeting of atoms or of the atomic mechanisms underlying the senses, were as promissory, and as unsatisfactory once pushed, as they were in Democritus. As mentioned above, even the collisions of macroscopic bodies became a phenomenon wanting a detailed atomic explanation in Epicurus's theory. Most of the problems that confronted Epicurus, over and above those that confronted Democritus, were occasioned by the introduction of minimum units of space and time as a response to Zeno. If I am right that Democritus did not attempt to adapt his atomism in response to Zeno then perhaps this is part of what earned him the nickname 'wisdom'.

Both Democritus and Empiricus needed to accommodate into their atomism properties of macroscopic bodies such as colour and degree of hotness, which are not properties possessed by individual atoms. As we have seen, Democritus denied that such properties exist in the world in reality. He interpreted them as impositions made on the world by us. Epicurus disagreed. For him, weights, colours, degrees of hotness and other properties predicated of macroscopic bodies are existent things. In fact they exist just as the senses reveal them to exist, although their existence, real as it is, must be distinguished from the mode of existence of the atoms which make up the macroscopic bodies possessing the properties in question.

There is much to be said for the claim that many properties of macroscopic bodies are real and not observer-dependent, notwithstanding the fact that those properties can be explained as resulting from the properties of the atoms composing them. Density provides a straightforward example. From an atomist's point of view wood is less dense than water because the ratio of space occupied by atoms to space occupied by void is less for the former than it is for the latter. Density, understood in this way, is not a property of atoms, but nor is it observer-dependent. Wood floated in water long before there were any humans around to observe it do so. A passage in Lucretius explicitly makes this kind of point with respect to the property that we would refer to as viscosity. With whatever facility wine flows through a sieve, 'olive oil, by contrast, is sluggish and takes its time', a fact that Lucretius presumes is to be explained in terms of the sizes and degree of entanglement of the component atoms of wine and oil respectively.[14] There is nothing here to motivate the view that viscosity is observer-dependent.

The point about the objective existence of properties of bodies is a valid one that is not threatened by the fact that those properties may arise from the atomic arrangements responsible for them. Epicurus and Lucretius do not make the point as forcefully as they might because of an excessive, and not totally appropriate, emphasis on colour. Whilst the properties of bodies responsible for the ways in which they reflect and transmit light are objective, there are grounds for saying that in some strong sense colour sensations are observer-dependent. It may well be that there is some sense in which there would be no colours, nor tastes nor smells, in a world without observers.

3.7 The Status and Role of the Evidence of the Senses

Epicurus was one of the majority of philosophers of his time who rejected total scepticism with respect to reliance on evidence provided by the senses which some followers of Democritus read into the latter's comparison of legitimate reason with the bastard senses.[15] The case for total scepticism with respect to the senses was incoherent insofar as evidence for their fallibility appealed to evidence provided by them, and, in any case, such total scepticism is pragmatically impossible as far as the practice of everyday life or philosophy is concerned. Attempts to give a positive account of sensory evidence that was philosophically respectable faced tricky problems that were tackled with considerable subtlety. Once some fallibility of the senses is recognised, some criterion would seem to be required to make it possible to discriminate between true and false testimony of the senses, but the nature and status of any such a criterion is problematic insofar as any attempt to justify it by appeal to sensory evidence would be question-begging. Another general philosophical problem associated with the senses is the fact that any instance of a claim justified by appeal to them seems to presuppose knowledge that goes beyond the evidence provided by the senses in that instance. For example, the claim that an apple is red seems to presuppose the concept redness with the aid of which it is possible to recognise the apple as red. Epicurus engaged in, and made significant contributions to, the philosophical debates on these issues, although it cannot be said that he arrived at a coherent position that solved all of the problems. The main points that I address concerning the status of Epicurean atomism do not depend on the philosophical subtleties and I will not pursue them.[16]

The view, attributed to Empiricus by Sextus Empiricus, that 'all sensibles are true, and that every impression is the product of something existent and like the thing which moves the sense' is too strong and difficult to defend. However, it is surely appropriate to concede to Empiricus that a wide range of significant truths can be established by a critical use of the senses. Whilst we have good reasons to doubt our judgements about the perceived shape of a distant tower we have no corresponding reasons for doubting the shape perceived in optimum conditions close-up. As far as the evidence bearing on atomism and competing philosophical accounts of reality are concerned, the assertions that the senses were looked on to provide evidence for were very course grained claims such as 'motion and change do occur', 'stones fall unless impeded' and 'the impact of two colliding stones results in a change in the motion of each of them'. There is no sensible reason to doubt that such claims are established by appeal to evidence provided by the senses.

One central argument employed by the Epicureans illustrates how course-grained truths established by the senses were considered adequate for establishing philosophical accounts of the ultimate nature of reality. It goes like this. Without void, motion is impossible. But the senses reveal that motion is possible. Therefore the void exists.[17] The argument is not sound because it does not take into account the possibility of circuital motions or the possibility of contraction and expansion. The Epicureans accepted a Parmenidean notion of being that ruled out these possibilities, but that notion of being was not accepted by opponents of atomism. The argument didn't work, but at least it is an example of the kind of argument that the Epicureans aspired to.

I have questioned the force of one of the key Epicurean arguments for atomism that appealed to evidence provided by the senses. However, let us not loose sight of the fact that the atomists, like all of their contemporary philosophers, made a distinction between the appearances and the reality behind the appearances, between the perceived growth of an olive tree, on the one hand, and the being of an olive tree that gives it its distinctive character as such, on the other. Even if there are no strong arguments from evidence of the senses to knowledge of ultimate reality, perhaps there are such arguments leading to knowledge of the appearances. Epicurus's own discussion of astronomical phenomena, or meteorological phenomena, to use his term, in his *Letter to Pythocles*, highlights a key problem for the claim that knowledge in this domain can be established by appeal to evidence of the senses. Epicurus there lists a number of competing possible explanations for phenomena such as eclipses of the sun and moon, the regular motions of the heavenly bodies, thunder, falling stars and so on. Epicurus repeatedly makes the point that there are a number of possible explanations for these phenomena consistent with the evidence of the senses, so that those who think they can pick out the correct explanations from amongst the possible ones 'have lost track of what it is possible for a man to understand'.[18]

A point that I wish to highlight is that, if it is the case that appeal to the senses is not capable of yielding knowledge serving to explain astronomical and meteorological phenomena, where many of the candidate causes can be observed, it can hardly be hoped that the atomic nature of reality, far removed from what can be observed, can be established by appeal to the senses. This was not Epicurus's view. Near the beginning of the *Letter to Pythocles* Epicurus contrasted the status of the basic claims of his atomism with that of meteorology. According to Epicurus, 'that the totality of things consists of bodies and intangible nature, and that the elements are atomic' are consistent with the observable phenomena in only one way, whereas in the case of meteorology, 'these phenomena admit of several different explanations for their coming to be and several different accounts of their existence which are consistent with our sense perceptions'.[19] Epicurus did not think that he had a method capable of yielding detailed knowledge of particular phenomena, but he did think he had one capable of establishing the truth of atomism.

3.8 Knowledge of Atoms: Getting Closer?

The general strategy exploited by the Epicureans to establish atomism did not differ from that utilised by Leucippus and Democritus to the same end. Certain ideas having plausibility in the domain of common sense were raised to the rank of fundamental principles and their consequences traced by rigorous reasoning. A philosophical account of the ultimate nature of reality developed in this way was subject to two types of constraint, logical consistency and compatibility with course-grained features of the world as revealed by the senses. We saw in the previous chapter that Democritus accepted the Parmenidean idea that there is just one kind of being. His identification of atoms with unchanging portions of being in fact likened these portions to idealised stones, in so far as they could collide and rebound, and to something like idealised hooks and eyes insofar as they could combine. Epicurus too accepted the Parmenidean notion of being and assumed his atoms to be unchanging portions of it. But, as we saw, he also took seriously Zeno's arguments to the effect that motion is impossible because paradoxical. He tried to adapt his atomism to cope with them. In so doing, and by introducing minimum magnitudes, he was led away from some of the intuitions about collisions and entanglement that had lent plausibility to the Democritean picture. To what extent did Epicurus progress beyond Democritus in making a case for the existence of atoms and for knowledge of their properties?

I suggest that there is a strong sense in which Epicurus failed to improve the case for atomism and that the details of the ways in which he failed serve to highlight problematic features of the enterprise. Like Democritus, Epicurus failed to make an adequate rational case for acceptance of the fundamental principles that underlay his system as opposed to other possibilities, such as those accepted by his Aristotelian or Stoic rivals. He also failed to improve on Democritus's far from convincing attempts to suggest atomic mechanisms for common phenomena such as gravity and sense perception. Indeed, his determination to respond to Zeno was to make matters worse in this respect.

Epicurus took the path to Zeno's paradoxes seriously. It makes common sense to claim that a continuous magnitude can be halved and, having been halved, can be halved again and so on. Epicurus blocked the path to the paradoxes that arise from assuming the halving to proceed to infinity by introducing indivisible magnitudes as constituting the parts of atoms and this forced him to adopt the view that the motion of atoms is in jerks, one unit of space per unit of time. This threatened to undermine the basic physical mechanism that had informed Democritean atomism, the collision and rebound of atoms. As we have seen, Epicurean atoms cannot make contact in the common sense because their edges can never occupy the same place, nor can they rebound in the common sense because their speeds can only change direction but not in magnitude as a result of collision. Epicurean atomism could not straightforwardly explain billiard-ball collisions!

The principles governing and the properties of atoms in Democritus and Epicurus involve selective abstractions and idealisations from the world of experience. The solidity of atoms was an idealised version of the solidity of stones. One problem, that arose with Democritus and persisted in Epicurus, concerns the criteria guiding the selection of principles or properties. Suppose the combination of copper and tin to form bronze is taken as paradigmatic of change rather than the collision of stones? Each portion of bronze, however small, has properties characteristic of it that differ from the properties of either of the constituent metals. Where does such a move leave the Parmenidean notion of being? Again, why not take the expansion and contraction of air as a primary mechanism and the impossibility of a vacuum as a fundamental principle, as can be done too such good effect in the context of siphons and lift pumps and so on? A natural response is to suggest that the best fundamental principles are those that can at least accommodate if not predict the widest range of common kinds of phenomena. Epicurus was worse off in this respect than Democritus insofar as the path from atoms to the paradigmatic properties of stones became problematic.

A key and important feature of Ancient atomism, which many then and now see as its main attraction, is the extent to which it dared to speculate about the nature of reality behind the appearances and sought to explain the appearances by reference to that hidden reality. Mythical, religious, teleological explanations of phenomena that appealed to scheming or designing gods were to be replaced by atomistic, causal ones. It is important to get this kind of assessment into perspective by recognising that many of the systems that rivalled Ancient atomism shared this basic feature. Aristotle, for example, wrote of reduction to first principles which 'must be not be derived from one another nor from anything else, while everything has to be derived from them'.[20] Aristotle then proceeds to discuss the various accounts of principles on offer. He does so in terms of contraries, the full and the empty in atomism, the hot and the cold, the wet and the dry underlying his own theory of elements, and so on. He recognises various levels of analysis, with the reduction to principles being the fundamental one. 'Some contraries are more primary than others, and some arise from others – for example sweet and bitter, white and black – whereas the principles must always remain principles'.[21] A general case for the inadequacy and incoherence of appeals to God as designer of the Universe and other teleological notions and the superiority of naturalistic, causal accounts of the functioning of nature of the universe, made in great detail by Lucretius, for example, does not hinge on the further assumption that the causal explanations be atomic ones. Long and Sedley (1999, p. 65) make the point that, as far as his opposition to teleology is concerned, Epicurus saw Aristotle as an ally.

Recognition of the fact that Ancient rivals of the atomists shared their goal of offering accounts of reality behind the appearances that would serve to explain them in a naturalistic, non-teleological way, then, invites the question of how strong the case was for the atomists' account of reality. The account of Epicureanism that I have offered indicates, I suggest, that that case was very weak, and can appear so much weaker once biological phenomena are brought to the fore in a way that I have not done. It is generally recognised that those phenomena find a much more secure home in Aristotle's philosophy than it does in atomism.

Epicurus's detailed defence of the simple life free of unnecessary fears and unrealisable and destructive aims constituted a major and consequential contribution to moral and social philosophy. In his physics he explored the possibility of dissolving Zeno's paradoxes by introducing indivisibles in a way that was ingenious and novel. His defence of the idea that explanation of the appearances by an underlying atomic reality does not render the world as it appears unreal, contrary to what can be read into some of Democritus's utterances, had philosophical merit. But he did not make significant headway in the task of establishing the existence and properties of atoms.

Notes

1. The main sources I have used are Long and Sedley (1999), Inwood and Gerson (1994) and Lucretius (1994).
2. Aristotle, *Physics*, 4, 6, 213b, 30–4, 9, 217b, 28.
3. For a helpful discussion of Aristotle's critique of the void on conceptual grounds see Pyle (1995, pp. 52–56).
4. According to Simplicius, as translated by Long and Sedley (p. 49), the Epicurians 'say that motion, magnitude and time have partless constituents, and that over the whole magnitude composed of partless constituents the moving object moves, but that each of the partless magnitudes contained in it does not move but *has* moved'.
5. The passage from Epicurus's *Letter to Herodotus* (pp. 56–59) that I paraphrase here is translated in Long and Sedley (1999, pp. 39–40).
6. This has been argued in Konstan (1979, pp. 403–407).
7. Lucretius (1994, pp. 599–615), Long and Sedley (1999, p. 40). The insertion of 'atom' in square brackets is my own. Long and Sedley (pp. 43–44) find in Epicurus's assertion that 'the process of combination out of minima with their own motion is an impossibility' an added reason for minima not having an existence separate from the atoms of which they are parts. Minima would be incapable of movement, they allege, for the same reason that Aristotle gave (*Physics*, 6, 10, 240b, 8–241a, 6) for points to have a motion of their own. In order to move from A to B, says Aristotle, a point must pass through a stage where it is part in A and part in B. But a point does not have parts. So it cannot move of itself. It moves only by participation in the motion of a parent body in which it is a point. So, according to Long and Sedlety, with

Epicurian minima. They can move only by participating in the motion of the atoms of which they are parts. This argument overlooks the fact that motion is not continuous for Epicurus. A minimum could move by jumping discontinuously from one minimum of space tc the next just as an atom must. I have stressed, rather, the extent to which minimal parts lack the features (shape, variety of sizes and edges) that enable atoms to play their role as physical entities.

8. *Physics*, 4, 8, 215a, 24–216a, 26.
9. Aristotle spells out this distinction in *Metaphysics*, 5, 4, 105a, 14–20 and *Physics*, I, 9, 192b, 1–38 and, in the context of animals, in *Physics*, 8, 4, 254b, 8–33.
10. Aristotle's position is spelt out in *On the Heavens*, 1, 8, 276a, 22–277a, 26 and 3, 2, 300a 20–23, 3, 302b, 9.
11. The importance of this mechanism in the atomists' conception of weight is stressed by Konstan (1979, Section III).
12. There are a number of conceptual conundrums that follow from the cinematographic account of motion, many of them raised by Sextus Empiricus. What happens for example, when two converging atoms separated by an odd number of space units reach the stage where they are separated by just one unit of space, and why? See Andrew Pyle (1995, p. 33).
13. The situation is complicated in Epicurean atomism because of the peculiar character of atomic collisions. However, the common-sense idea that the results of collision depend on the relative weights (degrees of unwieldiness) of colliding atoms is at least implicit in the writings of the Epicureans. Epicurus, for instance, remarks that an atom will continue to move in one direction at its unique speed ('as fast as thought') 'until it is in collision, either through an external cause or through its own weight in relation to the force of the impacting body' (*Letter to Herodotus*, 62, as translated by Long and Sedley (1999 p. 48)). The dependence of collisions on the unwieldiness of the colliding atoms, implicit here, is more explicit in Lucretius who writes that atoms 'as a result of their frequent, high-speed collisions, – recoil suddenly in opposite directions – not surprisingly, given that they are entirely hard with their solid weights and unobstructed by anything from behind'. (*De Rerum Natura*, pp. 86–88, as translated by Long and Sedley (1999, p. 46)).
14. Lucretius, *De Rerum Natura*, pp. 2, 390–397.
15. For the scepticism endorsed by Metrodorus of Chios and Anaxarchus, both followers of Democritus, see Long and Sedley (1999, p. 14).
16. Long and Sedley (1999, pp. 78–97) present a good and fairly detailed discussion of the relevant Epicurean texts.
17. For a version of this argument see Lucretius, *De Rerum Natura*, pp. 1, 328–345.
18. Epicurus, *Letter to Pythocles*, p. 98, as translated by Inwood and Gerson (1994, p. 27).
19. *Letter to Pythocles*, p. 86, Inwood and Gerson (1994, p. 19).
20. *Physics*, 1, 4, 188a, 27–28.
21. *Physics*, 1, 5, 189a, 18–19.

Chapter 4
Atomism in its Ancient Greek Perspective

Abstract Atomism was by no means the only account of the ultimate structure of matter put forward by the Ancient Greeks. One alternative, that of Aristotle, was the one that eventually won general support in the ensuing centuries leading up to the Scientific Revolution. Aristotle introduced form as a basic constituent of the world in addition to matter, and fashioned a notion of potential in addition to actual being which enabled him to overcome Parmenides' denial of change in a way that differed markedly from that of the atomists. The importance of the innovations of the Greeks for a history of atomism is not confined to their attempts to give accounts of the ultimate structure of matter. They developed knowledge that was much closer to experience, especially in medicine, biology and astronomy as well as in mathematics. Aristotle himself made contributions of that kind as exemplified in his classification of animals based on careful observations of their characteristics and his pioneering attempt to give a theory of the continuum. In his work there were a variety of suggestions that gross properties of substances are due to some underlying granular structure. These were destined to prove more significant as far as the path to atomic theories of the seventeenth century are concerned than the speculations of Democritus and Epicurus.

4.1 Philosophical Atomism Versus Less Ambitious Projects

The Ancient Greek atomists aimed to give an account of the deep structure of reality. The account was intended to apply, not just to a certain domain or a specific subject matter, but to reality in general. What is more, the account was intended to be ultimate in the sense that the principles appealed to were not intended to be in need of explanation at a deeper level. The aim was to give an account of the reality behind the appearances. That reality needed to be itself changeless since an ultimate account of change needed to grasp some persistent reality lying behind the change. Given the extreme nature of the objective that the atomists shared with other philosophers of Ancient Greece it should not be surprising that their bold claims to have reached it could not be adequately defended.

The speculations of the early philosophers, including the atomists, were not confined to those concerning the ultimate make up of reality in general. They often

A. Chalmers, *The Scientist's Atom and the Philosopher's Stone*, Boston Studies in the Philosophy of Science 279, DOI 10.1007/978-90-481-2362-9_4,

engaged in less ambitious projects designed to explain the appearances by invoking structures hidden behind them that were not, nor were they intended to be, ultimate. The wearing away of rocks was explained by assuming that pieces of rock too small to be seen were broken off and the passage of water through apparently solid rock was explained by appeal to invisible pores. The atomist Lucretius invoked such explanations, but so did a range of other philosophers, including Aristotle, as we shall see.[1] The fact that explanations of observable behaviour that appealed to some granular structure behind the appearances was not confined to the atomists should alert us to the danger of too readily identifying explanations of that kind as supporting philosophical atomism. The invisible particles detached from a ring as it wears away differ from the invisible parts detached from a wearing stone insofar as the former are particles of metal and the latter particles of rock. The explanations, if granted, fall short of explaining change by invoking unchanging portions of universal being in the void. The stories invoked by Lucretius involving washing drying on the line, metal rings wearing away in the passage of time and water seeping through invisible pores in rocks do not provide evidence for the ultimate account of reality of the Epicurean kind that his atomism was intended to provide. At best they provide analogies or conceptual possibilities designed to lend plausibility to the genuinely atomic, ultimate, mechanisms involved in philosophical atomism.

Invoking invisible granular structures to explain the appearances did not provide answers to the questions about the ultimate nature of reality that was of central concern to the Ancient Greek philosophers, then. Nor were such attempts peculiar to the atomists. Questions of the ultimate structure of reality apart, there remains the question of the status of hypotheses about structures behind the appearances. One kind of problem that stood in the way of making a case for them stemmed from the range of possibilities for explaining change touted by the Greek philosophers. Some involved rearrangements of particles of solid being, favoured by the atomists, others involved compression and rarefaction of a continuous medium as favoured by Anaximines, and yet others involved the fusing of one material with another to form a qualitatively new one, as copper combines with tin to form bronze. A further problem concerned the paucity of evidence for the hidden structures that, after all, did lie *behind* the appearances. The fact that some hypothesised hidden structure was a possible explanation of a phenomenon was insufficient to establish that it was the correct one. The ways in which these problems were eventually surmounted is what this book is about. At this stage of my story I stress that they were not solved by Democritus or Epicurus.

There was more to the pursuit of knowledge by the Ancient Greeks than a search for the ultimate structure, or even deep structure, of reality. Less ambitious objectives were involved, for example, in much of Hippocratic medicine, Ptolemaic astronomy, Euclidean geometry and craft skills such as metallurgy. Progress in those areas had to do with such features as the limited scope of the objectives involved, the availability and possibility of empirical support and the degree to which the resources of mathematics could be exploited.

The attempt by the atomists to explain material reality by appeal to nothing other than unchanging pieces of Parmenidean being in the void fell far short of winning universal appeal. As a matter of historical fact, neither Democritean nor Epicurean atomism won many adherents prior to their revival, albeit in a modified form, in the seventeenth century. The very different philosophical system constructed by Aristotle turned out to have more general appeal, to be more adaptable to a range of purposes and to be of greater historical significance than ancient atomism.

4.2 The Aristotelian Alternative

A basic problem with Greek atomism was the difficulty of reconciling the stark picture of inert lumps of matter, possessing shape and size only and moving chaotically in the void, with the range of activity and orderly behaviour characteristic of the observable world. Water flows and evaporates, members of living species develop, behave and reproduce in characteristic ways, foods and poisons reliably produce their effects, the sun regularly repeats its annual motions and radiates, and so on. Aristotle invoked form, in addition to the matter with which it was necessarily conjoined, to accommodate such order and activity. An entity in the world (other than God, the prime mover, who is pure form) is a complex of matter and form, with form being responsible for making that entity the kind of thing that it is and as responsible for its mode of activity. Aristotle adapted his notion of form to serve a number of overlapping functions resulting in a complex, detailed and integrated view of the world. Some general features of his system that bear on our history of atomism are summarised in the remainder of this section. In the following section I focus on some details that have a direct bearing on speculations about the granular structure of matter that were developed in medieval Europe.

Some general features of commonly occurring changes motivated Aristotle to introduce a distinction between two kinds of being, actual and potential being.[2] It is this distinction that gives him the scope to respond to Parmenides' denial of the possibility of change in a way that is quite different to that of Democritus. The distinction makes common sense in the context of a variety of natural changes, such as the melting of a block of ice or the growth of an olive seed into an olive tree. Part of what it is to be a block of ice is to have the capacity to melt at normal temperatures. That is one of the characteristics that render a block of ice distinct from a block of glass of similar size and shape. In changing to water the ice realises a potential that it possessed all along. In a similar way, an olive seed possesses the potential to become an olive tree and realises that potential when it does so. A crucial aspect of the being of an entity, that which makes it the kind of entity that it is, consists in the ways in which it is capable of acting and reacting and in what it is capable of doing or becoming. Parmenides' claim that change involves something coming from nothing is thereby confounded, Change involves actual existence coming from potential existence rather than something coming from nothing. Both the water and the olive tree are the actualisation of potentials that existed all along in the ice and olive tree respectively. (Aristotle exploited the distinction in his attempt to dissolve Zeno's paradoxes, arguing that a divisible magnitude is, in a sense, potentially but not actually infinitely divisible.)

When an olive seed grows into an olive tree a potential that existed in the seed all along is actualised. Such a change is an example of what Aristotle regarded as a natural change, a change that comes about as a result of the nature of the changing entity. Contrasted with this are forced or violent changes, such as the crushing of an olive seed into a powder. The growth of members of animal and plant species, the falling of a stone and the motion of a planet are all examples of natural changes. As such they are goal-directed. Growth of an organism is towards its maturation as a fully-developed member of the species, the falling of the stone is towards its natural place, the motion of a planet plays its part in maintaining the order of the heavens and so on. It is the form of a natural entity that is responsible for the goal-directed change, since it includes the potential being that makes that entity the kind of natural entity that it is.

Aristotle's notion of form makes most immediate sense in the area of biology, the area to which about one fifth of his writing was devoted and in which he conducted extensive empirical work. In Aristotle's view each member of an animal species is a member of that species because it possesses the appropriate form. It is the form that is responsible for the characteristic features and modes of behaviour and ability to mate with other members of the species and the characteristic progression of a member of a species from foetus to adulthood, The mother supplies the matter, the material cause, of an offspring whilst the father implants, or is the efficient cause of, the form. The form is the characteristic pattern, the formal cause, that renders the offspring a member of the species and it is the potentialities embodied in that form that guide that member towards its goal, that of a fully developed member of the species, the final cause. The presence or absence of the form is precisely what distinguishes a live member of a species from a dead one.

Forms, for Aristotle, have an objective existence, in conjunction with matter, as fundamental components of entities in the world. Aristotle inherited the notion of form from his teacher, Plato, but transformed it by removing Plato's idealism. Plato postulated an ideal realm populated by perfect forms such as the absolute good and the perfect triangle. Material triangular objects and human aspirations to instantiate the good are at best only rough copies of the ideal. In Plato's system, acquaintance with the real world could be attained via sense perception but knowledge of the ideal forms was to be acquired by mental activity, as exemplified in the knowledge of triangles provided by geometry. By contrast, for Aristotle, form exists only in individual worldly entities such as a triangular wooden set-square or an individual olive tree, and any knowledge we have of forms is to be gleaned from acquaintance with those individual entities.

All members of a species are, in a key sense, alike, which, for Aristotle, is the case because they possess the form of the species. But each member of a species differs in various ways from any other. No two olive trees are exactly alike. Here Aristotle distinguished between essential and accidental properties of individuals. The essential properties of an olive tree, such as the ability to bear olives, possessed by virtue of the appropriate form, are those that an olive tree must have if it is to be an olive tree. Other properties, such as a particular size or distribution of branches, are accidental ones.

The ability of a horse to contribute to the generation of horses as offspring, and the particular colour of a horse, are alike attributable to form in Aristotle's system. It would appear that there must be a range of forms arranged in some kind of hierarchy. A horse is a member of its species and behaves accordingly, by virtue of its possession of the form. The horse is more than an assembly of head, legs and body since a live horse needs to be distinguished from a dead one. The leg of a horse is such by virtue of the role it plays in the functioning of the horse as a whole. But it is made up of flesh, blood and bone. Each of these materials is, for Aristotle, made up of the four elements, air, earth, fire and water. For Aristotle, a particle of flesh must possess the form of flesh that serves to distinguish the flesh from a mere mixture of the elements. The elements themselves are composite, being reducible to pairings of the primary forms, the hot and the cold and the wet and the dry (water, for instance, being wet and cold). That is, the elements are prime matter plus form. So we have the form corresponding to horseness and the forms corresponding to the hot and cold and wet and dry and a range of forms in between, such as those of flesh and bone. The precise relationship holding between the various forms and, indeed, the question of whether they could co-exist was an issue on which Aristotle's writings are ambiguous. This provided room for disputes among his medieval commentators as we shall see.

All entities on earth are composed of the four elements air, earth, fire and water which are assigned natural places in Aristotle's earth-centred universe, earth at the centre, water above it, then air, then fire. The heavenly bodies from the moon outwards are made of a fifth element, aether. The earthly elements move by nature towards their natural place and aether moves by nature in circles centred on the earth. The four earthly elements perhaps map onto the three states of matter, solid, liquid and gas, with the addition of fire as differing from all three. But this was not the only grounds for Aristotle settling on the four elements. As indicated in the previous paragraph, the elements were themselves made up of combinations of the hot or the cold with the wet or the dry. As such, the elements can be transformed into one another. Water, the cold and the wet, can be transformed into air, the hot and the wet, by heating. I have stressed that Aristotle's theory was aimed at comprehending natural changes. There is plausibility in the claim that natural changes in the body of the earth are brought about through heating and cooling, and through wetting and drying, that is, between interactions between the hot and cold and the wet and dry. Aristotle suggested that other pairs of tangible properties, such as the hard and the soft and the viscous and brittle, could be reduced to, or explained by reference to, what he considered to be the primary ones, the hot and the cold, the wet and the dry (*Generation and Corruption*, 2, 1, 329a, 24–22, 330a, 29). (It wasn't only the atomists that were reductionists.)

For Aristotle, nature is an organised, integrated, reproducing whole. Asking questions of 'the why' of something calls for an answer that spells out how that something fits into the general scheme of things. The species are arranged in a hierarchy that reflects their mutual dependencies. They survive in an environment that has earth beneath, oceans on the surface and air above. The annual motion of the sun around the ecliptic gives rise to the seasons and so on. Aristotle sought to understand how things behave by nature, when left to themselves, thereby contributing to the overall scheme of things. Natural behaviour is distinct from the mode of behaviour that results from interference in the natural order, such as the crushing of a seed or the throwing of a stone. For this reason, technical knowledge of how the world behaves when interfered with, such as knowledge of projectile motion, levers or the working of metals, was a distinct, and lesser kind of knowledge.[3] Some of Aristotle's commentators interpreted him as holding that experiments that involved the disruption of the natural course of nature could not play a significant role in understanding it. However, some of Aristotle's more empirically orientated work suggests otherwise. The debate has an important bearing on our story and is discussed in some detail in the next chapter.

Aristotle was more concerned with the detailed comprehension of empirical phenomena than either Plato or the atomists. Against Plato, he insisted that forms exist only in concrete individuals and can be studied only by a scrutiny of those individuals. Knowledge is obtained by appropriate abstractions from what is observed, but the more abstract and universal it is, the more difficult it is to substantiate because further from the evidence of the senses (*Metaphysics*, 2, 2, 982a, 24). Aristotle's basic insistence on the world being explicable in terms of matter and form, the notion of prime matter in which forms inhere, the way in which the multiple uses of form fit together into a coherent system and his distinction between actual and potential being, were fundamentals of his system which were far from being directly observable, although they could be rendered plausible by analogy with observable features of the world. Aristotle could not establish the truth of the fundamentals of his theory involving matter and form any more than the atomists could. Like the latter, he could make a case for them only to the extent that they could make sense of, by accommodating, a wide range of observable features of the world in a way that was superior to other systems.

If Aristotle's various appeals to form are interpreted as serious attempts to offer explanations of specific observable phenomena then they can be criticised for being gratuitously fanciful, unsupported by evidence or circular and empty. Construing the male as the provider of form in procreation was totally unsupported by evidence whilst attributing the behaviour of horses to their possession of the horse form can appear circular. Aristotle's appeal to form makes most sense in the context of the problem highlighted by the Presocratics of characterising the ultimate nature of reality behind the appearances. Aristotle's forms made it possible to depict how ultimate reality could change in a way quite different from that of the atomists. At that level of abstraction there is a philosophical problem that is still with us. Contemporary philosophers are divided on the issue of the status of dispositional properties. Some insist that they need to be explained in terms of underlying 'categorical', that is non-dispositional, properties whilst others argue that change can only be adequately accommodated by admitting dispositional properties all the way down. The former have an affinity with the ancient atomists and the latter with Aristotle.

In evaluating speculations of the past we need to be sensitive to the level at which they were intended to operate. Much of what Aristotle claimed to know was based on careful observation and did not abstract or generalise all that far beyond such observation. The identification of species in the biological kingdom and the essential properties of the members of those species was something that needed to be accomplished by careful and extensive observation, as was the identification of the precise motions of the planets around the earth. Aristotle's philosophy left room for, and indeed required, extensive empirical research. It is also the case that pursuit of the various empirical programmes contributed to, and in some cases instigated by, Aristotle could be pursued independently of Aristotle's overall philosophy. So, whilst the scholastic philosophers of the medieval period were to spend much effort fine-tuning and interpreting Aristotle's fundamental philosophy, others were to find in his work a starting point for much more empirically oriented work. In the next section I identify some details of Aristotle's work that were to feed into issues very much related to our history of atomism. Early attempts to construct a matter theory that attributed a granular structure to matter had roots in Aristotle rather than the atomists!

4.3 Hints of a Granular Account of Matter in Aristotle

There are passages in Aristotle that had the potential to be developed into a theory of matter that involved attributing a granular structure to it. Insofar as this is the case, Aristotle sowed seeds for growth of theories that were 'atomic' in some weak sense of the term. His ideas were indeed developed in that direction by some medieval authors, as we shall see in the next chapter. Aristotle identified a special kind of change in a way that can be interpreted as recognition of what we would call chemical change as opposed to mere mixture. Elsewhere, he seems to suggest quite explicitly that substances have minimal parts. In yet another location he explores detailed conjectures about the sub-structure of terrestrial substances to explain their observable properties. Also of relevance to the future development of atomism was Aristotle's denial of the void, which put him at odds at least with the versions of atomism that originated with Democritus and Empiricus.

In *Generation and Corruption*[4] Aristotle classifies the various kinds of transformation that substances can undergo and identifies 'combination' as a significant one distinct from 'substantial change' and from 'alteration'. Alteration is a change in which there is an identifiable substratum that persists. A green leaf turning brown is an example, where the leaf itself is the persisting substratum. By contrast, in substantial change, there is no evident substratum. The decay of a leaf into dust or the evaporation of a puddle provide examples. What Aristotle calls 'combination' is distinct from substantial change and from alteration. The combination of tin and copper to form bronze is an example. Copper and tin are not identifiable as such in bronze, so, in this respect, combination is akin to substantial change. On the other hand, the copper and tin are in a sense in the bronze as components of it because they are recoverable from it, for 'it is evident that the combining constituents not only coalesce, having formerly existed in separation, but also can be separated out from the compound' (*Generation and Corruption*, 1, 10, 327b, 27–29).

'Combination', yielding a compound, is distinct from 'composition' that yields only a mixture. Copper and tin combine to form bronze whilst a mixture of wheat and barley is just that, a mixture of wheat and barley. According to Aristotle, a portion of bronze, however small, is still bronze, whereas a small sample taken from a mixture of wheat and barley may be a grain of wheat or a grain of barley. A compound is homogeneous in a way that a mixture is not. Aristotle used the term 'homoeomerous' to characterise those substances that are continuous 'all the way down' as it were, so that a portion of a homoeomerous substance, however small, remains a portion of that substance. As Aristotle explained in *Generation and Corruption* (1, 10, 328a, 10–14) 'if "combination" has taken place, the compound *must* be uniform in texture throughout, any part of such a compound being the same as the whole, just as any part of water is water; whereas if "combination" is "composition of the small particles", nothing of the kind will happen'.

To a modern reader, it looks as though Aristotle has singled out chemical change as distinct from mixture and has identified the problem of the sense in which elements can be said to exist in compounds, a problem that was not adequately solved until the twentieth century. However, any such interpretation should be approached with restraint. Aristotle's notion of combination involves the idea that constituents in a compound are recoverable. It is this fact that makes combination distinct from substantial change. Compounds from which the constituents could be recovered were few and far between in Aristotle's day and were to remain so for many centuries. Alloys provided him with rare examples of combination (and it is ironic that these are not strictly compounds from a modern point of view). Most of the processes known to Aristotle that we might today classify as 'chemical', involving, for instance, the extraction of metals from their ores, transformations brought about by fire, such as involved in cooking or the manufacture of red ochre from yellow ochre and the absorption of food by the body, were all irreversible as far as Aristotle was concerned, and appropriately classified by him as substantial change.

Aristotle's distinction between combination and mere mixing, where '"combination" is unification of the "combinables" resulting from their "alteration"' (*Generation and Corruption*, 1, 10, 328b, 24), whilst pointing in the direction of a key notion of chemical combination, points away from atomism insofar as it insists that compounds be strictly homoeomerous. In other locations Aristotle wrote in a way that medieval commentators were able to take as conducive to some form of atomism, as described below.

Aristotle entertains the idea that substances have least parts. A key passage comes early in the *Physics*, where Aristotle is surveying the various responses by others to Parmenides' denial of change, prior to offering his own solution. One of those responses was that of Anaxagoras. The latter's theory was, in a sense, the extreme opposite to Parmenides. Whereas for Parmenides being was one, leaving no option for it to change into anything else, for Anaxagoras being was infinitely many. Substances are divisible, in some sense infinitely divisible, and a sample of every substance contains within it portions of every other substance. Substances can be changed into others or abstracted from others because portions of them are there all along, waiting to be extracted. Water can be changed into flesh (as was presumed to take place in the bodies of animals and humans) because flesh particles are present in the water. Samples of flesh and water (and any other pairs of substances) contain an amount of each other, a sample appearing as flesh or water or whatever depending on which particles predominate. Change does not involve something coming from nothing but the extraction of portions already and ever present. A substance, which contains all other substances within it, owes its identity to the dominance or excess of particles of that particular substance.

Aristotle summarises Anaxagoras's position and responds to it as follows:

> According to the theory [of Anaxagoras] all such things are already present in one another and do not come into being but are constituents which are separated out, and a thing receives its designation from its chief constituent. Further, anything may come out of anything – water by segregation from flesh and flesh from water. Hence, since every finite body is exhausted by the repeated extraction of a finite body, it seems obviously to follow that everything cannot subsist in everything else. For let flesh be extracted from water and again more flesh be produced from the remainder by repeating the process of separation: then, even though the quantity separated out will continually decrease, still it will not fall below a certain magnitude. If, therefore, the process comes to an end, everything will not be in everything else (for there will be no flesh in the remaining water); if on the other hand it does not, and further extraction is always possible, there will be an infinite multitude of finite equal particles in a finite quantity – which is impossible. (*Physics*, 1, 4, 187b, 23–35)

Our discussion of Zeno's paradoxes and the response to them by Aristotle and Epicurus indicated that in talking of division it is important to be clear whether it is actual or potential division at issue, a distinction made explicit by Aristotle, and whether it is physical or metrical division at issue, a distinction that Aristotle rarely made explicit but which is necessary to make coherent sense of his arguments. Since Aristotle is discussing processes of transformation that are presumed to take place in nature, the parts or particles invoked in the context of the appraisal of Anaxagoras's theory need to be actual, physical parts and the divisions actual physical divisions. Aristotle made this more or less explicit a few lines prior to the above quotation stipulating that 'by "parts" I mean components into which a whole can be divided and which are actually present in it' (*Physics*, 1, 4, 187b, 15). With this understood, Aristotle is able to rule out the possibility of infinite division because such a division would lead to an actual infinity of magnitudes, all of the same size, at the presumed limit of geometrical division. This is a physical impossibility for Aristotle for a range of reasons which he invokes elsewhere. An actual infinity of parts runs foul of Zeno's paradox of division. It will not be possible for the particles resulting from infinite division to be next to each other or to touch and so on. Given this, physical extraction of flesh from water must come to an end when an actual minimum of flesh is reached which is finite in size but not capable of further physical division.

Aristotle has other arguments for the existence of these 'least parts' of substances. One involves the scale-variance evident in nature (*Generation and Corruption*, 1, 4, 187b, 14–21). The sizes of the members of animal and plant species vary only within limits. This is readily understandable if they are made up from building blocks of a definite size. On this view, making up a minutely small mouse from its component least parts is no more possible than constructing a tiny doll's house from house bricks. By contrast, if members of a species are made up from an infinity of infinitely small parts there is no obvious reason why the results should not show an unlimited variation in size.

Another thought that can be read into Aristotle's discussion of Anaxagoras is that, insofar as combination of substances requires that they be in contact, natural division into parts facilitates such contact (*Generation and Corruption*, 1, 10, 328b, 1–5). Combination can often be facilitated by grinding one or more of the components into a powder.

The 'least parts' of substances foreshadowed here by Aristotle are atoms insofar as they are least parts. But they are importantly different from Democritean atoms in ways that Aristotle alluded to. Copper and bronze may well have least parts, but this need not entail that the least part of copper is a geometrical part of the least part of bronze. For Aristotle, particles of a substance can be altered by being transformed into another substance by way of a change of form, rather than by being physically divided or physically joined to another particle. 'Water and air are, and are generated "from" each other, but not in the way in which bricks come "from" a house and again a house "from" bricks' (*Physics*, 1, 4, 188a, 15–17).[5] Least parts of copper and tin become a least part of bronze not by being adjoined but by being jointly transformed into a single entity, a least part of bronze. We will see in the next chapter how these remarks of Aristotle's were built into the beginnings of a theory of natural minima by some of his medieval successors.

Another source of 'atomistic' ideas that later writers found in Aristotle's writings was his work *Meteorology*. That work, as Aristotle explains at the outset, is concerned with explanation of natural, terrestrial, events that are less orderly than celestial ones. The first of the four books comprising the work is concerned with phenomena assumed to take place in the upper regions of the terrestrial region, from comets and meteors down to clouds and winds. The second book is concerned with phenomena associated with the sea and with the weather whilst the third deals with optical phenomena such as the rainbow and halos around the sun and moon. The fourth book concerns the properties of naturally occurring materials, such as combustibility, fragility, degree of elasticity and viscosity.[6] The discussion of the whole book is notable for its attention to a wealth of empirical data, concerning such things as wind currents, temperature distribution in the sea, and the physical and chemical properties of a range of substances, although the main intent of the book is to incorporate that data into an account of matter based on the principles of change, the hot and the cold and the wet and the dry, together with the four elements, air, earth fire and water, that they inform.

A number of passages lend themselves to development in 'atomistic' or corpuscularian terms insofar as they involve explanations of the properties of macroscopically homogeneous substances by reference to underlying pores and particles.

> A thing is viscous when it is ductile as well as being liquid or soft. And this characteristic belongs to all bodies with interlocking parts, whose composition is like that of chains; for they admit of considerable extension and contraction. Bodies which have not this characteristic are friable. (*Meteorology*, 4, 9, 387a, 11–15)[7]

> Some things are combustible, some incombustible; for example, wood is combustible and wool and bone, while stone and ice are incombustible. All things are combustible which

> have pores which fire can penetrate and which contain in their longitudinal pores too little moisture to overcome the fire. (4, 9, 387a, 17–21)

Aristotle also talks of solubility in water in terms of its having pores able to admit particles of solute (385a, 28–30).

I have already brought attention to the fact that Aristotle insisted that the pores he invoked to explain various properties of materials were not voids. He objected to the void on a range of grounds.[8] Some of the more abstract ones were tied up with the idea that only material bodies (and hence not empty space) can have dimensions. Only individual material objects have shape and size on Aristotle's view. The shapes and sizes can be abstracted, in thought, from their material basis, and theorising about them constitutes the science of geometry.[9] A material cube thus constitutes a space rather than occupies it. For Aristotle, unoccupied space is unintelligible.

Other arguments involve motion. Whereas the atomists had introduced void to make motion possible by giving atoms a space to move in, Aristotle argued almost the reverse. The space in which objects move, for Aristotle, can make room for an object to move in by becoming rarefied, rather than by offering emptiness. In addition, motion in a medium is possible through the circulation of the medium, exemplified in the way that water can circulate in making way for the fish.

In an infinite void such as that posited by the atomists there can be no special places or directions. For Aristotle, this flies in the face of the fact that heavy objects fall towards, and light ones rise away from, the centre of the earth. In an infinite void a moving object would move indefinitely, there being no reason for it to do otherwise, a conclusion that Aristotle clearly regarded as an absurdity. Since, says Aristotle, objects fall at speeds inversely proportional to the resistance offered by the medium, objects falling in a void, which would offer no resistance, could have speeds that bear no possible ratio to their speeds in a medium. That is, they would fall at an infinite speed.

I have identified two themes in Aristotle's writings that promised development in the direction of some kind of atomic theory, the discussion of natural minima in the *Physics* and speculations about granular structure of matter in *Meteorology*. Both themes were developed by medieval authors, as we shall see in the next chapter. However, I have stressed that the materials constituting natural minima and the corpuscles surrounding pores were characteristic of the substances they composed and quite unlike the universal matter of Democritian atomism. Nor did the Arsitotelian position require a commitment to void.

4.4 Granular Versus Ultimate Structures

Aristotle's explanations of the phenomena by appeal to structures behind the appearances work at various levels. This can be brought out by considering two differing ways in which he distinguishes between homoeomerous and non-homoeomerous substances. In *Generation and Corruption* homoeomerous substances are construed as absolutely continuous, retaining their identity all the way down, as it were. Substances having an atomic structure are not homoeomerous on this view. By contrast, in *Meteorology*, Aristotle attributes a granular structure to substances he describes as homoeomerous. For instance, in *Meteorology* (4, 10, 388a, 10–20), he gives metals, stone, bone and flesh as examples of homoeomerous substances but also attributes various properties of those substances to the existence of pores and variously shaped corpuscles. Substances that owe their properties to hidden corpuscles and pores cannot be homoeomerous in the strict sense at work in *Generation and Corruption*. What are we to make of this apparent tension?

I suggest the answer lies in the difference between the kinds of project in which Aristotle is engaged in the two works. In *Generation and Corruption* Aristotle is concerned with the Presocratic quest to give an account of the ultimate structure of reality. He outlines the theories of his predecessors on this issue, criticises them and attempts to do better. Since these philosophies are meant to capture the reality behind the appearances they are not of a kind that can be straightforwardly confirmed by appeal to the appearances. The empirical examples invoked by Aristotle should be interpreted as illustrating conceptual possibilities or as offering analogies rather than as providing empirical evidence. This is not the case for Aristotle's project in *Meteorology*, where the granular structures invoked by him are clearly intended to be taken literally as proposed explanations of various empirical phenomena. Neither Aristotle's designation of materials as 'homoeomerous' nor his suggested explanations of their properties by appeal to particles and pores in *Meteorology* are intended to be ultimate in the way that the matter theories considered in *Generation and Corruption* were.[10]

Aristotle's speculations about microstructure in *Meteorology* are reductionist in a rough and ready sense. The combustibility of wool and bone is explained by proposing that those two substances contain pores that allow fire to penetrate and which do not contain sufficient moisture to extinguish the fire (4, 9, 387a, 18–22). The reductions involved are not ultimate reductions. The material that surrounds the pores in wool, on Aristotle's account, is not itself wool because it is not combustible like wool is. But nor is it akin to the universal matter of Democritus's theory because it must differ from the material that separates the pores in bone. Wool is a structure of woolstuff separated by pores and bone is a structure of bonestuff separated by pores. Nor should the pores themselves be interpreted as Democritian void. Aristotle (386b, 5) makes it quite clear that the pores are full, but full of some substance other than the material they are pores in.

Because Aristotle's speculations about granular structure were not ultimate explanations they can plausibly be construed as conjectures more open to empirical investigation than theories about the ultimate structure of matter. However, this aspect of *Meteorology* should not be exaggerated. Aristotle's speculations were much conditioned by his desire to reduce terrestrial phenomena to the action of the four elements, these in turn owing their character to the hot and the cold and the wet and the dry. In the next chapter we will be taking a close and critical look at the way in which Aristotle's conjectures were exploited and extended in the Middle Ages, especially by alchemists.

The distinction I have introduced in the context of an interpretation of Aristotle's works, between ultimate matter theory and non-ultimate granular or corpuscular structures, will prove to be an important one for understanding the history of atomism. As I indicated in the opening section of this chapter, the distinction can usefully be invoked in the context of Greek atomism also. Explanations of clothes drying on the line and metal rings wearing away, invoked by Lucretius, had a plausibility that was quite independent of the quest to reduce material reality to portions of universal being possessing only shape and size.

4.5 Greek 'Science'

Parmenides, Democritus, Plato and Aristotle were typical Ancient Greek philosophers insofar as they sought to characterise the ultimate nature of reality and it should not be an occasion for surprise to note that they lacked the resources to make significant progress. But ultimate knowledge of the kind sought by philosophers was not the only kind of knowledge constructed by the Ancient Greeks. They made progress in constructing and defending a wide range of knowledge claims that I will categorise under the title 'science', without intending to imply too much by that term other than to signify that it was a kind of knowledge distinct from, and less ambitious than, philosophy and on somewhat more secure ground.

Some Greek science was firmly grounded in observation and amounted to little more than generalisations from it, which is not to say that it lacked discernment or sophistication. Some of this knowledge was established by way of experiment, that is, by way of exploratory interventions in nature. Dissections of the kind Aristotle conducted and his tracing of the development of embryos by breaking open birds eggs are examples. Equally impressive were the range of mechanical or pneumatic devices constructed by Hero of Alexandria in the first century AD, such as the flask made to rotate by an escaping jet of steam and the device which opened a temple door as a result of expansion caused by a fire near a hollow alter.[11] Significant progress was made in medicine, including, for instance, the identification of the optic nerve as responsible for carrying optical information from eye to brain and the curative power of drugs extracted from plants, including the recognition of the development of immunities. There was also detailed documentation of astronomical and biological phenomena, knowledge about metallic ores and how to extract metals from them and so on. The vast and impressive array of generalisations based on observation accumulated by the Ancients in the millennium following Parmenides was not totally unproblematic. For instance, Aristotle's beliefs that there is a blood vessel connecting the liver with the right arm and that the heart has three chambers and that comets and meteors are located in the upper atmosphere are indications that establishing significant generalisations by way of observation was not straightforward and free of pitfalls. Nevertheless, there is no doubt that the Ancients were able to make significant progress in establishing a wealth of knowledge through careful observation and generalisations from it.[12]

Another highly significant and impressive component of ancient science took the form of mathematical generalisations involving abstractions from observable data. Euclidean geometry is the classic example. The status accorded to geometry varied from one philosophy to another. Platonists took it as being true of ideal geometrical shapes but not of the shapes of objects in the real world. Aristotle took geometry to be true of the shapes of real physical objects abstracted from their physical nature. For him, spheres, for example, cannot exist in themselves as geometrical objects but only as the shapes of physical orbs or balls or whatever. But the shape can be abstracted, in thought, from the physical being of an object, and geometry constitutes truths about shapes so abstracted. Whatever philosophical nuance is put on the status of geometry, it could be, and was, interpreted as constituting truths based on self evident axioms that could be safely applied in surveying, statics, architecture, optics and astronomy. Problems of philosophical interpretation only came to the fore when geometry was pushed beyond its roots in common observation to the domain of the very large or very small. We have already discussed problems posed by the possibility of infinite division implicit in the geometrical notion of continuity, whilst the question of whether the infinite space of Euclidean geometry corresponds to anything in reality was another source of philosophical dispute.

Ancient mathematical 'science' was a possibility in areas where it was feasible to formulate abstractions that could be plausibly construed as self-evidently true whilst being applicable to the domains they were abstractions from. The adequacy of these sciences became questionable once they were extended in ways that called into question the self-evidence or the empirical applicability. Geometrical optics provides an instructive illustration. Here the necessary conceptual abstraction, initiated by Euclid about 300 BC, came in the form of a ray of light. Some such notion is implied in the idea of a line of sight as indicative of the location of the object sighted. The law of reflection for rays of light was a geometrical formulation whose status as a self-evident truth was reinforced by Hero's demonstration, some three and a half centuries after Euclid, that the path from object to eye via a reflecting surface is a minimum just when the angle of incidence at the reflecting surface equals the angle of reflection. In the second century AD Ptolemy reported experiments designed to illustrate the truth of the law of reflection, but these were probably regarded as about as necessary as the empirical vindication of Pythagoras's theorem by measuring the sides of triangles. The situation is much trickier when it comes to refraction. Ptolemy reported the results of measurements of angles of incidence and refraction too. He made it clear that he thought refraction must be governed by some 'law', but was unable to formulate that law. In the main, his results are in conformity with a law. The angles of refraction that he records for angles of incidence taken in ten degree intervals conform to the stipulation that the second differences between the series of angles of refraction be constant. This law is false, and the departures from it for large angles of incidence can easily be detected by measurements of the kind described by Ptolemy. Ptolemy clearly adjusted his readings so that they conformed to the 'law' he has presupposed. It is quite clear that Ptolemy's experiments were not designed to rigorously test the adequacy of mathematically formulated laws in the way that was to become typical of mathematical physics.[13]

Greek geometrical science was a possibility in areas where appropriate abstractions could plausibly be construed as self-evident and empirically adequate. It ran into trouble when the criterion of self-evidence became problematic, when empirical data failed to conform, or when appropriate abstractions were difficult to come by. The latter problem was evident in the difficulties experienced in mechanics, where the move from statics to dynamics proved problematic.

Ancient empirical and mathematical science can be construed in some sense as abstracted from or based on generalisations from experience. This is not the case with other aspects of Greek science which clearly went beyond observation in postulating hypothetical explanations of them. Examples are the postulation, from Hypocrates on, of the balance of four hypothetical liquids, the humours, to account for the health of the body, the postulation of solid crystalline spheres by Eudoxus and Aristotle to account for the motions of the heavenly bodies and understanding vision in terms of emanations passing from the eye to an observed object. These hypotheses were akin to the claims of abstract philosophy insofar as they purported to explain observable phenomena by reference to what lay beyond or behind them, but were unlike abstract philosophy insofar as the hypotheses were more limited, being directed at explaining a delimited range of phenomena rather than reality in general. Some of the hypotheses invoked were 'atomistic' in some sense of the term, although a preferable term is 'corpuscular' because it is more neutral and does not invite an identification of the particles involved with atoms of the kind specified by Democritus and Epicurus. We have already encountered Aristotle's speculations involving corpuscles in *Meteorology*. A number of authors presumed that corpuscles interspersed with void were necessary to explain the transmission of light through transparent solids, while Hero assumed the same with respect to the compressibility of air. These hypotheses were adequate or inadequate to the extent that they offered convincing explanations of the phenomena but all lacked support insofar as they were not borne out by evidence independent of the phenomena they were designed to explain. Hypotheses about atoms or minute corpuscles in this kind of context were at best unsubstantiated speculations.

Notes

1. For instance, Lucretius (1994, pp. 36–37) describes the wearing away of rings and the seeping of water through rocks in this way and Aristotle appeals to pores and particles of various shapes and sizes to explain a range of phenomena in *Meteorology IV*. Aristotle (*Physics*, 8, 3, 253b, 13–20) also refers to the wearing away of stones by dripping water in terms of the breaking off of imperceptible pieces.
2. *Generation and Corruption*, Book 1, chapter 9. See also *Physics*, Book 3, chapters 1–3.
3. For the distinction between knowledge of how the world behaves naturally and how it behaves when interfered with see Collingwood (1960, pp. 80–92) and the opening of the Aristotlelian work 'Mechanical Problems' translated in Hett (1936, pp. 331–411).
4. The key source is Book 1 of *Generation and Corruption*, especially chapter 10.
5. A similar point is made in *Generation and Corruption*, 1, 9, 327a, 5–25.
6. The suggestion that Book IV of *Meteorology* was not by Aristotle has been discredited. See, for example, D. Furley (1983, pp. 73–93) and W. Newman (2001, p. 307).

7. The translations of Aristotle's *Meteorology* are from Aristotle (1962).
8. For a succinct summary of Aristotle's case against the void see A. J. Pyle (1995, pp. 49–63).
9. See Lear (1982).
10. Newman (2006, p. 25) has stressed the extent to which there is a corpuscular tradition having its roots in Aristotle. He names both *Generation and Corruption* and *Meteorology* as sources of it, referring to those to works as 'more empirically oriented' than *Physics* and *De caelo*. My position entails that it is wrong to identify *Generation and Corruption* as empirically oriented. Doing so blocks the way to my solution of the problem of the differing uses of 'homoeomerous' by Aristotle.
11. See Boas (1949) for an accessible summary of Hero's achievements and influence.
12. A good taste of the extent of the achievements of Greek science, and references for further reading, can be obtained from a glance at the original sources collected in Irby-Massie and Keyser (2002).
13. An instructive analysis of the status of Greek geometrical optics is A. Mark Smith (1981, pp. 73–99) as is his account of Ptolemy's study of the law of refraction in Smith (1982, pp. 221–240). See also my critique of Ptolemy's experiments in Chalmers (1990, pp. 126–132). These works make it clear that it cannot be maintained that the fundamentals of Greek mathematical physics were borne out by experimentally testing them in anything like the modern sense.

Chapter 5
From the Ancient Greeks to the Dawn of Science

Abstract Those scholastics that were hard line Aristotelians were more concerned with the logical cohesion of Aristotle's system than with matching it to the world. More liberal Aristotelians were very concerned with such a match and included experiment as a means to attain it. Alchemy was central to their efforts. In attempting to absorb that practice into their philosophy, Aristotle's minima constituting the limits of division and his speculations in *Meteorology IV* on the granular structure of matter were gradually transformed into least parts of bulk substances pre-existing any division. These trends reached a well-developed state in the natural minima theory of Daniel Sennert. He likened his minima to the atoms of Democritus. However, there were crucial differences. Sennert's minima had properties other than shape and size, characteristic of the substances they were least parts of and, in true Aristotelian spirit, owed those properties to a substantial form. Sennert's atomism was to have a major influence on Robert Boyle.

5.1 Introduction

This chapter concerns developments that took place over a period of two millennia. Consequently, it cannot be considered to be history in any rigorous sense. I draw attention to deliberations and practices engaged in during the period that provided raw materials for seventeenth-century philosophers and scientists to employ and build on without pretending to give a detailed account of the context and circumstances in which those deliberations and practices arose.

As I have already noted, Ancient Greek atomism in the form articulated by Democritus and Epicurus did not attract many adherents prior to its revival in the seventeenth century. It was mainly the philosophical system of Aristotle that was taken over and developed, first by the Arabs, who became the proprietors of Greek philosophy after the fall of the Roman Empire, and then by philosophers in the West as knowledge of Greek philosophy reached Western Europe during the medieval period. From the thirteenth century on Aristotelian philosophy became integrated with Christianity, through the work of the likes of Albert the Great and Thomas Aquinas, to form the orthodoxy against which Renaissance and seventeenth-century philosophers were to increasingly react.

A. Chalmers, *The Scientist's Atom and the Philosopher's Stone*, Boston Studies in the Philosophy of Science 279, DOI 10.1007/978-90-481-2362-9_5,

Democritean and Epicurean atoms were of an extreme kind. They were passive, stone-like entities possessing only shape, size and solidity and were totally permanent and changeless. Atoms were capable of motion in the void, this motion, and the impacts to which it could give rise being the only source of activity in nature. There was little prospect of developing this stark picture to provide convincing explanations of natural phenomena supported by evidence. However, the view that the properties and behaviour of material systems be attributed to the properties and behaviour of the tiny particles of which those systems are composed need not be interpreted in the extreme way involved in Democritean and Epicurean atomism. The tiny particles can be assumed to have a wider and richer array of properties than those allowed by the Greek atomists, and they need not be regarded as themselves changeless. Suppose we take atomism to involve simply the assumption that natural systems and materials are made up of tiny invisible particles possessing properties capable of accounting for the properties and behaviour of the systems and materials of which they are part. Then we find various themes along these lines developed in medieval and Renaissance philosophy that have their origins, not in Democritus or Epicurus, but in the Aristotelian texts that we discussed in the previous chapter.

5.2 Natural Minima

We have seen how Aristotle, in his *Physics*, came to propose that natural materials such as flesh and water have least parts as a response to Anaxagoras's claim that everything exists in everything else in the form of indefinitely divisible particles. Many medieval and early Renaissance philosophers elaborated on these views, sometimes drawing together the relevant passages in the *Physics* and texts drawn from other of Aristotle's works in a way that Aristotle himself did not do. In the main this treatment of natural minima, as the least parts conjectured by Aristotle came to be known, was an exercise in Aristotelian philosophy. The aim was to develop a coherent account of minima that fitted in with Aristotelian philosophy in a way that was free of paradox. It was not motivated by a desire to extend the theory of natural minima towards one better able to guide and be borne out by observation and experiment. The discussion of whether flesh and water have least parts, for instance, was not motivated by an attempt to better understand how water is converted into flesh by the body. We must beware of too readily reading something like the modern atom into the natural minima as characterised by the medieval commentators on Aristotle.

The medieval discussions of natural minima draw on and exploit distinctions that have at least their seeds in Aristotle's writings. One such distinction is the one I have referred to as the distinction between metrical and physical division. More in keeping with the medieval terminology, this distinction can be expressed as one between mathematical division and natural division. Once we follow Aristotle and accept that space and time are continuous, every extended body is infinitely divisible by virtue of being extended. But what of the division of a natural body insofar as it is natural? Natural minima were proposed as a limit to the possibility of natural division of a body of that kind. This is not the same as saying that the minimum is physically indivisible, but it is to say that the minimum, if divided, will no longer be parts of the original natural substance. A minimum of flesh will no longer be flesh if it is divided.

A second, related, distinction is that between actual and potential division. We have seen how Aristotle exploited the distinction to help him circumvent Zeno's paradoxes. A piece of flesh that is perfectly continuous, and so not actually divided, is potentially divisible into parts. The question arises of whether such divisions, if actualised, can proceed indefinitely. The defenders of natural minima followed Aristotle in holding that the division cannot proceed indefinitely without ultimately destroying the flesh as flesh.

When applied to the division of natural bodies, the distinctions between metrical and physical (or natural) division and between actual and potential division lead to considerations that bring out just how inappropriate it is to attribute to the medieval defenders of natural minima the idea that bodies are composed of a collection of natural minima in a way analogous to the modern view that they are made of a collection of atoms. An extended piece of flesh can potentially be divided at any place. What is more, flesh can be regarded as infinitely divisible precisely because it is continuous. Exceedingly small yet potentially divisible parts of flesh in a sense exist as parts of the flesh so long as they remain parts of the whole. However, once natural divisions are actualised, and parts of flesh are severed from the whole, a stage is reached where the part is too small to exist as a natural portion of flesh. Early in the fourteenth century John of Jandum explicitly made the point that a homogeneous natural substance is in a sense infinitely divisible and yet cannot be physically divided into parts having properties characteristic of the whole.

> There is no minimum of magnitude for a continuous natural substance, as long as the parts remain united with the whole; there is no natural minimum for these parts except insofar as they are separated from the whole.[1]

There are hints in Aristotle that a portion of substance cannot be less than that of its natural minimum for otherwise it would not be able to resist the corrupting influence of the surrounding medium. Aristotle had expressed the view that a small quantity of wine added to a large volume of water is transformed into water, whereas a sufficiently large quantity of wine can retain its identity as wine (*Generation and Corruption*, 1, 10, 328a, 27–29). It takes a minimum amount of wine to resist being dominated by and transformed into water. This presumably raises the possibility of the magnitude of the minimum of a substance being relative to the medium by which it is surrounded.

There is no doubt that there was extensive attention paid to the notion of natural minima by the medieval and Renaissance commentators on Aristotle. John Murdoch (2001) lists well over 30 authors who made contributions in that period. Natural minima were incorporated into significant atomic theories of matter in the late sixteenth and early seventeenth centuries, as we shall see. But such moves had not taken place prior to that, and authors such as van Melsen (1960) and Emerton (1984) read much more of the atomistic view into the medieval literature than is warranted as Murdoch (2001, pp. 93–95 and 130) has observed.

5.3 Hard Line Versus Liberal Interpretations of Aristotle

The so-called 'scholastics' who elaborated on Aristotle's philosophy in medieval Europe found the works of Aristotle sufficiently rich and ambiguous to find plenty of ground for dispute over just what that philosophy amounted to. In this section I elaborate on some general issues over which there was disagreement and which have an important bearing on our history of atomism. One of them concerns whether or not more than one form is present in a complex entity such as a horse or a lump of metal. Another concerns the related issue of the extent to which nature can be appropriately understood by way of artificial interventions in it. I will distinguish what I will call a hard line approach, typified by the philosophy of Thomas Aquinas, and more liberal approaches. It was the latter that offered scope for developments able to accommodate atomic or corpuscular theories of matter.

According to the hard line stance on form, natural entities in the world are what they are by virtue of the 'substantial form' that makes them what they are. On the extreme version of this view, natural entities as such cannot possess more than one form. A horse is, in a sense, composed of flesh, blood and bone and so on, but it is the substantial form of horseness that is responsible for coordinating these parts into a horse. It is the presence of the substantial form that is responsible for the difference between a live horse and a dead one. Natural wholes are more than the sum of their parts, and substantial forms are responsible for making them the wholes that they are. According to the extreme version of this view, flesh and bone do not exist as such in a live horse. To admit that they do would be to admit more than one form to be involved in the composition of a horse, the forms responsible for fleshness and boneness as well as the form of horseness. On the hard line view, the substantial forms of flesh and bone as such are created and the substantial form of horse destroyed on the death of a horse. After death all organisms alike are presumed to naturally decompose into the four elements, air, earth, fire and water. But even the elements exist only potentially, as opposed to actually, in a live organism such as a horse according to the hard line view. It is not only living organisms that are what they are by virtue of a substantial form. Naturally occurring substances such as metals or salts are presumed to posses them too.

On the hard line view, there is a clear distinction between natural things and human artefacts, only the former possessing a substantial form making them the wholes that they are. Aristotle's distinction between natural and forced motions or changes was invoked to justify the distinction. Stones fall naturally to the ground and acorns grow into oak trees because of the substantial forms that make them what they are. Throwing a stone or grinding an acorn into dust is to impose unnatural states on them. The parts of a live horse form a natural whole in the way that the parts of a bed made of oak does not. As Aristotle pointed out, if a bed made of freshly hewn oak were to take root it would grow into a tree not a bed. An artefact like a bed will fall to the ground by virtue of being a heavy object but will not behave as a whole in a way characteristic of beds analogous to the behaviour of a natural whole such as an oak tree or a horse. Artefacts lack the substantial forms that make natural wholes the kinds of wholes that they are.

Substantial forms had theological ramifications for the scholastics. They are the creations of God not humans. Humans can construct beds but only God can make a tree. It is no co-incidence that Thomas Aquinas was among those that took a hard line on the centrality of substantial forms.

The hardline stance on substantial forms meshed with a second hard line position that barred artificial experimentation as an appropriate device for learning about nature. We cannot expect to learn what is distinctive of natural things by artificial interventions that disrupt their natural mode of behaviour. Burning or carving wood is no way to find out how trees behave as trees. '*Techne*', knowledge of the way the world behaves when interfered with, has some pragmatic value but it is distinct from '*scientia*', knowledge of the world as it is by nature. In modern terminology, technology involves a lowlier kind of knowledge than science, and experiment bears only on the former.

The distinction between natural and artificial substances meshed with another aspect of the hard line view, namely, that substances undergo sequences of transformations that are one-way. The substantial form of a horse guides its development from conception through birth and growth into a fully developed animal, and, upon death, corresponding to the loss of the substantial form, the horse decays into air, earth, fire and water. Non-living natural substances such as metals are also formed in the earth in a way governed by substantial forms and they too degenerate into air, earth fire and water. Substances such as horses or metals can be formed from the four elements only by the natural processes engendered by substantial forms. To believe that humans can artificially construct natural substances from components is to believe that they can play God, bringing substantial forms into existence and driving nature in reverse.

The hard line view that I have sketched in broad terms did not go unchallenged. Even its supporters needed to reconcile the Aristotelian idea that substances are what they are by virtue of form and the equally basic Aristotelian tenet that all material things are composed of the four elements. One way of coping with the problem, the one most easily accommodated by the hard liners, was to view the elements as existing only potentially rather than actually in the wholes they composed. An alternative, was to assume that the forms of the elements persist in wholes along with the substantial forms of those wholes, but with the forms of the elements playing some subservient role, dominated by and subservient to the substantial form.

The various lines on the way in which elements exist in substances could, without too much difficulty, be adapted to accommodate more complex forms informing the parts of wholes along with the substantial form of the whole but organised or dominated by them. There is a common sense in which natural things as well as artefacts are composed of parts and have properties that derive from the properties of the parts. Insofar as a horse is composed partly of flesh and the properties of the horse depend on the properties of flesh, and insofar as an Aristotelian must understand the properties of flesh to stem from the form of flesh, then surely there is a sense in which the form of flesh, as well as of bone, blood and so on, must be presumed to exist in the horse. The same issue arises in the case of inanimate objects. The recovery of copper and tin from bronze suggests that the forms of copper and tin persist in the bronze in some sense. Liberal Aristotelians found ways of accommodating these realisations to their philosophy by assuming that forms of parts persist in wholes along with substantial forms, perhaps in the way that individual notes exist in a harmonious chord.

Liberal scholastics were also able to find room for a role for experiment as a tool for shedding light on knowledge of natural things, and were able to find support for their position in Aristotle's work. Aristotle had suggested that 'art', that is, artificial intervention can aid and even perfect nature. Gardening can readily be construed as doing just such a thing. What is more, reproducing natural processes in artificial situations can shed light on those natural processes themselves. In *Meteorology* (3, 4 374a 35–374b 5) Aristotle discussed the formation of the rainbow invoking the spectra of colours induced in artificially produced sprays to support his analysis. He also conducted dissections in biology and famously traced the development of a chicken from its embryo by breaking open eggs at various stages of incubation.

An area of practical knowledge that posed a problem for hard-line Aristotelianism was alchemy. The remainder of this chapter is concerned with the way in which liberal Aristotelians rose to that challenge. Some of them were led by this route to formulate atomic theories of matter.

5.4 Aristotelianism and Alchemy

Alchemy had flourished in Alexandria in the Hellenistic period that marked the end of the Greek era. Alchemical tracts were translated into Arabic from the eighth century AD and became a focus of attention for western philosophers in the thirteenth century. By that time there existed a well-developed practical alchemical tradition. It involved experimental manipulation of materials, preparing them and breaking them down into components by a range of techniques such as roasting, distilling, subliming and dissolving. The pragmatic aim was to produce materials of use or value, especially, but by no means exclusively, gold. The transformation of materials, corresponding to what we now call chemical transformations, posed a philosophical puzzle that had been pinpointed by Aristotle, as we have seen (*Generation and Corruption*, Book 1, Chapter 10). When two constituents combine to form a compound, the result is a substance whose properties differ from those of its constituents. Nevertheless, there is a sense in which the constituents persist in the compound insofar as they can be recovered. Aristotle does not give an example when making this point, but the composition of bronze from copper and tin would have served his purpose. The transformations of materials that alchemists brought about posed a fundamental challenge to Aristotelians. In particular, the idea that materials with characteristic modes of behaviour could be artificially built up from components could be seen as involving the manufacture of substantial forms and driving nature backwards in just the way that was ruled out by the hard line Aristotelians.

A large body of alchemical writings influential in the late thirteenth century were attributed to Jabir ibn Hayyam, believed then to have been an Arabic author living in the eighth century. He became known as Geber, a Latinised version of Jabir. A key text from amongst those writings is *Summa perfectionis* (Sum of perfection) containing alchemical matter theory plus accounts of the experimental basis for it. William Newman, the scholar who has done most to unearth the details of the incorporation of alchemy into medieval Aristotelianism, has made a strong case that the author of that text was in fact Paul of Taranto, a Franciscan monk from Southern Italy writing in the late thirteenth century.[2] Another work by that author, under his own name, had the revealing title *Theorica et practica*. Those two works between them spell out the alchemical theory that is the object of our discussion. I will, for convenience, refer to it as 'Geber's theory'.

As we noted in the previous section, Thomas Aquinas took the strong line that only one form, the substantial form, can exist in natural substances such as a horse or a metal. When such substances decay they decay into the four elements of which they are composed. As Newman reports, Paul of Taranto explicitly invokes this Thomistic stand in *Theorica et practica* and takes issue with it. Is it not highly implausible to deny that flesh and bone exist as such as parts of a living horse, as the Thomistic view would seem to require? More significant for our purposes are the alchemical examples. Components of compounds are not destroyed when they enter into combination to the extent that they can be recovered. A metal can be transformed into a calx by intense heat and can then be recovered from the calx. If copper and lead are alike reduced to the four elements by intense heat, then why is it that lead, but not copper, can be recovered from the calx of lead and copper, but not lead, recovered from the calx of copper? The combinations and recoveries attained in the experimental practice of the alchemists seems to require the notion of intermediate substances more complex than the four elements but less complex than the substances they combine to form. For an Aristotelian, this meant that the existence of some plurality of forms in substances had to be accommodated.

There is a remarkable passage in *Theorica et practica* highlighted by Newman (2006, p. 41) that illustrates Paul of Toranto's stand on the issue. It reads as follows:

> This [the existence of intermediate principles] is expressly proven by certain experiments of this art, for all metals and minerals are incinerated and calcined in their own ways, as if by the resolution of their substance they are reduced to the nature of earth. But then they are resolved by techniques of art into a water, then into air through vapour and smoke, and presently through the resolution of their smoke they are reduced to the nature of water; having been fused by a strong fire, they return to their own original nature of whatever mineral body or metal. But if there were a complete resolution to the simple elements and not to certain mineral or metallic principles which are nearer than the first simple bodies, the metal or such and such a body would no more return from them upon [its exposure]

to fire than anything else made up of the simple elements, and gold would no more return from gold than would stone or wood [return from gold], especially since fire is a common agent, behaving alike towards all and each. But since these [metals and minerals] return just the same as before, it is manifest that they were only resolved to certain components of theirs and not to the simple elements or to the prime matter, as those foresaid [philosophers] mistakenly assert.

This passage makes it clear that Paul of Toranto is conversant with and intent on comprehending the detailed experimental practice of the alchemists. But he is also an Aristotelian philosopher, addressing the question of the existence of forms intermediate between substantial forms and the four elements. The transformations and reproductions involved in the experimental practice of the alchemists convince him of the necessity to postulate a hierarchy of intermediate forms. A hard line Aristotelian might well have been impressed by the fact that burning wood gives off fumes and flames and bubbling water leaving ash, taking this as evidence for reduction of wood to the four elements. But Paul of Toranto is aware that the transformation of substances results in products that are characteristic of the substances transformed and that sometimes the original substances can be recovered.

Paul of Toranto (alias Geber) assumed a hierarchy of forms. At the base of the hierarchy are the forms of the four elements basic to Aristotelian philosophy. At the next level of complexity are the principles mercury and sulphur, assumed by alchemists to be the components of metals. The grounds for this assumption of the alchemists stemmed from empirical considerations rather than Aristotelian, philosophical ones. Sulphur is given off when metallic ores or impure metals are heated and metals combine with mercury to form amalgams, acquiring in the process a sheen that is an enhancement of any sheen they already possessed. The forms of the metals themselves are more complex still. Paul of Toranto is taking the theory and the practice of the alchemists seriously and modifying Aristotelian philosophy to accommodate them. The details of the way he did so are to be found in the works he wrote under the name of Geber. They involve appeal to a granular structure of matter that Newman, for one, takes as important beginnings of a corpuscular theory.

We have seen that, according to hard line Aristotelianism, the formation of substances and their subsequent decay can take place only naturally under the guidance of God given substantial forms. Natural cycles are one way. The ability of alchemists to break down substances into and build them up from their components would appear to fly in the face of this. But Aristotle's own distinction between combination and mere mixture provides a way of avoiding this conclusion. The very fact that alchemists could recover the components of their artificial productions could be seen as evidence that those productions were mixtures of those components, akin to mixtures of wheat and barley, rather than genuine substances.

In the next section, drawing heavily on the work of Newman, I investigate the way in which these issues led Paul of Toranto in the direction of an atomic theory.

5.5 Geber's 'Atomism'

Geber was an Aristotelian. As such, he believed all terrestrial substances to be composed of the four Aristotelian elements. He was also conversant with the practical achievements of the alchemists and familiar with their understanding of metals as composed of the principles mercury and sulphur. Geber reconciled these two viewpoints by way of a particle theory. He assumed the four elements to be particulate and that the parts of the elements combine together in stable clusters to form the parts of mercury and sulphur. These parts of mercury and sulphur are 'of a very strong composition'. That is, they resist dissociation by intense heating or any of the other treatments from the alchemist's experimental repertoire that they might be subjected to. Geber offered experimental evidence for the particulate structure of sulphur and mercury involving their sublimation. When they are sublimed, mercury is deposited in tiny droplets and sulphur in a fine powder which 'seems to reveal their particulate structure to the naked eye' (Newman, 2006, p. 29). Further, they are sublimed as a whole leaving no deposit. This is taken to be evidence of the degree to which the parts of mercury and sulphur are bound together as wholes and resist separation.

Geber drew connections between ease of sublimation and particle size. Components of a substance made of small particles are more readily driven off by heat than those made of large ones. Separation can be achieved, first by a relatively gentle heating, which drives off substances made up of parts of small size, and then by progressively more severe stages of heating which will drive off substances with progressively larger parts. This explains, for example, how excess sulphur can be driven from metal ores, leaving the metal behind. Geber also attributed density of materials to the various degrees to which the particles composing them are closely packed.

As I have construed it thus far, Geber's matter theory is particulate in the rough and ready way in which Aristotle's conjectures in *Meteorology* were. However, Geber's *Summa perfectionis* gives some grounds for an interpretation that is more strictly atomic and Newman takes advantage of them.[3] On this interpretation, the parts of the elements that cohere to form parts of mercury and sulphur, and the parts of the latter that cohere to form parts of minerals and metals are least parts (natural minima). A further assumption is that the least parts of a single material, such as earth or mercury or gold, are all alike. Insofar as materials such as these are homoeomerous, they are not so in the strict sense of Aristotle's definition in *Generation and Corruption* since they will lose their identity as a material of their kind once they are divided beyond the level of their least parts. When Geber describes a substance such as mercury as homoeomerous he must mean it either in the rough and ready sense that Aristotle employed in *Meteorology IV* or in the (novel) technical sense that the least parts of a homoeomerous substance contain parts that are identical, strongly bound together, a condition not met by a mere mixture of wheat and barley.

Interpreting Geber's theory as particulate only in a rough and ready, as opposed to a strict atomic or corpuscular sense has the advantage of avoiding problematic features of the theory that arise if the latter course is adopted. The explanation of density in terms of the degree to which minima are closely packed is a case in point. Newman interprets Geber as follows:

> Here 'Geber' tells us that gold is made of the smallest possible particles of mercury and sulphur (subtillissima substantia, subtiles partes). Because these particles are so tiny, they can be pushed together without leaving much interstitial space. Therefore the resulting metal, Gold, will be heavy.
>
> The author's [Geber's] reasoning is absolutely clear. The particles making up silver are larger than those of gold. Therefore, the interstitial spaces between such particles are also greater than those of gold: as a result, silver will be lighter than gold. (Newman, 1991, p. 145)
>
> Even more clearly here than in the former passage, the author explains that a higher specific weight is due to a smaller size of particles, since this allows them to be closely packed. (Newman, 1991, p. 146)

One problem here is that the first of the passages cited above attributes the density of gold to the size and close-packing of mercury and sulphur particles that make up a gold particle whereas the second passage attributes it to the size of the gold particles themselves.[4] If the measured density of gold is to be traced back to an agglomeration of particles, as Newman interprets Geber as doing, then there are a number of variables that need to be specified. We need to know the size and density of the mercury and sulphur particles making up a minimum of gold, the degree to which these particles are closely packed and also the degree to which the minima of gold are closely packed in a bulk sample of gold whose density can be measured. No such specifications are given by Geber. Incidentally, it is difficult to see how a density for gold greater than that of its components, mercury and sulphur, can be reconciled with Geber's account of density as interpreted by Newman.

Each of the passages quoted above assume that small particles can be packed more closely than large ones. However plausible this may seem, and whether Geber believed it to be true or not, it is false. The ratio of matter to space for closely packed particles of the same shape and size does not vary with particle size. If a photograph of closely packed particles is enlarged, the empty and filled volumes are magnified alike, yielding larger particles filling space to exactly the same degree as the original smaller ones. It would seem that particles can be more effectively packed if they differ in size, with the smaller ones insinuating themselves in the gaps between the larger ones, but there is no evidence that Geber exploited such an assumption.

There seems to be an ambiguity in Newman's construal of Geber's position on the question of whether least parts of a substance can vary in size. In places it is made clear that the minima of a given substance, such as mercury or sulphur, are identical to each other. As Newman (1991, p. 147) puts it 'each particle of mercury (or sulphur, as the case may be) is itself composed of the four elements, and is identical to every other particle of mercury'. But Newman (1991, p. 151) also writes that 'the homogeneity of natural mercury is only relative, since it still has particles of different size' so that 'if the particles of mercury can only be increased somewhat in size, with their uniformity, coherence and strong mixture remaining, the substance will become impervious to fire'. Sulphur too, 'can be composed of particles varying in size' (p. 157). If corpuscles of sulphur or mercury can be divided so long as the proportion of air, earth, fire and water in them remains the same, then the idea that there is a minimum part or atom of sulphur or mercury becomes unnecessary or redundant.

I certainly believe that Newman has more work to do if he is to adequately substantiate his claim that 'it is not too much to view his notion of a *fortissimo composition* joining discrete corpuscles as having a kinship with the chemical bond of contemporary chemistry' (2006, p. 29).

I contend that it is possible to make most coherent sense of Geber's alchemical matter theory if it is interpreted as corpuscular only in a rough and ready as opposed to a strict sense. On this interpretation Geber's '*partes*' is to be translated as 'parts' rather than 'particles'. Further, '*subtiles partes*' are components of a material that are chemically active because they are able to penetrate the pores in solids with which they interact. A substance that results from a 'strong composition' is simply one that is able to resist decomposition by heat or other agents. Geber's reference to substances combining 'per minima' is interpreted as referring to the fact that combination is made possible or facilitated either by preparing them in solution or in liquid form so that they are more easily divided and more penetrative, or by grinding them into a powder, as Aristotle had noted long before (*Generation and Corruption,* 1, 10, 328a, 33–36).[5]

My interpretation of Geber is more conservative that Newman's and, as a consequence, is more readily construed as being borne out by experiment. Newman (2006, p. 29) takes the fact that mercury is deposited as tint droplets and sulphur as a powder when sublimed 'to reveal their particulate structure to the naked eye'. This cannot be sustained if 'particulate' is intended in some strict corpuscularian sense. The mercury droplets are droplets that can be divided, yielding smaller droplets of mercury in keeping with their homoeomerity in Aristotle's sense. Mercury droplets change size when pressed together and coalesce. The sulphur particles too can be divided into smaller particles of sulphur. Neither the droplets nor the powder qualify as the non-homoeomerous particles attributed to Geber by Newman. The fact that pure mercury and sulphur sublime without leaving a residue is a sign both of their purity and of their ability to resist dissociation by heat. The observed phenomena give no warrant for the stronger claims about their corpuscular or atomic structure.

5.6 The Status and Fate of Geber's Integration of Alchemy and Aristotle

In the works of Geber, alias Paul of Taranto, we find Aristotelian matter theory and experiment linked to an unprecedented degree. However, my critique of Newman's attribution of an atomic theory to Geber should make us wary of seeing in Geber's work the beginnings of a successful attempt to make experimental contact with atoms. Even the interpretation of Geber as invoking a particulate structure in a rough and ready, as opposed to an atomic or corpuscular, sense, should not be too readily seen as a successful bringing together of experiment and matter theory. It is clear that Geber (Paul of Taranto) was intent on incorporating alchemy into Aristotlelian philosophy, and using the former to bolster the latter. So, for instance, his identification and defence of substances intermediate between the four elements, on the one hand, and horses or alloys on the other, was centrally concerned with scholastic debates about whether one or a plurality of forms can exist in complex bodies or substances. The four elements themselves were the legacy of Aristotle rather than objects of experimental manipulation. It is true that Geber appealed to detailed experimental findings, such as the recovery of metals from their calces to defend his own interpretation of Aristotle. The philosophical and experimental dimensions are both present and intertwined. But I question the extent to which they form an integrated whole, with the philosophical theory guiding experiment and the experiment confirming the philosophy. The centrality of Aristotelian forms and the four elements was presupposed by Geber, whilst the experimental results were practical results appropriated by Geber rather than being fruits of his theorising.

My stance here is borne out by the subsequent fate of Geber's work. The experimental part did not find a place in the commentaries on and extension of Aristotle that dominated the agenda in medieval and early Rennaisance universities. The existence of blood and bone in the body or copper and tin in bronze were sufficient to fuel the debate about the possibility of the co-existence of forms in a substance. The novelties uncovered by the alchemists added little of significance to that debate. Attempts by late thirteenth-century liberal interpreters of Aristotle to include alchemy in the university curriculum failed.[6] As for experimental alchemy, this progressed in the hands of artisans outside of the university context without being concerned with or guided by systematic philosophy and was sufficiently progressive to spawn the crafts of metallurgy and apothecary. Insofar as the practice was informed by theoretical assumptions these took the form of not particularly coherent or logically worked out metaphorical appeals to vitalistic notions such as sexual attraction or analogies between for instance, the generation of metals in the earth and the growth of trees, or the extraction of potent extracts from a substance to the extraction of the kernel from a nut. None of these devices made coherent sense if taken too literally. In the sixteenth century Paracelsus moulded such modes of thinking into a non-Aristotelian world view that lacked the clarity and logical rigour of scholastic philosophy but which was influential with practitioners, including medical practitioners outside of the university context, nevertheless.

5.7 Currents of Thought Leading to Sennert's Atomism

I have expressed reservations about the assumption that the *Summa perfectionis* and *Theorica et practica* expounded an Aristotelian atomic theory. There can be no such doubt about the theory that Daniel Sennert had come to articulate by the time of his death in 1637.[7]

Sennert was a professor of medicine at Wittenburg. The philosophical framework in which he worked was explicitly Aristotelian, but that fact did not prevent him articulating a matter theory that was an atomic theory in key respects. Developments of the medieval theory of natural minima and of Geber's theory were key sources for Sennert. There are also references in his work to Democritus as a source, although Sennert's theory differed both in substance and in intent from that of his Ancient Greek precursors, Democritus and Epicurus.[8]

We saw earlier in this chapter that medieval authors read into Aristotle an account of natural minima where those minima were conceived of as limits to division or as the least portion of a substance that could resist the corrupting effect of its surroundings. The minima were not conceived of as particles composing a bulk substance. If v is the volume of a natural minimum of a substance then interpretation of that minimum as a pre-existing particle waiting to be divided, as it were, would rule out volumes of that substance lying in between v and 2v, whereas interpreting the minimum as a limit of division will allow a size of 7/4 v or any other size of a portion of the substance between v and 2v. There is evidence that during the Renaissance a number of authors came to interpret the natural minima as parts of the substances they composed and existing prior to, and independently of, division. The trend is quite explicit in the writings of Julius Caesar Scaliger in the mid sixteenth century which were explicitly invoked by Sennert.

In Scaliger's view, natural minima have an autonomous existence as parts of substances, and he appealed to them to explain a range of physical and chemical phenomena, at least in a notional way. Chemical combination takes place 'per minima' where this is taken to mean that minima of the combining substances come together to form a minimum of the compound. Scaliger attributed a range of physical properties to arrangements of minima. It was in this way that he explained the difference between rain, hail and snow, all presumed to be made up of the same minima of water.[9]

Scaliger interpreted the chemical union of minima to form the minimum of a compound in Aristotleian rather than Democritean terms. The component minima are not merely next to each other in the compound minimum, but are in some way fused to form a new unit, the minimum of the compound. Yet the union was such that the combining minima retained their identity to the extent that they could be recovered from the compound. The form of the minima of the components in some sense persists in the compound, but in a way that is subservient to the newly generated form of the compound.

Sennert adapted Scaliger's theory to his own purpose. Although Sennert shared Scaliger's concern to construct a theory that was thoroughly Aristotelian, he was also driven by more down to earth empirical or practical concerns stemming from his practice of medicine, and especially the chemistry relevant to that enterprise. As far as this empirical dimension was concerned, Sennert drew heavily and explicitly on Geber's *Summa perfectionis*.

5.8 Sennert's Atomic Theory

In the theories of his maturity Sennert endorsed the Democritean view that 'all things are made of atoms' explicitly citing Democritus in this connection.[10] However, as we shall see, the natural minima that constituted Sennert's atoms differed markedly from the atoms of Democritus, which they needed to do to enable Sennert to be both an Aristotelian and an atomist.

Sennert's natural minima formed a hierarchy. Most basic were the least parts of the four elements, air, earth, fire and water. Those atoms combined in stable clusters to form least parts of more complicated minima of mercury, sulphur and salt. Here Sennert followed the alchemy of his day in adding salt to sulphur and mercury as constituting the principles of metals. Minima of those three principles could combine to form relatively stable clusters that were the minima of metals. Such minima could themselves combine to form yet more complicated clusters, for instance, making up the least parts of the mixt (compound) precipitated when salt of tartar is added to a solution of silver in aqua fortis.

The clusters forming the natural minima of a substance other than the four elements and the principles salt, mercury and sulphur were only relatively stable, persisting through some changes and being broken down into less complex minima in others. Sennert made much of 'reduction to the pristine state' to illustrate the former circumstance. The recovery of silver after dissolution in aqua fortis (nitric acid) utilised by Sennert has been re-enacted by Newman (2006, pp. 81–82) and illustrated by clear colour photographs in his book. Silver is dissolved in aqua fortis whereupon it disappears into a solution which can be filtered without leaving a residue. When a solution of salt of tartar (potassium carbonate) is added, a curdled precipitate (of silver carbonate) is formed. The precipitate is filtered, washed and heated in a crucible, whereupon metallic silver is recovered. Sennert presents this as evidence that natural minima of silver persist throughout the reaction. However, in Sennert's view, the stability of the silver minima is only relative because the silver can degenerate into a calx.[11]

Sennert used the terms *synkrisis* and *diakrisis* to describe the combinations and dissociations involved in the transformation of natural minima. Those terms are precisely the ones used in the Aristotelian tradition to describe the combining and separating of atoms in Democritean atomism. Indeed, as observed above, Sennert himself likened his minima to the atoms of Democritus. However, there are fundamental differences. Democritus was intent on reducing the world to the shapes, sizes and motions of portions of being (or matter). Sennert's natural minima, by contrast, had properties peculiar to the substances they were minima of, most important of these being their affinity for and propensity to combine with or their repulsion for, and their propensity not to combine with, minima of other substances. Sennert's atoms were 'minima of their own genus'. There is no suggestion, in Sennert's theory, that those properties be reduced to more fundamental ones. In Sennert's theory, there is a sense in which complex minima are composed of less complex ones right down to the minima of air, earth, fire and water. However, the properties of complex minima cannot be reduced to the properties of those fundamental ones. Minima of mercury, for example, have components which are minima of air, earth, fire and water, but the properties of mercury are other than the sum of the properties of those minima. After all, compounds (mixts) need to be distinguished from mere mixtures. The yellow, curdled solid precipitated when a solution of salt of tartar is added to a solution of silver in aqua fortis possesses distinct properties differing markedly from the properties of the solutions from which it is formed.

The sense in which constituents exist in mixts posed a fundamental problem for the Aristotelian notion of form, as we have seen, and it was one with which Sennert was compelled to grapple. Basics of Sennert's chemistry, especially 'reductions to the pristine state' as described above, convinced him that components really do exist in the compounds that they form, their subsequent recovery being witness to that. This implied, for Sennert, that natural minima persist as components of the more complex minima resulting from their combination. On the other hand, he needed to accommodate the fact that compounds possess properties qualitatively different from those of their components. The details of Sennert's response to the problem were not all the same, but it is clear that he needed to be committed to some hierarchy of forms. The form of a minimum of mercury, for instance, was presumed to persist in some way in the natural minimum of gold of which it formed a component. Superimposed on the forms of the mercury, salt and sulphur composing the gold is a superior form, the substantial form of gold, conferring on the gold its characteristic properties. This mirrors the way in which the substantial form of a species is present in a living member of that species and makes it more than a collection of flesh bone and blood, notwithstanding the fact that flesh, bone and blood are components of the organism. A living organism is more than the appropriate mixture of its parts, signified by the fact that a recently dead member of a species would contain the same parts but would be a qualitatively different entity nevertheless. In like manner, a portion of a compound is more than an assembly of its constituents. Something like the hierarchy of forms adopted here by Sennert would seem to be required by Aristotle's own recognition that there exist compounds whose properties are qualitatively different from those of their components and from which the components can be recovered (*Generation and Corruption*, 1, 10, 327b, 24–31). The elements arise from the imposition of form on prime matter. The elements can then be seen as the 'matter' on which the appropriate form can be imposed to constitute mercury, sulphur and salt. These in turn constitute the 'matter' that can be informed to yield metals and so on.

Insofar as he admitted a plurality of forms and defended his matter theory by appealing to experiment, Sennert was very much a liberal as opposed to a hard line Aristotelian. As such he needed to respond to criticisms of his hard line opponents epitomised by Thomas Erastus, a professor of medicine in Heidelberg writing in the second half of the sixteenth century.[12] Erastus stressed the one-way character of nature's cycles. Substances are generated from the elements and then decay back into the elements under the guidance of the unique, God given substantial forms that make them what they are. On this view, the substances artificially produced by chemists in the laboratory are artefacts rather than genuine substances and lack a substantial form. They are mere mixtures that lack the unity of natural substances that the latter owe to their substantial form. Reductions to the pristine state made much of by chemists such as Sennert was construed by Erastus as evidence that the productions of the chemists were indeed mere mixtures as opposed to genuine substances. Artificial tampering with substances in the laboratory or workshop could not yield knowledge of natural substances in Erastus's view, and to presume that such practices could produce the substantial forms responsible for making substances what they are is to presume that the experimenter can play God.

Sennert made explicit his sophisticated response to these kinds of criticisms.[13] He insisted that the processes involved in chemical experimentation and the products it yielded were in a strong sense natural, in spite of the artificial and contrived nature of the experimental set-up. By their artificial interventions the chemists can bring substances together and apply heat, for example, but what eventuates as a result does so naturally and not at the whim of the experimenter. Consequently, the substantial form, that Sennert presumed to be responsible for making a chemical product the substance that it is, is a natural outcome depending on the natures of the components as determined by their substantial forms. In this respect chemistry is like gardening or farming and a manufactured chemical is no more an artefact than a grafted tree or chicks raised with the aid of incubation. The chemist can only exploit the God given ways in which substances can be combined and broken down.

Substances in experimental situations behave in the natural way dictated by their substantial forms just as they do in nature, then. However, Sennert makes the point that naturally occurring situations are typically highly complicated, so much so that it is difficult to learn anything precise from them. The problem can be tackled by artificially contriving situations that are sufficiently simple and contained to make it possible to learn something. Here is how Sennert put it in the *De chymicorum* of 1629, as cited by Newman (2004, pp. 252–253).

> One must not, therefore, make such superficial and incidental judgements about the works of nature. Rather they must be inspected a bit more deeply. And what is not presented to us casually must be sought from art and labor [industria]. In judging the unrestricted [*externis*] resolutions of nature, many impediments arise: the resolved parts do not present themselves for examination, so that it might be known whether the elements are pure or of another type, instead they are spread out in vapors and dissipate. But there is greater certitude in the works of art, where nothing is lost, but all the materials are treated in sealed vessels, and the resolved parts are collected within, and the heterogeneities are separated from the homogeneities, so that a correct judgement may be made of all.

To Erastus's claim that the products of chemical experimentation are mixtures rather than genuine substances just because they can be broken down into as well as made up out of their components Sennert could reply as follows. There is more to the distinction between a mixt, such as bronze or nitre, and a mixture, such as that of wheat and barley, than the fact that division eventually separates the mixture in a way that it does not separate the genuine mixt (compound). There is the additional fact that, in the case of the mixt, the resulting substance possesses properties that are other than and not a mere average of, the properties of the components, as is the case with a mere mixture. The yellow sludge (silver carbonate) that is filtered off after adding potassium carbonate to the solution resulting from adding silver to nitric acid, in Sennert's classic reduction to the pristine state, has properties quite other than properties of silver, or nitric acid or potassium carbonate. For Sennert, an Aristotelian, albeit a liberal one, this means that the silver nitrate must possess a substantial form that makes it the kind of substance that it is distinct from its components, notwithstanding the fact that silver can be recovered from it by heating.

There were a number of inter-related ways in which Sennert sought to explain the properties and behaviour of materials by appeal to atoms. Some properties could be explained structurally, that is, by appeal to the ways in which atoms are arranged. Explanations of density by appeal to the degree to which atoms are closely packed, of the change from liquid to solid state by the degree to which atoms are bound together, and the explanation of the differences between ice, snow and frost by appeal to different arrangements of atoms of water are straightforward examples. Newman (2006, pp. 230ff.) notes that Sennert came to use the word 'immutation' to describe the aggregations and dissociations of particles involved in this type of explanation. Another type of explanation appealing to atomism involves the way in which atoms of components combine to yield 'atoms', that is, least parts, of a compound. This is Sennert's version of the *synkrisis* and *diakrisis* involved in Democritian atomism. The views of Sennert and Democritus are similar to the extent that atoms combine to form wholes in which they remain as recoverable parts. But for Sennert the complex whole is not a mere juxtaposition of atoms. It possesses distinctive properties that are qualitatively different from those of the component atoms. The least parts of substances possess the properties that they do by virtue of their substantial forms. Sennert was an Aristotelian and did not assimilate the strict reductionism of Democritus and Epicurus in the ways that the mechanical philosophers that came after him attempted to do.

5.9 The Status of Sennert's Atomism

Sennert's theory was the most detailed and influential of various versions of atomism that became popular in learned circles in the early decades of the seventeenth century.[14] My critical assessment of its status in this section takes the views of William Newman as my foil, notwithstanding the fact that, had it not been for his scholarship, I would have scant knowledge of the atomic theory that I am about to appraise.

To what extent can Sennert's atomic theory be seen as at least the beginnings of experimental knowledge of atoms? Newman clearly is of the opinion that this was indeed the case. In his view (2006, p. 96) 'Sennert has an operational atomism that relies on the analytical tools of the laboratory to have the final say in determining the permanence of substances'. Sennert, as Newman construes him, is a cautious empiricist concerned not to claim more in his matter theory than is warranted by observation and experiment. Sennert takes seriously the dictum 'the things into which composition can be dissolved are the things out of which they are made' (based on Aristotle, *De Caelo*, 3, 302a, 15–18) and interprets this, according to Newman

(2006, p. 97), in a way that anticipates Lavoisier's notion of a chemical element as a substance that cannot be broken down by laboratory techniques. Sennert the empiricist, unlike the mechanists who construed atoms as variously shaped portions of universal matter in strict Democritean fashion, refrained from speculating about the shapes and sizes of atoms. 'Thus instead of going down the path of a Descartes or a Lemery, Sennert argues again from the phenomena' (2006, p. 131). Knowledge of how the substantial forms responsible for the properties of natural minima brought about their effects have to be accepted as unachievable insofar as it lay beyond the bounds of experimental investigation. As Newman (2006, p. 139) puts it, the 'unknowable nature of Sennert's substantial form is an Aristotelian empiricist's statement of nescience'.

It needs to be acknowledged that Sennert was significantly progressive insofar as he took the results of experimental investigation very seriously indeed and defended its role in the quest for knowledge of the natural world in a sophisticated way as we have seen. As a professor of medicine he was interested in experimental knowledge for its practical implications, for example for the manufacture of drugs and for the understanding of the spread of contagious diseases. But he also saw the need to incorporate the new experimental knowledge into his Aristotelian view of the world generally, including theology. In particular, he aimed to incorporate chemical knowledge into his general philosophy in a way that the hard line Aristotelians could not. Sennert had more success in finding a place for chemistry in the university than his thirteenth century precursors such as Roger Bacon had been when attempting a similar thing.

Newman portrays Sennert as an advocate of a cautious empiricism exemplified by Lavoisier's stand on the chemical elements, construed as substances that could not be broken down further by chemical means. There are hints in Sennert of the idea that there are limits to what one can hope to know by experimental means. He stressed that substantial forms are unknowable, that the causes of manifest phenomena such as magnetism or the harmful effect of a poison are occult (or hidden) in the sense that they lie beyond the bounds of our knowledge, and suggested it was futile to attempt to experimentally reduce all qualities to the four fundamental ones, the hot and the cold and the wet and the dry.[15] Nevertheless, there were two central aspects of Sennert's position that put it beyond the bounds of what can be supported by the resources available to a cautious empiricist, namely, his Aristotelianism and his atomism.

The general Aristotelian view on matter and form including the reduction of the material realm to the four elements was presupposed by Sennert. In his articulation of chemistry Newman may well be right to claim that Sennert treated his natural minima as 'black boxes' the cause of whose properties could not be experimentally investigated. But this did not prevent him assuming that the minima were ultimately composed of the four elements. Nor did it prevent him from pursuing and debating the issue of whether more than one form could exist in a substance, and just how a plurality of forms could be articulated. The assumption that substantial forms were in some sense God-given rather than man-made was fundamental to his theology.

Sennert's philosophy generally was constrained by experiment but went beyond what could be warranted by it in significant respects.

Sennert's atomism, also, was accommodated to experimental phenomena rather than being confirmed or supported by them.[16] Experimental practice had revealed a wealth of knowledge about how substances can be combined or decomposed to yield well-defined products with characteristic properties. Some of the more novel examples involved reactions that were reversible insofar as ingredients could be recovered form the compounds that they combined to form, made much of by Sennert and his 'reductions to the pristine state' as we have seen. To what extent can this knowledge be taken as evidence for atomism? To what extent, for example, can the recovery of silver after dissolution in nitric acid, graphically illustrated by the photographs in Newman's book, be taken as evidence that silver is composed of atoms or natural minima of that substance? Experiment does show that the silver is in some sense in the liquid that results from adding nitric acid to it insofar as the silver can be recovered from it. On the other hand, that liquid (a solution of silver nitrate) does not have either the chemical or physical properties of silver, so the silver is not present in any straightforward sense. The atomistic 'explanation' involves the assumption that silver and nitric acid are composed of minima that combine, per minima, to form minima of the compound. Silver can be recovered because in some sense the minima of silver are 'in' the minima of the compound ready to be extracted. But in what sense are minima of silver 'in' the minima of the compound. They cannot be in the compound minima simply by residing next to a minimum of nitric acid, because this will be incapable of explaining why the compound has qualitatively different properties from those of silver. That is, the problem of understanding the sense in which ingredients are 'in' the compounds from which they are recoverable is not solved by invoking minima. Rather, it merely shifts the problem from the level of combining substances to the level of combining minima. If Sennert really were the cautious empiricist that Newman paints him as, one would expect him to have admitted ignorance about the processes underlying chemical change rather than confidently attributing it to combining atoms or minima.

Apart from qualifications concerning the degree to which Sennert's Aristotelian philosophy and his atomism can be said to have been supported by, rather than accommodated to the available experimental evidence, there is the question of how productive the relation between Sennert's theory and experiment was. The degree to which Sennert's theory involved accommodations of experimental phenomena rather than their prediction tells against its productiveness. Sennert's attempts to devise a theory of subordinate forms that could accommodate chemistry and his views on natural minima as the bearers of substantial forms did not have an internal dynamic of their own. Provided they could be rendered compatible with the phenomena there was not much to choose between one articulation rather than another. The advance of experimental knowledge came mainly through the efforts of artisans rather than the learned working in universities. Christoph Meinel (1988, p. 72) makes this point quite forcefully.

> The experience of practical men, separated from the mainstream of learning by educational and social barriers, had become more influential since the Renaissance. By the very nature of their crafts they treated matter in a nonphilosophical, purposeful way. For obvious reasons, metallurgists, assayers, chemists and apothecaries were more concerned with the properties of the products than with the theory of the processes. . . . Alchemists and practical men, on the other hand, knew a great deal about [metals and minerals] and they knew how to handle and study them experimentally.

As Klein (2007) has pointed out, the recovery of silver in its pristine state was not an experiment that was the fruit of Sennert's or any other theorist's philosophy. It was an embellishment of techniques common in sixteenth-century metallurgy and described in metallurgical booklets. The fruitful integration of theory and practice that was to become a hallmark of science is not to be found in the work of Sennert or other matter theorists of the time who invoked atoms.

It is not my attention to admonish Sennert for not inventing modern science. At the time he wrote it was not at all apparent that experimental research unguided by philosophy could yield knowledge of the inner structure of matter. Those who chose to speculate about such matters needed to develop philosophies that far transcended what empirical research of the time could offer. This book aims to explore the way in which an atomic theory of matter that was highly general, experimentally supported and independent of philosophy became possible. The point my appraisal of Sennert's atomism is meant to make is that our story is far from having reached its end. Making experimental contact with atoms was very much a task for the future in the middle of the seventeenth century. In the next three chapters I argue that the mechanical philosophers who came after Sennert did not do much to further the cause.

Notes

1. As cited by Duhem (1985, p. 41).
2. See Newman (1985).
3. Newman's translation of Geber's 'partes' as 'particle' invites an atomistic reading of Geber's text more strongly than Geber's Latin necessarily implies.
4. The same ambiguity persists in Newman (2006, pp. 32–33), as Ursula Klein (2007) has observed.
5. Newman (2006, p. 34) reads Geber in a way that lends itself to this interpretation. '*Subtiles partes*, "subtle particles", are small, volatile, and capable of penetrating deeply into narrow pores. *Grosse partes*, "gross particles", are larger (though still perhaps imperceptibly small), "fixed" or non-volatile, and far less penetrative than their subtle counterparts'. Materials can be rendered more reactive by dividing them into parts, by dissolving them in solvents or by grinding them to a powder. On my reading, Newman's 'particles' are to be interpreted simply as small parts.
6. See Newman (1991, Chapter 1)
7. Sennert (1629) and Sennert (1636). Once again, Newman is my main source of information, especially Newman (2006). See also Newman (1996), Meinel (1988), and Emily Michael (2001).
8. Lucretius's poem *De Rerum Natura* was rediscovered in the west in 1417 and republished in 1473. Latin translations of Leucippus, Democritus and Epicurus appeared in 1473. Initial

interest in these theories was centred on the social philosophy and the problem of free will. It was not until late in the sixteenth century that serious attention was given to the atomic theory of matter contained in these sources. See Hilary Gatti (2001, pp. 163–164) and Meinel (1988, pp. 70–71).

9. See Clericuzio (2000, p. 12).
10. See Newton (2006, p. 94)
11. There is a basic problem here that was not to be adequately resolved until the time of Lavoisier. It concerns the issue of whether chemical reactions involve the building up or the breaking down of substances. Arguments of the type employed by Sennert could be used to support the claim that calces (oxides) rather than metals persist through chemical change. The calx of silver can be dissolved in nitric acid, silver carbonate precipitated by adding salt of tartar, silver generated by gentle heating and then, finally, by intense heating, the calx can be recovered 'in its pristine state'. Adequate resolution of the ambiguities here required attention to the changes in weight involved in reactions and the recognition of the chemical role of gases.
12. For details of Erastus's position see Newman (2006, pp. 45–65).
13. See, especially, Newman (2004, pp. 250–256).
14. Other atomic theories were formulated, for instance, by Sebastian Basso and Etienne de Clave in France and Joachim Jungius in Germany.
15. See Newman (2006, pp. 138–144) for documentation.
16. My low estimate of the degree to which atomism in the early seventeenth century had experimental support is shared by Meinel (1988).

Chapter 6
Atomism, Experiment and the Mechanical Philosophy: The Work of Robert Boyle

Abstract Boyle articulated and defended a strict version of the mechanical philosophy, a theory about the ultimate structure of matter. According to that philosophy, the material world is made up of corpuscles of the one impenetrable matter possessing a definite shape and size and capable of motion. Boyle was also a pioneer of experimental science, best exemplified in his pneumatics. Boyle himself distinguished between the two forms of knowledge, arguing that the 'intermediate causes' involved in his experimental science, such as the weight and spring of air, were empirically accessible in a way that the ultimate mechanical corpuscles were not. As a consequence, Boyle's experimental science could not be fruitfully guided by the mechanical philosophy and the success of his experimentation did not constitute significant support for it. This is at least implicit in some of Boyle's own remarks. What was scientific about the scientific revolution, in my view, was the emergence of experimental science as distinct from philosophical theories about the ultimate structure of matter.

6.1 What Was Scientific About the Scientific Revolution?

Our discussion of Sennert's atomism in the previous chapter gives a hint of one aspect of the scientific revolution that took place in the seventeenth century, the increased emphasis on experiment as the source of and grounds for scientific knowledge. The utilisation of artificial experiments to throw light on nature that had been marginalised in the work of the Aristotelian alchemists became a new focus of attention. The writings of Francis Bacon were emblematic of the new move.

Another aspect of that revolution, which was already making its presence felt in the latter part of Sennert's lifetime in academic circles, was the replacement of the Aristotelian world-view by the mechanical one. Those mechanical philosophers who were atomists recast atomism in a form that reconciled it with the new philosophy and, in doing so, constructed a version of atomism that had marked similarities with that of Democritus and Epicurus and which, consequently, but as we shall see, largely mistakenly, can easily be construed as a revival of the ancient theories.

This chapter involves an investigation of the relationship between the switch from an Aristotelian to a mechanical matter theory, on the one hand, and the rise

A. Chalmers, *The Scientist's Atom and the Philosopher's Stone*, Boston Studies in the Philosophy of Science 279, DOI 10.1007/978-90-481-2362-9_6,

of experimental science on the other. I argue that these two changes were not as closely connected as is typically supposed. What was scientific about the scientific revolution, in my view, was the emergence of experimental science as distinct from deep theories about the ultimate structure of matter, mechanical or otherwise. What was scientific about the scientific revolution was the emergence of science as distinct from philosophy. In the seventeenth century, atomism retained the speculative, philosophical status that it had possessed in the hands of Democritus and Epicurus. It did not feed productively into the new experimental science.

René Descartes constructed the first systematic version of the view that came to be known as the mechanical philosophy.[1] For Descartes the material world is full of inert matter (which he identified with extension) and the only cause of changes in motion is the pushing of matter against matter. Descartes was not an atomist, insofar as he regarded the world to be a plenum and he regarded matter to be indefinitely divisible. Nevertheless, he did appeal to particles of matter in his construction of mechanical explanations of a range of chemical, magnetic, optical and other phenomena. Already present in Descartes is the division between the material realm, composed of inert matter, and the spiritual realm of souls, minds and angels. This division was adopted in some form or other by all the seventeenth-century mechanical philosophers and constituted the main difference between their world-view and that of Democritus and Epicurus.

Descartes' French compatriot, Pierre Gassendi, was a mechanical philosopher who was an atomist who explicitly presented his view as a revised version of that of Epicurus. However, my main focus will be on the version of atomism developed in the context of the mechanical philosophy by Robert Boyle. There are several reasons for this. Boyle was one of the most articulate expositors and defenders of the new philosophy and did much to popularise it.[2] However, articulation of the mechanical philosophy is by no means Boyle's only claim to fame. He is also famous as an experimentalist. His voluminous and wide-ranging writings are full of detailed descriptions of many experiments, most notably in chemistry (and alchemy) and in pneumatics. Indeed, he is known mostly through his experimental discovery of the law that bears his name. If we are to find a version of the mechanical philosophy in the seventeenth century that was intimately linked with and borne out by experiment, then it is in the works of Boyle that we are most likely to find it.

Boyle articulated an atomistic version of the mechanical philosophy. It involved the total elimination of Aristotelian forms and sought to reduce the material world to the arrangements and motions of particles of universal matter possessing an unchanging shape and size. He also conducted experiments in pneumatics, many of them involving his newly-devised air pumps. The claims supported by those experiments involved appeal to the weight, spring and pressure of air. Consequently, they did not qualify as 'mechanical' in the strict sense, as Boyle openly acknowledged. Implicit in Boyle's practice, and also made quite explicit by him, is a distinction between 'matters of fact' established by experiment and the fundamental matter-theory called the mechanical philosophy.

The mechanical philosophy was 'philosophical' insofar as it offered an ultimate theory of matter in general. By contrast, Boyle's pneumatics was subject-specific and the explanations it made possible were not ultimate insofar as weight, spring and pressure remained unexplained. I urge that the emergence of the latter kind of knowledge as distinct from the former is what renders the name 'scientific revolution' appropriate. In the remainder of this chapter I aim to give a detailed articulation and defence of this view in the context of Boyle's work.[3] I am particularly concerned to defend myself against the charge of anachronism, along the lines that I am illegitimately imposing a modern distinction between science and philosophy on the seventeenth century. Boyle was intent on defending his mechanical philosophy as well as his experimentation and he explored and insisted on a detailed relationship between the two in a way that does not mirror the contemporary separation of science and philosophy as distinct practices assigned to different university faculties. Nevertheless, a distinction was there in practice in Boyle's work and sometimes made explicit by him.

Newman (2006, p. 2) sees the scientific revolution to which Boyle contributed as 'the great disjunction between the common view of matter-theory before and after the mid-seventeenth century', the former involving immaterial forms and the latter arrangements of minute, robust corpuscles. I, by contrast, construe the revolution in terms of the emergence of experimental science as distinct from philosophical matter theories. In my view, Boyle was one of the important pioneers of experimental science as opposed to philosophy but not a successful defender of an atomistic version of the mechanical philosophy that was experimentally based.

6.2 Boyle's Version of the Mechanical Philosophy

In the early 1650s Boyle was already an advocate of and participant in 'Baconian science', that is, science based on experimentation. He became a participant in this tradition through his association with the German expatriate Samuel Hartlib and his circle. We know Boyle was knowledgeable about experimental chemistry as practiced on the continent by the likes of Rudolph Glauber and that he was tutored in the practice of alchemy by the American émigré, George Starkey.[4] Boyle moved to Oxford late in 1655 and there began extensive experimentation, especially in chemistry and pneumatics. It was in that period that Boyle constructed his air pump and thereby revolutionised pneumatics. It was also in that period that he began a careful formulation of his mechanical philosophy. In this section I am concerned with the latter.

The key sources for Boyle's articulation of the mechanical philosophy are 'The origin of forms and qualities according to the corpuscular philosophy', written in the late 1650s and published in 1666 and 'About the excellency and grounds of the mechanical hypothesis' published in 1674.[5] According to Boyle there is one universal matter characterised by its impenetrability. The world is composed of particles of this matter that are too small to be detected by the senses and which remain undivided in physical processes, although they are divisible mentally or by divine omnipotence. Boyle refers to these as *minima naturalia* or *prima naturalia*.[6] As portions of matter, the natural minima have a distinctive and permanent shape and size and they can move. These are the only properties they possess in addition to the impenetrability characteristic of all pieces of matter. They are the 'primary affections'. Natural minima can combine into relatively stable clusters, which clusters will themselves have distinctive shapes, sizes and motions, and which may well remain undivided through various kinds of change. The main tenet of the mechanical philosophy is that all the phenomena of the material world are to be explained in terms of, traced back to, or reduced to, the motions and arrangements of portions of matter characterised in terms of their primary affections only. Boyle referred to the arrangements and motions of the invisible particles assumed to be responsible for a body's observable properties as its 'texture'.

The reductionist character of Boyle's mechanical characterisation of the world required that he explain how observable bodies come to have a range of properties, some detectable by the senses, such as colours and smells, and others, such as temperature or degree of elasticity, determining how bodies interact with each other. Perceptible properties ('sensible qualities' in Boyle's terminology) are explained as responses in us due to the impact of mechanical particles (that is, particles characterised solely in terms of their shape, size and motion) on our sense organs. Visible objects have the colours that they have in given circumstances as a result of the interaction of their corpuscular structures with the structures that constitute light and the interaction of that light with our eyes (themselves made up of a characteristic arrangement of mechanical particles) in those circumstances. Properties such as temperature or chemical properties, which, as Boyle recognises, are possessed by bodies independently of whether they are perceived by humans or not, are to be explained in terms of mechanical particles and their motions. Temperature of a body arises from the relative vigour of the motions of the particles composing it whilst the ability of gold to be dissolved by nitric acid is to be attributed to the relationship between and interaction of the shapes and motions of the mechanical particles making up those two substances.

So far, a marked similarity between Boyle's mechanical philosophy and ancient atomism should be obvious. Major differences enter in when it comes to the non-material world of souls and minds and the related issue of the role of God. Unlike the Ancient Greeks, Boyle and other mechanical atomists (as I shall refer to those mechanical philosophers who were atomists) restricted their mechanical, atomistic explanations to the material realm. Boyle (2000, Vol. 5, p. 300) made it quite clear that he rejected the atheistic implications of Ancient atomism and insisted that the 'Reasonable Soule, that is said to inform the humane Body' is immune from mechanical reduction and he explicitly invoked God as actively involved in the construction and operation of the material world in a number of ways. Firstly, Boyle saw God as the author of both matter and any motion that it possesses. Secondly, he regarded it as necessary that God break matter into pieces and arrange those pieces into the interconnected structure that constitutes our world. Thirdly, God is the author of the laws of motion that govern the regular behaviour, that is, motion and collision, of particles of matter and is constantly at work in the world to ensure that those laws are obeyed. Here is how Boyle (2000, Vol. 5, p. 306) summarised the role of God in the mechanical world in the 'Origin of forms and qualities':

> I shall not scruple to say ... That the Origine of Motion in Matter is from God; and not onley so, but that thinking it very unfit to be believ'd that Matter, barely put into Motion and then left to itself, should Casually constitute this beautiful and orderly World: I think also further that the wise Author of Things did, by establishing the laws of Motion among Bodies, and by guiding the first Motions of the small parts of Matter, bring them to convene after the manner requisite to compose the World, and especially did contrive those curious and elaborate Engines, the bodies of living Creatures, endowing most of them with the power of propagating their Species.

I will refer to the mechanical philosophy as I have summarised it in this section as the 'mechanical philosophy in the strict sense'. Boyle and other mechanical philosophers often used the term 'mechanical' in senses weaker than the strict sense. These weaker senses will be characterised and their applicability discussed in later sections of this chapter. Given the range of usages, it is important to be clear about which sense of mechanical is at issue when discussing the character and status of the 'mechanical philosophy' as critics of earlier work of mine have been quick to point out.[7]

There is one aspect of Boyle's deliberations that do not fit into my characterisation of the strict version of his philosophy. I refer to his appeal to 'seminal principles' to explain phenomena, especially biological phenomena such as reproduction. Clericuzio (1990) gives considerable emphasis to this aspect of Boyle's philosophy but I am more inclined to see his resort to seminal principles as what Newman (2006, p. 215) describes as 'a sort of rearguard action intended to evade certain explanatory difficulties resulting inevitably from the postulation of a purely mechanical universe'.[8] Since my focus in this book is the introduction of the atom into the physical sciences rather than biology I bypass this debate.

6.3 Boyle's Case for the Mechanical Philosophy

The main general arguments for the mechanical philosophy offered by Boyle appealed to its clarity, intelligibility and simplicity. The notions of matter and the sizes, shapes and motions of portions of it are clear, easily understood and clearly not in need of any further explanation at a deeper level. Further, explanations that appeal to the motions and interactions of portions of matter are unproblematic and accepted, where they can be upheld, even by opponents of the mechanical philosophy. Boyle contrasted the clarity and intelligibility of mechanical explanations with the obscurities and ambiguities of explanations offered by his opponents, especially the alchemists and the scholastics. There is no doubt that there were plenty of obscure if not unintelligible texts for Boyle to target in this respect and Boyle's writing is clear (if somewhat prolix). But not all of Boyle's opponents can be accused of lack of clarity, and it is worth pointing out that Boyle found arguments of Daniel Sennert, who accepted versions of Aristotelianism and alchemy, sufficiently clear to virtually reproduce them in his own work, albeit without acknowledgement.[9]

Boyle, along with other mechanical philosophers, saw a key merit of his scheme to be an avoidance of the obscure and empty nature of the scholastic appeal to the

Aristotelian notion of form. The title of his key work 'The origin of forms and qualities according to the corpuscular philosophy' signals his intent to give an account of form and qualities as an alternative to the Aristotelian one. One of Boyle's targets was the notion of substantial form. As we have seen, these forms were claimed to inhere in complex bodies of a particular kind and to confer on those bodies the properties of the kind. A sample of gold is what it is by virtue of possessing the substantial form of gold, just as a horse is a horse through possessing the substantial form responsible for horseness. Boyle's response was to argue that properties of wholes could be explained in terms of the properties and mode of combination of their parts without the need for the addition of anything over and above those parts. He exemplified his position by reference to the workings of a watch understood in terms of the inter-relationships between its component parts. He also pointed out that specific knowledge of the substantial forms and their mode of operation was beyond our access, which rendered appeal to them empty. One of the strongest points in favour of the Aristotelian position involved the difference between a live human and a recently dead one, where it could plausibly be argued that the two were composed alike of the same parts. The substantial form in the living being was to account for the difference. But here Boyle (2000, Vol. 5, p. 300) did not disagree, insofar as he accepted the crucial presence of the soul in the living body and himself likened it to a substantial form.

Another argument of Boyle's against a scholastic view concerning forms built on what we have located in the work of Sennert, although, again, there is no acknowledgement of Sennert by Boyle in this respect. A common scholastic assumption was that substantial forms exist in natural bodies but not in artefacts (so that these scholastics would have been unimpressed by Boyle's analogy with watches). The case against this involved arguing for the identity of chemical substances formed either naturally in the earth or artificially in the laboratory, with the latter often being deliberately composed from parts, that is, from component substances, in the laboratory.

As well as appealing to substantial forms to account for the being of a body as a whole, scholastics invoked forms that they called 'real qualities' to account for individual properties of a body, such as the whiteness of snow or the fusibility of a metal. Boyle's response was that little is gained by way of understanding by attributing properties such as whiteness or fusibility to the presence of otherwise unspecified real qualities of whiteness and fusibility. In any case, argued Boyle, it is very unclear what the mode of existence of this quality is that is added to snow to make it white, or to metals to make them fusible, since qualities were not considered by the scholastics to be a material part of the subjects they were qualities of. 'Nor could I ever find it intelligibly made out', Boyle (2000, Vol. 5, p. 309) wrote, 'what these real Qualities may be, that they [the scholastics] deny to be either Matter, or modes of Matter, or immaterial Substances.' Boyle advanced his case by showing how properties could be changed or accounted for mechanically, removing a need to invoke real qualities. For instance, glass can be changed from transparent to white merely by grinding it, where the ensuing whiteness can be explained in terms of the multiple reflections from the many surfaces of the resulting pieces of glass and the resulting interaction of the reflected light with our eyes, with no need to attribute the whiteness to some form or quality added to the glass (Boyle, 2000, Vol 5, p. 320). Likewise, the capacity of a polished sphere to form images can be explained by appeal to the reflecting power of its smooth surface and the poisoning effect of food containing ground glass can be explained by the cutting action of the pieces of glass (Boyle, 2000, Vol. 5, pp. 311–312). No 'real qualities' are needed.

The only properties that matter can have primitively are those of shape, size and motion, the primary affections, together with the impenetrability characteristic of all matter. These are the properties that must be possessed by any portion of matter by virtue of being such. Other properties need to be explained away as arising from the primary affections. The idea that a portion of matter should possess primitively some property other than the primary ones was considered by Boyle to be unintelligible.

6.4 Boyle's Use of the Macroscopic/Microscopic Analogy

Boyle's mechanical explanations, insofar as they appeal to the shapes, sizes and motions of unobservable particles, necessarily go beyond what can be justified by direct appeal to the observable. The problem of how to gain knowledge of the unobservable micro-realm by appeal to observations of the macro-realm has been highlighted by Maurice Mandelbaum (1964, pp. 88–112) and called by him 'the problem of transdiction'. Mandelbaum (1964, pp. 107 and 110–111) attributes to Boyle a response to this problem which he labels 'the extension of sense knowledge by analogy' and 'the translation of explanatory principles from the observed to the unobserved'. Madelbaum's position has recently been endorsed and amplified by William Newman (2006, pp. 203–208). I will first outline the position Mandelbaum and Newman attribute to Boyle. Then I will argue that the arguments are grossly inadequate as a way of defending Boyle's mechanical atomism and have been given more credibility than they warrant.

Boyle, in a variety of places, attempted to render plausible claims about the micro-realm by drawing analogies with the observable macro-realm. The observed compression of snow into a snowball is an effect that can be assumed to apply at the corpuscular level also, and so account for the effect of pressure on firmness of materials (Mandelbaum, 1964, p. 107). Objects of various sizes, such as collections of apples, walnuts, filberts, wheat, sand and flour, when poured from a sack more closely resemble fluids the smaller the individual objects are. This makes it reasonable to suppose that the fluid-like behaviour of a molten metal is due to the fact that its minute, component particles have been separated by the heat (Newman, 2006, pp. 204–205). The way in which two highly polished sheets of glass cohere suggests that the corpuscles composing a solid cohere in the same way (Mandelbaum, 1964, p. 111).

Another move by Boyle that certainly has a superficial plausibility at least, and which Mandelbaum and Newman invoke, is the idea that laws that apply at the observational level and which do so independently of size can be assumed to apply at the micro-level also. Boyle (2000, Vol. 8, pp. 107–108) put the case in 'Excellency of the mechanical hypothesis' as follows:

> [F]or both the Mechanical affections of Matter are to be found, and the Laws of Motion take place, not only in the great masses and the middle-siz'd Lumps, but in the smallest Fragments of Matter; and a lesser portion of it, being as well a Body as a greater, must, as necessarily as it, have its determinate Bulk and Figure. And he that looks upon Sand in a good Microscope will easily perceive, that each minute Grain of it has as well its own size and shape, as a Rock or a Mountain. And when we let fall a great stone and a pebble from the top of a high Building, we find not but that the latter as well as the former moves conformably to the Laws of acceleration in heavy Bodies descending. And the Rules of Motion are observ'd not only in Cannon Bullets, but in Small Shot; and the one strikes down a Bird according to the same Laws that the other batters down a Wall. ... And therefore, to say that, though in Natural Bodies whose bulk is manifest and their structure visible the Mechanical Principles may be usefully admitted, they are not to be extended to such portions of matter whose parts and Texture are invisible, may perhaps look to some as if a man should allow that the Laws of Mechanism may take place in a town clock, but cannot in a Pocket-Watch.

What are we to make of these extrapolations from the observable to the unobservable? I acknowledge that they have some force. In many respects bodies do behave independently of their size, and, in such cases, it is reasonable to conjecture that observed behaviour carries over into the unobservable micro-realm if there is no evidence to the contrary. However, whilst I can tentatively accept such arguments, a mechanical atomist such as Boyle cannot afford to give too much scope to them because *they run counter to and undermine the main tenets of that philosophy*. If the mechanical philosophy is true then the world of interacting minima is qualitatively different from the observable world. In the latter, bodies have a range of properties. They have shape and size and a degree of motion or rest, to be sure, but they also have colours, are rigid and elastic to some degree, are hot or cold, have a taste and so on. Boyle's atoms, his natural minima, are quite unlike this insofar as they lack all such properties. 'And if we should conceive that all the rest of the Universe were annihilated', wrote Boyle (2000, Vol. 5, p. 315) in the 'Origin of forms and qualities', 'it is hard to say what could be attributed to it besides Matter, Motion (or Rest), Bulk, and Shape' Whatever the case for this claim is, it cannot be based directly on observation because no such body has ever been observed. Insofar as Boyle's natural minima are perfectly rigid and impenetrable and lack all properties but the primitive affections, they are quite unlike the bodies of our experience.

With this general point in mind, we can see the problematic character of the arguments from Boyle invoked by Mandlebaum and Newman as giving support to mechanical atomism. The projection of the compressibility of snowballs onto small unobservable portions of it cannot proceed as far as natural minima because they are incompressible. If the adhesion between two polished surfaces is attributed to air pressure, as Boyle came to do, then it becomes a problematic explanation of the alleged cohesion of the minima of a solid. A mechanical philosopher cannot afford to assume that the fluidity of water can be extended, by analogy, to the corpuscles composing it. As for the law of fall, whilst it is true that this law is scale invariant as far as experiments on heavy bodies are concerned, it cannot be assumed to apply to Boyle's minima because they lack weight. For the mechanical philosophers, the latter property is one that required explanation by reference to the motions and impacts of corpuscles characterised solely in terms of the primary affections as Boyle himself openly admitted on more than one occasion. For a mechanical philosopher, no law involving properties other than the primary affections can be carried over to the level of minima however scale-invariant they may appear at the level of observation. As for the approach of the behaviour of poured powders to that of liquids the smaller the size of the particle, taken as an argument for the particulate character of liquids, this argument flounders because of crucial ways in which the analogy breaks down The powders form a pile in a way that poured liquids do not, while pressure applied to liquids is transmitted isotropically through them in a way that is not the case for powders.

I do not here wish to cast doubt on some efficacy for arguments from analogy, nor do I wish to discredit the idea that laws apparently scale-invariant should be assumed to hold generally until there is evidence to the contrary. What I do claim is that these arguments cannot help Boyle defend his mechanical atomism, for the simple reason that that philosophy implies a radical lack of analogy between the observable world and the world of atoms.

Faced with the task of defending the mechanical philosophy in the strict sense, Boyle was in a position similar to that of the Ancient Greek philosophers. He wished to give an account of the ultimate nature of material reality, the reality behind the appearances. Like the Ancients, he drew on the knowledge of the observable world available to him, abstracted aspects of it and turned them into fundamental principles. He argued for the intelligibility of those principles and used analogies with known phenomena to render their general applicability plausible. But there was plenty of opportunity for philosophers to disagree with Boyle's selection of principles. It could be assumed that there is more than one kind of matter, contrary to the fundamental assumption of the Ancient atomists shared by Boyle and the mechanical philosophers generally, or it could be argued that Boyle's 'primary affections' were insufficient to capture the degree of variety and activity evident in the world. If a reality behind the appearances is to be specified then there is a fundamental question of whether properties operative at that level are a subset of observable properties and if so, which subset. Arguments involving some macroscopic-microscopic analogy of the kind invoked by Boyle were not up to the task and I believe that champions of them such as Mandlebaum and Newman have underestimated their problematic character.

There is more to be said about the way in which Boyle marshalled knowledge of the phenomena to defend his mechanical matter theory. As Newman rightly insists, the main area in which Boyle looked for empirical support for his mechanical philosophy was chemistry. I delay most of my discussion of that particular topic to Chapter 8, which is devoted to the emergence of modern chemistry. Empirical support from areas other than chemistry is discussed later in this chapter. Before exploring that issue further, it is necessary for me to investigate in some detail the character of the knowledge of the phenomena that Boyle was able to invoke. That knowledge typically took the form of experimental knowledge, much of it of Boyle's own devising.

6.5 Boyle's Experimental Science as Distinct from the Mechanical Philosophy

From the late 1650s onwards Boyle conducted experiments extensively and published the results of his efforts in tracts that he called 'histories', 'experimental essays' and 'new experiments'. He referred to the knowledge acquired by means of his experiments as 'matters of fact', 'physiology' and as knowledge of the phenomena.[10] I will refer to this body of work as Boyle's experimental science. In his articulation, appraisal and defence of that work Boyle distinguished between experimental science and the mechanical philosophy. In this section I focus on the character and status of the body of knowledge that Boyle considered to have established by experiments as Boyle himself construed them.

In 1661 Boyle published a collection of 'experimental essays' based on experimental work he had conducted at Oxford during the previous 5 years. In the first of these, 'A proemial essay . . . with some considerations touching experimental essays in general', Boyle (2000, Vol. 2, pp. 9–34) spelt out the character of the knowledge he believed himself to have produced through his experimental work. There are also scattered references bearing on the issue in his unpublished papers. Especially significant is an unfinished 'Essay of various degrees or kinds of ye knowledge of natural things'.[11] From these sources and others it is possible to construct a picture of Boyle's view on the status of knowledge established by experiment as distinct from the matter theory codified in his mechanical philosophy.

According to Boyle (2000, Vol. 2, p. 21) there is a 'scale, or series of causes' and corresponding 'degrees of explication'.[12] Highest in the scale of natural causes are the truly mechanical causes stemming from the motions of minima or atoms of matter, characterised in terms of the primary affections. The most fundamental explanations are those that invoke such causes. Lowest in the scale of causes are the most readily accessible causes, such as the weight of a stone invoked to explain its fall. An exemplification of Boyle's scale of causes is the following. We can explain the mercury level in a barometer by appealing to air pressure. We then move up the scale to explain pressure by invoking the elasticity (the 'spring') and weight of air. Perhaps the elasticity of air can be explained by appeal to the elasticity of its component particles and so on until we reach the highest level, where elasticity of the particles is explained by appeal to the shapes, sizes and motions of portions of universal matter.

Boyle acknowledged that the highest level of causes and the fundamental explanations were difficult, if not impossible, to arrive at, so that 'we may aspire to, but must not always require or expect, such a knowledge of things as is immediately derived from first principles'.[13] Consequently, explanations in experimental science typically involve appeal to 'subordinate principles' and 'intermediate causes'.

And, indeed, there are oftentimes so many subordinate Causes between particular Effects and the most General Causes of things that there is left a large field, wherein to exercise Men's Industry and Reason, if they will but solidly enough deduce the Properties of things from more general and familiar Qualities, and also intermediate Causes (if I may so call them) from one another. (Boyle, 2000, Vol. 2, p. 23)

Boyle listed gravity, fermentation, springiness and magnetism amongst the subordinate principles. Indeed, he made it plain that most of the science of his day, including his own chemistry and pneumatics, was to be regarded as offering knowledge of intermediate rather than first, that is mechanical, causes.

Of the subordinate or intermediate causes or theories of natural things, there may be many: some more and some less remote from the First Principles and yet each of them capable to afford a just delight and useful instruction to the mind. And these we may call for distinction sake the Cosmographical, the Hydrostatical, the Anatomical, the Magnetical, the Chymical and other causes or reasons of Phenomena as those which are more immediate (in our way of estimating things) than ye general and Primordial causes of natural effects.[14]

Boyle acknowledged that it is 'very fit and highly useful, that some speculative wits, well versed in mathematical principles and mechanical contrivance' should speculate about fundamental mechanical explanations of phenomena, but he also expressed his reservations about those involved in such an enterprise since 'they oftentimes give forced and unnatural accounts of things rather than not to be thought to have deriv'd them immediately from the highest principles'. What is worse, 'they despise and perhaps too condemn or censure all yt knowledge of the works of nature yt Physicians, Chymists, and others pretend to, because they cannot be clearly and easily deduc'd from ye doctrines of Atoms, or ye Catholick Laws of motion'.[15] In Boyle's view, explanations that appealed to intermediate causes and were substantiated by experiment were not to be despised. They might fall short of fundamental or ultimate explanations, but they were genuine and useful explanations for all that.

He gives some Reason, why Stones and Iron, and all other heavy bodies, will swim in Quick-silver, except Gold, which will sink in it; that teaches that all those other bodies are *in specie* (as they speak) or bulk for bulk, lighter than Quick-silver, whereas Gold is heavier. He, I say, may be allow'd to have render'd a Reason of a thing proposed, that thus refers the Phaenomenon to that known Affection of almost all Bodies here below, which we call Gravity, though he does not deduce the Phaenomenon from Atoms, nor give us the cause of Gravity; as indeed scarce any Philosopher has yet given us a satisfactory Account of it. So if it is demanded, why, if the sides of a blown Bladder be somewhat squeez'd betwixt one's hands, they will, upon the removal of that which compress'd them, fly our again, and restore the Bladder to its former figure and dimensions; it is not saying nothing to the purpose, to say, that this happens from the spring of those Aerial Particles, wherewith the Bladder is fill'd, though he, that says this, be not perhaps able to declare, whence proceeds the Motion of Restitution, either in a Particle of compress'd Air, or any other bent spring. (Boyle, 2000, Vol. 2, p. 22)

The aim to establish intermediate explanations by way of experiment, then, is legitimate useful and productive. What is more, the strong implication is that the search for fundamental atomic explanations is typically futile and unproductive. When Boyle wrote that 'the most useful notions we have, both in Physics, mechanics, Chymistry, and ye medicinal art, are not deriv'd from ye first principles, but from intermediate Theories, notions and rules' he was in effect saying that experimental science is able to proceed productively independently of the dictates of the mechanical philosophy.[16] He insisted that it would be 'backward to reject or despise all explications that are not immediately deduced from the shape, bigness and motion of atoms or other insensible particles of matter' and urged that those that persist with that mechanical programme 'undertake a harder task than they imagine'.[17]

Boyle employed experiments to support matters of fact and was quite explicit about what could be achieved by doing so. I cannot improve on the account of Boyle's reflections on the capabilities of experiment for establishing matters of fact given by Rose-Mary Sargent (1995, pp. 159–180). I summarise key points she attributes to Boyle as follows: The normal course of nature involves a complicated combination of multiple causes producing their effects in ways that are difficult to fathom. The artificial conditions of an experiment enable the situation to be simplified, individual causes isolated and their effects investigated.[18] Chemical changes, for example, are best investigated by preparing pure substances and engineering their combination in simple, controlled conditions. The purity of chemicals is best accomplished by a range of tests rather than a single one. Indeed, it is always advisable to support matters of fact by a range of tests preferably involving a variety of instruments. The ability to purposefully reproduce an experimental effect is important, but of great significance is the ability to vary the conditions under which causes can be made to yield their effects. An experiment presupposes knowledge of the experimental situation. The purification of a chemical substance requires knowledge of chemistry sufficient to dictate the procedures to be carried out and the tests to be administered. Just because experimental inferences presuppose knowledge they are fallible. Matters of fact are never final, although a sound experimental case is necessary to revoke a claim that has been supported by a range of independent tests.[19] An experimental claim is supported by a range of independent evidence in a way similar to that in which a claim in a court of law is substantiated by evidence from independent witnesses. In either case, to the deny the claim supported is to accept unexplained coincidences

Boyle's pneumatics had good claims to the status of matters of fact established by experiment. Boyle claimed that air has a pressure arising from its weight and its 'spring' and appealed to it to explain a range of phenomena involving barometers, syringes and the like. Boyle supported his claims by a range of evidence, the most novel and striking involving use of his air pump. Over a decade or more Boyle improved and modified his air-pump experiments and augmented them with other experiments in response to alternative explanations of his results offered by critics. Acknowledging the challenge to his 'grand hypotheses' concerning the weight and spring of the air Boyle proceeded to defend them, and the case he made was a detailed experimental one whose status he compared to the case made by Harvey for the circulation of the blood.[20]

The status of Boyle's pneumatics as an example of experimental knowledge, as distinct from the kind of knowledge that the mechanical philosophy involved and also from what Boyle termed metaphysics, is brought out in Boyle's dispute with Hobbes. The latter insisted that the receiver evacuated by Boyle's air pump was occupied by some subtle matter rather than coarse air, and attributed pressure to the circulation of that subtle matter. He also criticised Boyle for attributing elasticity to air without being able to explain it. According to Hobbes, the mechanical philosopher, attributing elasticity to air was tantamount to admitting that it could move itself.[21] Boyle responded in several ways, all of which involved retreating to what he regarded as 'matters of fact' that could be established by experiment. He refrained from taking a stand on the possibility of a vacuum and on whether his evacuated receiver or the space above the mercury in a barometer constituted one. He regarded such an issue as 'metaphysical' because not susceptible to experimental investigation.[22] He did claim that his evacuated receiver was relatively free of air and was able to give a range of experimental evidence for that claim. He freely admitted he had not given a mechanical explanation of the 'spring' of the air but insisted that he had shown experimentally that air has a spring and that it can be appealed to in order to explain the behaviour of barometers and various phenomena revealed by use of his air pump and otherwise.[23]

Another instructive dispute involved Spinoza's challenge to Boyle's interpretation of his experiments on nitre (potassium nitrate). The dispute was carried on in correspondence between Spinoza and Oldenburg, with the latter's exposition of Boyle's view constructed in consultation with Boyle.[24] In those experiments Boyle transformed nitre into 'fixed nitre' (potassium carbonate) by plunging red-hot carbon into it and then recovered the nitre by adding spirit of nitre (nitric acid). Spinoza criticised Boyle for not substantiating a truly mechanical account of the process and attempted to repair the deficiency. He claimed that nitre and spirit of nitre are in fact composed of the same matter, the difference lying in the rapid motion of the underlying matter in the case of spirit of nitre. Boyle declined to speculate about a precise mechanism. He claimed to have shown that nitre can be broken into and built up from fixed nitre and spirit of nitre and urged that this posed problems for the appeal to substantial forms. (The latter point did not impress Spinoza, who already took the dismissal of substantial forms for granted.)

Boyle's understanding of experimental knowledge makes it quite different from what would nowadays be described as some extreme positivist or empiricist ideal, according to which facts are given and able to speak for themselves in some straightforward way. Establishing experimental facts involves purposeful activity, guided by knowledge of the situation investigated. Boyle (2000, Vol. 2, p. 14) made it clear in his 'Proemial essay' that such work often needed to be guided by hypotheses and that appropriate concepts (which Boyle referred to as 'notions') necessary for the framing of such hypotheses needed to be fashioned (2000, Vol. 2, p. 20). He referred to the spring and weight he attributed to air as 'grand hypotheses' but by the end of the series of experiments involved in resolving disputes with critic such as Hobbes he was claiming them as 'matters of fact'.

Boyle advocated and put into practice means for establishing experimental matters of fact by putting them to a range of experimental tests and by ruling out alternatives and possible sources of error. He put the idea into practice to great effect in his pneumatics. Experimental knowledge is not established infallibly because there may remain sources of error unknown and hence untested for. However, from the new perspective, an objection to a piece of experimental knowledge should take the form of one that opens up the possibility of some new experimental test. A generalised point about the fallibility of experimental reasoning will not do, nor will an appeal to untestable possibilities (such as positing an ethereal medium in the receiver of Boyle's pump with no testable consequences). In evaluating experimental claims, the results of tests are primary and the consent of the community is secondary. The aim is to produce knowledge rather than consensus about knowledge.[25]

Experimental knowledge can be strongly established, although not infallibly, as matter of fact. At the other extreme Boyle placed metaphysics that could not be tested by experiment at all. This leaves the question of the status of the mechanical philosophy. I have given ample evidence that Boyle distinguished it from experimental knowledge. Nevertheless, Boyle did not leave room for doubt that he considered the mechanical philosophy to be empirically supported in some way. He distinguished between it and metaphysics. Boyle's view on the empirical status of the mechanical philosophy is the issue discussed in the next section.

6.6 Empirical Support for the Mechanical Philosophy

The characterisation I have offered of Boyle's mechanical philosophy and his experimental science suggests a distinct lack of fit between the two. Mechanical atomism was remote from what could be tested experimentally whilst Boyle's experimental science was required to pass stringent experimental tests. Boyle acknowledged and exploited a distinction between untestable metaphysics and his experimental science but did not conclude that his mechanical philosophy constituted metaphysics. Rather, he claimed that his mechanical atomism was susceptible to and in need of experimental support. This is reflected in the fact that he referred to the mechanical philosophy as a hypothesis. Early in the 'Origin of forms and qualities' Boyle (2000, Vol. 5, p. 296) explicitly made the point that 'by the Lovers of real Learning it is very much wish'd, that the Doctrines of the new Philosophy (as 'tis called) were back'd by particular Experiments, the want of which I have endeavour'd to supply'.

Not only did Boyle construe the mechanical philosophy as in need of experimental support, but also he claimed to have supplied a good deal of such support. He repeatedly claimed that the experiments that he appealed to in his chemistry, such as his experiments on nitre, lent support to his mechanical philosophy. Towards the end of his 'Excellency of the mechanical hypothesis' Boyle (2000, Vol. 8, p. 114) clearly expressed the view that experimental support for the mechanical philosophy was on the increase.

> [T]he Sagacity and Industry of modern Naturalists and Mathematicians, having happily applied them [mechanical principles and explications] on several of those difficult *Phenomena* (in *Hydrostaticks*, the practical part of *Opticks, Gunnery, &c.*) then before were referr'd to occult Qualities, 'tis probable that, when this Philosophy is deeplier searched into and further improv'd, it will be found applicable to the solution of more and more of the *Phenomena of Nature*.

> And therefore, if the Mechanical Philosophy go on to explicate things Corporeal at the rate it has of recent years proceeded at, 'tis scarce to be doubted, but that in time unprejudic'd persons will think it sufficiently recommended by its consistency with it self, and its applicableness to so many Phenomena of Nature.

The general problem with claims such as these, implicit in the discussion of the previous section, is that the scientific successes that Boyle invokes in support of the mechanical philosophy do not involve mechanical explanations in the strict sense because they involve appeal to such things as weight, elasticity, the reflecting and refracting properties of materials and so on rather than involving reductions to the primary affections only. This is precisely the point Boyle makes when he invokes the scale of causes, with experimentally accessible ones near the bottom of the scale and remote mechanical ones at the top. It is exemplified in Boyle's response to criticisms from Hobbes and Spinoza that we have described above. If Boyle's pneumatics and his experiments on the analysis and synthesis of nitre do not provide strict mechanical explanations then how can those examples of experimental knowledge be invoked to provide empirical support for the mechanical philosophy in the strict sense?

I mention one more instance of the problematic character of Boyle's insistence that his experimental matters of fact provided support for the mechanical philosophy before offering a solution to the puzzle. In an essay 'Of the imperfections of the chymist's doctrine of qualities' Boyle criticises attempts by chemists to reduce the object of their study to the action of the three principles, mercury, salt and sulphur. One of the key objections raised by Boyle (2000, Vol. 8, p. 166) is to the effect that, even if their attempts were in some sense empirically successful, the theory of the chemists would be inadequate because it leaves the properties of the three principles themselves, such as the fusibility and inflammability of sulphur or the weight and solidity of salt, unexplained. 'For even when the explications seem to come home to the phenomena, they are not primary and, if I may so speak, *fontal* enough'. This stand of Boyle seems to be in conflict, for example, with his stand on the status of his pneumatics, where he insists that his explanations have merit in spite of the fact that weight, spring and pressure of air remain unexplained. The properties invoked by Boyle in his experimental science are no more 'fontal' than the principles of the chemists.

I suggest the way to dissolve these apparent tensions in Boyle's writings is to respect his distinction between the mechanical philosophy and his experimental science and to recognise that, whilst he sought empirical support for the mechanical philosophy, it was of a different, and weaker, kind than the stringent kind of support he demanded of experimental knowledge. In the previous section I documented Boyle's distinction between the mechanical philosophy implicated at the top of the scale of causes and experimental knowledge of causes lower down the scale. It remains for me to locate in Boyle's writings the idea that the kind of empirical support sought for the mechanical philosophy was different in kind from that demanded of experimental knowledge.

Boyle did not claim that his mechanical philosophy qualified as a matter of fact. In the context of his pneumatics, for instance, he was quite explicit on the point that to show that air has weight and a spring and to explain a range of phenomena by appeal to it was one thing, whereas to give an explanation of the weight and spring of air by appeal to the configurations and motions of underlying corpuscles was another. He claimed to have done the former while acknowledging his inability to do the latter. The whole tenor of his 'Proemial essay' in which Boyle (2000, Vol. 2, p. 14) introduced and explained the purpose of his early empirical work is that there is a distinction to be drawn between matters of fact and philosophical systems such as those articulated by Gassendi, Descartes and Aristotle. Some 15 years later, in a tract on the 'mechanical origin or production of divers particular qualities' Boyle commented specifically on the relation between matters of fact and his own mechanical philosophy. In his view the 'corpuscular doctrine' could be backed up by appeal to experiment to the extent that possible corpuscular mechanisms could be proposed that served to explain the phenomena or render them compatible with the mechanical philosophy. The case could be further strengthened by pointing to the difficulty of reconciling the phenomena with alternative hypotheses such as those involving appeal to Aristotelian substantial forms or real qualities. The more matters of fact that could be accommodated by the mechanical philosophy, and the greater the problems posed by them for alternative hypotheses, the stronger the case for it. However, Boyle did not consider the case to be strong enough to qualify that philosophy as a matter of fact. Showing that corpuscular mechanisms could be contrived capable of explaining the phenomena was not sufficient to establish that the contrived mechanisms corresponded to the true ones.[26] I quote Boyle's own words on the distinction, central to my argument, between the status of matters of fact and the mechanical philosophy.

> There is yet another way of arguing in favour of the *Corpuscular* Doctrine of Qualities, which, though it do not afford direct proofs of its being the best *Hypothesis*, yet it may much strengthen the Arguments drawn from other Topicks, and thereby serve to recommend the Doctrine it self. For, the use of an *Hypothesis* being to render an intelligible account of the Causes of the Effects or Phaenomena propos'd, without crossing the Laws of Nature or other Phaenomena, the more numerous and the more various the Particulars are, whereof some are *explicable* by the assign'd Hypothesis, and some are *agreeable* to it, or at least are nor dissonant from it, the more valuable is the Hypothesis, and the more likely to be true. For 'tis much more difficult, to find an *Hypothesis* that is not true which will suit with *many* Phaenomena, especially if they be of various kinds, than but with *few*. And for this Reason, I have set down among the Instances belonging to particular Qualities some such experiments and observations, as we are now speaking of, since, although they may be not direct proofs of the preferableness of our Doctrine, yet they may serve for Confirmation of it. (Boyle, 2000, Vol. 8, p. 325)

Boyle proceeds to illustrate his point drawing an analogy with clocks. Boyle raises the (no doubt apocryphal) story of the Chinese who, when first confronted with a working clock, presumed it to be a kind of animal. Boyle explains that offering an explanation of the clock's behaviour by contriving some possible mechanism involving wheels, springs and so on is sufficient to dispel the belief that the clock must be animated, even though the postulated mechanism may not correspond to that in place in the clock in question. In like manner, contriving a possible mechanism that serves to explain some phenomenon provides some grounds that the phenomenon is at bottom mechanical. The proposed mechanism may not be the true one, but the fact that it is a possible one at least shows that invoking substantial forms is not necessary.[27]

Boyle is here responding to the claim that the mechanical explanations, whilst they can be effectively applied to mechanisms such as clocks and watches, cannot possibly be extended to explanations of all the phenomena of nature. 'To remove therefore this grand Prejudice and Objection', Boyle (2000, Vol. 8, pp. 326–327) writes, 'which seems to be the chief thing that has kept off Rational Inquirers from closing with the Mechanical Philosophy, it may be very conducive, if not sufficient, to propose some Mechanical accounts of Particular Qualities themselves as are intelligible and possible, and are agreeable to the *Phenomena* whereto they are applied'. Given the phenomena, the mechanical philosophy can be made plausible and even probable to the extent that hypothetical mechanisms can be devised that are sufficient to account for those phenomena. Such a mode of argument does not constitute 'direct proof' of the mechanical philosophy, but it does support that hypothesis to some degree and at least makes appeal to such things as sympathies, antipathies and substantial forms unnecessary.

When Boyle raised the issue of empirical support for the mechanical philosophy in 'Excellency of the mechanical hypothesis' he drew an analogy with code-breaking. Just as one can claim to have found the correct key to a cipher by showing how application of that key makes sense of a variety of encrypted messages, so is the mechanical philosophy supported by being rendered compatible with a variety of phenomena. Immediately following this analogy comes the passages referred to above in which Boyle claims empirical support for the mechanical philosophy.

The distinction between the strong mode of support implicit in Boyle's experimental science and the weaker notion he invoked in the context of his mechanical philosophy can be summed up by recognising that the former involves confirmation by empirical evidence whereas the latter involves mere accommodation to evidence. Arguments from coincidence work for the former in a way they do not for the latter. An experimentalist can hope to provide a range of evidence for a claim to the extent that denying that claim involves admitting a remarkable coincidence. Such arguments lose their force in the case of mechanisms that have been contrived to fit phenomena established by other means. It is no coincidence that a range of mechanisms fit the phenomena if they have been contrived to do so. Here, Boyle's analogy with code-breaking is deceptive. The case that the correct key to a cipher has been found rests on the fact that that one key naturally applied to a range of cases yields intelligible messages. The case would lose its force if the key needed adjusting or amplifying to meet the needs of each individual case.

Another mark of the distinction between accommodation and confirmation that can be discerned in Boyle's work is the fact that he is content to offer more than one, mutually inconsistent, mechanical accommodations of a phenomenon. For instance, as we shall have occasion to discuss in more detail in Chapter 8, Boyle (2000, Vol. 8, p. 470) suggests that the transformation in the properties of mercury sublimate when it is combined with more mercury can be explained, either by the sharp edges he attributes to particles of sublimate be sheathed as a result of the combination with mercury or that the corpuscles of sublimate be arranged in bundles with their sharp edges pointing inwards. This contrasts with Boyle's attitude to alternative explanations in the context of his experimental science. In his pneumatics, for example, he goes to great pains to conduct experiments to counter alternative explanations to his own. Two accommodations are better than one so long as mere accommodation is the aim. A stronger sense of empirical support is required of experimental knowledge.

With the distinction between the confirmation of and accommodation to the evidence in hand, there is scope for removing some of the apparent tensions in Boyle's writings that I have pinpointed earlier in this section. When Boyle criticises the chymists for employing their principles in a way that is not sufficiently 'fontal' he is speaking as a mechanical philosopher who is seeking ultimate explanations at the top of the series of causes. When he defends his appeal to weight, spring and pressure in his pneumatics he is speaking as an experimental scientist. When Boyle, in the 'Excellency of the mechanical hypothesis' claims that that hypothesis had empirical support that was increasing, his claim makes most sense if the support in question is of the weak kind appropriate for supporting ultimate claims about the structure of matter.

I have done my best to free Boyle's remarks about empirical support for the mechanical philosophy from inconsistency by highlighting the distinction Boyle himself draws between that philosophy and experimental knowledge and the mode of support appropriate in the two domains. However, I insist that, if the mechanical philosophy is interpreted in the strict sense summarised in Section 6.2 of this chapter, then support for it, even in the weak sense identified in the previous section, was scant. Boyle acknowledged that neither he nor any other of his fellow mechanical philosophers were able to devise mechanisms capable of explaining common properties such as weight and elasticity. Boyle cannot claim advances in hydrostatics, optics and gunnery as support for the mechanical philosophy in the strict sense, as he seems to do in the passage from the 'Excellency of the mechanical hypothesis' quoted early in this section until he has constructed strict, albeit hypothetical, mechanical mechanisms able to account for the properties, such as weight, involved in those sciences. Not even colliding billiard balls, an archetypal example of a mechanism, can be accommodated by the mechanical philosophy in the strict sense once it is realised that the colliding balls must be elastic to some degree.[28] Some of my critics insist that I am constructing a problem of my own devising here by imposing on Boyle's mechanical philosophy a sense of mechanical that is inappropriately strict. I take up this issue in the next section but one.

6.7 The Lack of Fertility of the Mechanical Philosophy

Suppose my claims about the lack of experimental support for the mechanical philosophy in the strict sense is accepted. There remains the possibility that that philosophy proved its worth by usefully guiding experimental science. Was Boyle's mechanical atomism fertile insofar as it encouraged lines of experimental enquiry that bore fruit? In this section I argue that it was not, nor was its general character conducive to its being so.

As we have seen, Boyle's mechanical philosophy was to be supported by contriving mechanisms able to account for the phenomena. This implies that the phenomena be known prior to the construction of the mechanisms and hence independent of them. A theme in the opening pages of Boyle's characterisation and promotion of experimental knowledge in his 'Proemial essay' is that most attempts at constructing grand systems of natural philosophy are deficient because based on inadequate knowledge of the phenomena. This implies that the latter knowledge is independent of and prior to the mechanical explanation. Later in the essay Boyle makes the point that research must necessarily start at the bottom of the scale of causes and work up.

> And though it must not be denied, that it is an advantage as well as a satisfaction, to know in general, how the qualities of things are deducible from the primitive affections of the smallest parts of matter; yet whether we know that or no, if we know the qualities of this or that body they compose, and how it is disposed to work upon other bodies, or be brought on by them, we may, *without ascending to the top of the series of causes, perform things of great moment*, and such as, without the diligent examination of particular bodies, would, I fear, never have been found out *a priori*, even by the most profound contemplator.[29]

Boyle (2000, Vol. 2, p. 12) does, nevertheless, raise the possibility that a philosophical system, of which I take his own mechanical philosophy to be an example, might guide and inspire the search for new experimental knowledge in its support.

> For such kind of Writings [compleat Bodies or Systems of Physiology], if their Authors be (as for the most part they are) subtle and inquisitive men, there may be very good use, not so much by their gratifying the Intellect with the plausible account of some of Nature's Mysteries; as because on the other side their Writers, to make good their new Opinions, must either bring New Experiments and Observations, or else must consider those that are known from a new Manner, and thereby make us take notice of something in them unheeded before; and on the other side, the curiosity of Readers, whether they like or disapprove the Hypothesis propos'd, is wont to be thereby excited to make trial of several things, which seeming to be Consequences of his new Doctrine, may by their proving agreeable or repugnant to experiment either establish or overthrow it.

There is a key feature of Boyle's mechanical atomism that stands in the way of interpreting it as capable of guiding experiment in the fruitful way depicted in the above passage. It is a feature that Boyle himself stresses when responding to the charge that the stark world of impenetrable atoms possessing only shape, size and motion is incapable of accounting for the vast variety of observable phenomena. The response that Boyle (2000, Vol. 8, p. 113) has to the objection is that the mechanical philosophy is so flexible, insofar as it is free to postulate atoms with innumerable shapes, sizes, arrangements and motions, that it is possible to accommodate it to phenomena of 'as great a variety as need be wish'd for, and indeed a greater than can easily be so much as imagin'd'. I stress the flexibility that Boyle presents as a positive attribute of his matter theory by quoting at length from his 'History of Particular Qualities' where he is responding to the 'Grand difficulty' that a great variety of phenomena should be derived simply from particles of matter in motion. A large part of his response stresses the extent to which his mechanical atomism is 'fertile' and 'comprehensive' on account of the vast variety of shapes, sizes, arrangements and motions that the mechanical philosopher is free to invoke.

> And so though Figure be one of the most simple modes of Matter; yet it is capable, *partly* in regard of the surface of the figur'd Corpuscle (which may consist of Triangles, Squares, Pentagons &c.) and *partly* in regard of the shape of the body itselfe, which may be either flat like a cheese, or Lozenge; or Spherical like a Bullet; or Elliptical, almost like an Egge; or Cubical like a Dye; or Cylindical like a rolling-stone; or Pointed like a Pyramid, or Sugar-Loaf; Figure I say, though but a simple mode, is upon those and other scores, capable of so great a multitude of differences, that is concerning Them, and their Affections, that *Euclid, Apollonius, Archimedes, Theodosius, Clavius*; and later writers then he, have demonstrated so many Propositions. And yet all the hitherto nam'd Figures are almost nothing to those irregular Shapes, such as are to be met with among Rubbish, and among hooked and branched Particles &c. that are to be met with among Corpuscles and Bodies; most of which have no particular Appellations; their Multitude and their Variety having kept men from enumerating them, and much more from particularly naming them.
>
> To which let me add, that these Varieties of Figure, and Shape, do also serve to modifie the Motion, and other Affections of the Corpuscles endowed with them, and of the compounded Body whereof it makes a part. (Boyle, 2000, Vol. 6, p. 276)

Boyle proceeds to document the wide diversity of possible motions at even greater length.

There are two points I wish to make about this emphasis by Boyle on the flexibility of his mechanical philosophy. The first is that this flexibility detracts from the merit and significance of the ability of the mechanical philosophers to adapt their system to the phenomena. Boyle (2000, Vol. 8, p. 114) explicitly claims that the mechanical principles 'being so general and pregnant that among things corporeal there is nothing *real* (and I meddle not with *Chymerical* beings, such as some of *Paracelsus*'s) that may not be deriv'd from, or be brought to a subordination to, such comprehensive Principles'. If Boyle is able to specify in advance of the discovery of phenomena that the mechanical philosophy will be able to accommodate them, then instances of such adaptations can hardly be seen as significant confirmations of that philosophy. A theory that is adaptable to everything explains nothing. (Here Popperians prick up their ears.) Incidentally, the flexibility of the mechanical philosophy notwithstanding, its supporters were not able to devise mechanisms to account for gravity and elasticity at their own admission.

Closely allied with the foregoing point is the observation that just because of the freedom a mechanical philosopher like Boyle had to attribute shapes, sizes, arrangements and motions to atoms to suit the purpose in hand, the matter theory in its generality was not capable of predicting any phenomena and so not capable of guiding experimental investigation. This reinforces the point made above that knowledge of the phenomena needed to be in place before mechanical explanations of them could be contrived. Knowledge of Boyle's intermediate and subsidiary causes was needed prior to the possibility of reducing them to mechanical causes by contriving mechanisms. Claims to the effect that Boyle's mechanical philosophy in the strict sense was fertile and fruitful were unfounded as far as the production of experimental knowledge is concerned.

Newman (2006) might well object to the above assessment on the grounds that it neglects the main area in which there was a fruitful inter-relation between the mechanical philosophy and experiment, namely, chemistry. On this point he would have had the support of Boyle. I re-iterate my intention of treating the case of chemistry in detail, in Chapter 8.

6.8 The Various Senses of 'Mechanical'

In the foregoing I have attributed to Boyle an adherence to a matter theory that I have referred to as 'the mechanical philosophy in the strict sense' and have insisted on a distinction between it and Boyle's experimental science. There are senses of 'mechanical' other than the strict sense. Indeed, there are common usages of 'mechanical' that have stronger claims to the appellation than the strict sense, which is 'mechanical' in a technical sense that is strained and artificial compared with the common sense. As we have seen, according to the strict sense of 'mechanical' a clock is not a mechanism insofar as the weight of the pendulum bob that drives it has not been reduced to the shapes, sizes, arrangements and motions of particles of impenetrable matter. Is it not an unusual usage of 'mechanical' that denies the right of a clock to be called such? Surely a clock is an archetypal mechanism as far as common usage is concerned. Given this, the possibility has to be entertained that my key claims about the distinction between the mechanical philosophy and experimental science is an artificial consequence of the very strict and technical way I interpret the mechanical philosophy to be 'mechanical'. If the mechanical philosophy is interpreted in some common, as opposed to the strict, sense, then perhaps that philosophy and the new experimental science championed by the likes of Boyle form a coherent whole just as Boyle implied that it did. My critics mentioned in Footnote 7 argue along these lines.

There certainly are common senses of 'mechanical' that differ from the strict sense. Clocks, watches and levers, and machines or engines generally, are mechanisms in the common sense. A feature of such mechanisms is that their behaviour is accounted for by reference to the inter-relations of their parts. Insofar as the workings of a clock involves the transmission of motion from that of the pendulum bob to the hands via inter-locking gear wheels, no appeal to something over and above the inter-related parts, such as substantial forms, is called for. As Newman (2006, p.186) points out, there is a long tradition of such mechanical explanations going back to the Hellenistic engineers of Antiquity. Boyle (2000, Vol. 2, p. 87) did use 'mechanical' in this common sense, and even cited the analogy between the explanations involved in his matter theory and those to explain the working of mechanical engines as the justification for the term 'mechanical philosophy'. He also employed a common sense of mechanical when he described the spring and weight of the air as 'mechanical affections'.

Another sense of mechanical, again with a long history, translates roughly as 'artisanal'. The mechanical arts were taken to include, not only the manufacture of mechanical devices, but also alchemy. According to this usage, the new experimental science championed by Boyle could be said to be mechanical insofar as it involved and encouraged artificial intervention in nature as a means of understanding it.

So why do I not heed my critics, drop my insistence on interpreting Boyle's mechanical philosophy in the strict sense and substitute for it a common sense of mechanical which would enable the new experimental science to be embraced as part and parcel of the mechanical philosophy just as its proponents often seemed to imply and many historians since then have assumed?

My main reason is this. When Boyle articulates and defends the mechanical philosophy, then, as we have seen, it is the strict sense of mechanical that is involved. What is more, most of the arguments for that philosophy, that the primary affections are simple, clear and intelligible and not in need of further explanation, do not work if mechanical is interpreted in the common sense. Size and shape as applied to pieces of universal matter may be perfectly clear and in need of no further explanation, but the mechanical philosophers, including Boyle, did not think this to be the case for weight and elasticity.

Boyle's mechanical philosophy was designed to eliminate, not just some extreme scholastic version of substantial form, but Aristotelian and scholastic form in general. His major essay outlining the mechanical philosophy aimed to give the 'origin of forms and qualities' generally, not substantial forms only. Anstey (2000, pp. 153–154)) rightly insists, with the approval of Newman (2006, p. 178, fn.), that Boyle's mechanical philosophy was, first and foremost, a theory of qualities. Boyle made it abundantly clear that the aim of his mechanical philosophy was to reduce qualities to the 'primitive affections' possessed by portions of matter that they necessarily possess by virtue of being such. In an essay devoted to the 'history of particular qualities', Boyle (2000, Vol. 6, p. 267) distinguished the primitive affections from all other qualities, requiring that the former be reduced to the latter.

> And there are some other Attributes, namely *Size, Shape, Motion* and *Rest*, that are won't to be reckon'd among Qualities, which may more conveniently be esteemed the *Primary Modes* of the parts of Matter; since from these simple Attributes, or Primordial Affections, all the Qualities are deriv'd.

As I have already noted in my summary of Boyle's case for his mechanical philosophy in Section 6.3 of this chapter, Boyle could not accept the intelligibility of the idea that qualities, such as the whiteness of a wall or any other quality other than the 'primary affections', which were something other than 'Matter or modes of Matter or immaterial substances', could be added to matter to confer on it the property in question. Boyle (2000, Vol. 5, p. 308) was intent on avoiding the assumption that 'there are in Natural Bodies store of *real Qualities* and other *real Accidents*, which not onley are no Moods of Matter, but are real Entities distinct from it'. Intermediate causes and subordinate principles involving properties such as the spring and weight of the air that played a crucial part in Boyle's experimental science are mechanical in a common sense. But to admit them into the mechanical philosophy in the strict sense is to undermine Boyle's main argument concerning the intelligibility of the latter.

The importance of the distinction between the common and strict senses of 'mechanical' and the importance of eliminating qualities other than the 'primitive affections' of matter from the mechanical philosophy in the strict sense can be brought out by attending to a general feature of the structural explanations that are involved in mechanical explanations in the common sense. As we have noted, such explanations involve explaining wholes in terms of the properties and inter-relations of their parts. Whether they are assumed implicitly or invoked explicitly, such explanations need to ascribe properties to the parts. Explanations of the workings of a clock that appeal to its mechanical structure presuppose the rigidity of gear-wheals and the weight of the pendulum bob. How are such properties to be regarded? From the point of view of the competing matter theories of the seventeenth century those properties can be attributed to the presence of forms or real qualities or they can be reduced to the primary affections of universal matter. As a mechanical philosopher Boyle was committed to the latter position. A third possibility is to recognise the need for appeal to non-ultimate 'subordinate causes' in science and to simply ignore the issue of an ultimate explanation of them. This is the course implicit in Boyle's experimental science.

Boyle's articulation of arguments for the mechanical philosophy does not make sense if mechanical is taken in the common rather than the strict sense. But there is more to it than that. The use to which Boyle put the mechanical philosophy outside of experimental science, especially in a theological context, required the strict rather than common sense of mechanical. For instance, the necessity of God's active intervention in nature can be persuasively argued for if nature is composed of inert pieces of matter possessing only shape, size and motion and becomes weakened by addition of properties such as weight and elasticity which, if attributed to matter primitively, has the consequence that matter can move itself, without the need for God's intervention. In putting his case for the active role of a deity in the workings of nature Boyle in effect makes explicit that his view requires mechanism in the strict rather than any weaker sense. Here are Boyle's own words:

> I shall next take notice, That Philosophers, who scorn to ascribe anything to God do often deceive themselves in thinking they have sufficiently satisfied our Enquiries, when they have given us the nearest and most immediate causes of some things; whereas oftentimes the assignment of those Causes is but the manifesting that such and such Effects may be deduc'd from the more Catholick affections of things, though these be not unfrequently as abstruse as the *Phenomena* explicated by them, as having onely their Effects more obvious, not their Nature better understood: As when, for instance, an account is demanded of that strange supposed Sympathy betwixt Quick-silver and Gold; in that we finde that whereas all other Bodies swim upon Quicksilver, it will readily swallow up Gold and hide it in its Bosom. This pretended Sympathy the Naturalist may explicate by saying, That Gold being the only Body heavier than Quick-silver of the same bulk, the known Laws of *Hydrostatikcs* make it necessary that Gold should sink in it and all lighter Bodies swim on it: But though the cause of this Effect be thus plausibly assign'd, by deducing it from so known and obvious affection of Bodies as Gravity, which every man is apt to think he sufficiently understands; yet will not this put a satisfactorie period to a severe Inquirer's Curiositie, who will perchance be apt to alledge, That though the Effects of Gravity indeed be very obvious, yet the Cause and

> Nature of it are as obscure as those of almost any *Phenomenon* it can be brought to explicate, and that therefore he that desires no further account desists too soon from his Enquiries, and acquiesces long before he comes to his Journey's end. And indeed, the investigation of the true nature and adequate cause of gravity is a task of that difficulty that, in spite of aught I have hitherto seen or read, I must yet retain great doubts whether they have been clearly and solidly made out by any Man. And sure, *Pyrophilus*, these are divers Effects in Nature, of which, though the immediate Cause may be plausibly assigned, yet if we further enquire into the Causes of those Causes, and desist not from ascending in the Scale of Causes till we are arriv'd at the top of it, we shall perhaps finde the more Catholick and Primary causes of Things to be either certain primitive, general and fix'd Laws of Nature (or rules of Action and Passion among the parcels of the Universal Matter); or else the Shape, Size, Motion, and other primary Affections of the smallest parts of Matter, and of their first Coalitions or Clusters, especially those endowed with seminal Faculties or properties; or (to dispatch) the admirable conspiring of the several parts of the Universe to the production of particular Effects; of all which it will be difficult to give a satisfactory Account, without acknowledging an intelligent Author or Disposer of things.[30]

Boyle here, in effect, makes explicit the point that experimental science employing mechanisms in the common sense falls short of explanations in the strict mechanical sense, and then proceeds to make use of mechanism in the strict sense to make his theological point. I defy my critics to make sense of this passage by sticking to an interpretation of the mechanical philosophy that interprets mechanical only in the common sense.

A further reason why it is inappropriate to interpret Boyle's mechanical philosophy only in the common sense, or to confuse that sense with the strict sense, is that allegiance to the common sense of mechanical is insufficient to distinguish Boyle from his opponents in the way that he clearly wished to be. None of the Aristotelians or chemists with whom Boyle took issue need have had a problem with structural explanation of levers that is mechanical in a common sense. Explanation of the properties of chemicals in terms of their components was central to an alchemical tradition embraced by Aristotelians such as Daniel Sennert, as we have seen. The funicular hypothesis proposed by the Aristotelian Franciscus Linus as an alternative to Boyle's explanation of pneumatical phenomena had as strong a claim as Boyle's appeal to the spring of the air to be mechanical in the common sense. As we have seen, the artisanal sense of mechanical involved in the mechanical intervention into nature as a means of understanding it had been pioneered by Aristotelian alchemists from the thirteenth century up until Sennert and beyond. If there was a sense of mechanical that distinguished Boyle from his opponents it involved the radical rejection of form involved in his mechanical philosophy in the strict sense.

There are two senses of mechanical, and both are evident in the work of Boyle. One is the fundamental matter theory that I refer to as the mechanical philosophy in the strict sense. The other involves a number of inter-related notions that I have grouped together under the term 'mechanical philosophy in the common sense'. There are ways in which the experimental science that blossomed in the seventeenth century can be construed as mechanical in the common sense. But if we are to construe the scientific revolution in terms of the emergence of this kind of knowledge then we must distinguish that change from the transformation in matter theory from an Aristotelian to a mechanical view. I claim that that change was distinct from the emergence of experimental science, was not supported by it and did not contribute significantly to it.

6.9 Boyle's Mechanical Philosophy and Experimental Support for Atoms

The attempt I have made to distinguish between Boyle's mechanical philosophy in the strict sense and his experimental science, insofar as it is successful, undermines any claim to the effect that Boyle's theory constituted important progress towards an atomic theory of matter that was supported by experiment. To the extent that Boyle's mechanical philosophy assumed permanent particles of universal matter with unchanging shape and size as the building blocks of the material world, it was an atomic theory. But I have argued that that philosophy was far removed from what could be tested and supported experimentally. Boyle did make significant contributions to experimental science to be sure, with his pneumatics constituting an outstanding example. But the claims Boyle established as experimental matters of fact did not invoke or imply atoms.[31]

Notes

1. The main source is Descartes (1983).
2. Peter Anstey (2000, pp. 153–154) suggests that the term 'mechanical philosophy' was first introduced by Henry More in the context of Descartes' philosophy. But there is no doubt that Boyle was largely responsible for bringing the term into wide currency.
3. In focussing on Boyle I do not intend to imply that he brought about the scientific revolution single-handed. My focus is due to the fact the Boyle was both a key articulator of the mechanical philosophy and a prolific experimenter. These two aspects of his work are inter-related in important and instructive ways, and Boyle himself explicitly discussed that relationship.
4. See Hunter (1995), Newman (1994, pp. 54–91) and Newman and Principe (2002).
5. Boyle (2000, Vol. 5, pp. 281–442 and Vol. 8, pp. 99–117). These two essays are reproduced in Stewart (1979, pp. 1–96 and 138–154) along with other key papers by Boyle concerned with an exposition and defence of the mechanical philosophy. For a detailed exposition of Boyle's mechanical philosophy see Anstey (2000)
6. Boyle's use of the term natural minima here is misleading, since Boyle's natural minima have more in common with the atoms of the Ancients than with the natural minima of the medieval Aristotelians. The latter were minima of the substance they were minima of, whereas Boyle's minima, like those of the Greeks, are composed of the one universal matter.
7. Anstey (2002a), Pyle (2002) and Newman (2006, pp. 175–189) criticised the specification and critique of Boyle's mechanical philosophy in Chalmers (1993) on the grounds that I give an inappropriately strict account of 'mechanical' ignoring the less strict usages at work in Boyle. I have defended myself against Anstey and Pyle in my (2002b). I revisit the issue and take Newman's critique on board in some detail later in this chapter.
8. In similar vein, Anstey (2002b, p. 627) writes that Boyle's seminal principles 'were required to explain those phenomena that appeared beyond the capabilities of the corpuscular hypothesis'.
9. On Boyle's debt to Sennert see Newman (1996).

10. These three terms occur in the opening pages of 'A proemial essay' (Boyle, 2000, Vol. 2, pp. 9–35) written as an introduction to a variety of 'experimental essays' or 'physiological essays' that followed.
11. Boyle (1990, Vol. vii, f184, reel 5, frame 189). I have consulted the microfilm version of Boyle's papers at the Royal Society and have included the reel and frame numbers in my references.
12. The quotation is from Boyle's 'Proemial essay'. The scale of causes is also invoked in Boyle's 'Essay containing a requisite digression, concerning those that would exclude the deity from intermeddling with matter', in Boyle (2000, Vol. 3, p. 245). Something like Boyle's scale is found in Francis Bacon's *Novum organum*, Book 1, CIV.
13. Boyle (1990, Vol. viii, f184, reel 5, frame 189), underlined in the original.
14. Boyle (1990, Vol. ix, f40–41, reel 5, frame 250).
15. The quotations are from Boyle (2000, Vol. 2, pp. 21–22) and Boyle (1990, Vol. viii, f184, reel 5, frame 189) respectively.
16. Boyle (1990, Vol. ix, f40, reel 5, frame 250).
17. Boyle (1990, Vol. viii, f166, reel 5, frame 168).
18. We noted in the previous chapter that Sennert had made a similar point.
19. When I prepared the section on experiment for the revised edition of my introductory text, *What is this thing called science?* (Chalmers, 1999, pp. 27–40) I was insufficiently aware that most of the points I was making had already been made explicit by Boyle.
20. For the use of the expression 'grand hypotheses' see Boyle (2000, Vol. 3, p. 125).
21. Hobbes, *Dialogus physicus*, translated in Shapin and Schaffer (1985, pp. 254–255).
22. For Boyle's categorisation of issues, such as the possibility of a vacuum or the infinite divisibility of matter, see 'Some specimens of an attempt to make chymical experiments useful to illustrate the notion of the corpuscular philosophy' (Boyle, 2000, vol. 2, p. 87).
23. The dispute between Boyle and Hobbes, and also other adversaries of Boyle, namely More and Linus, is dealt with in detail in Shapin and Schaffer (1985). Much of their discussion is conducive to my interpretation of the status of Boyle's experimentation.
24. For the relevant correspondence and discussion by the translators see Hall and Hall (1965, pp. 458–470). The dispute is discussed by McKeon (1928, pp. 137–152) and, more recently, by Antonio Clericuzio (1990, pp. 561–589).
25. Here I distance myself from contemporary sociologist of science and social constructivists who see the need to bring wide-ranging social elements involving consent into the evaluation of scientific knowledge, so that knowledge has the status of a social convention, a line that forms part of Shapin and Shaffer's study of Boyle's experimentation.
26. Here and elsewhere, Boyle (2000, Vol. 8, p. 327) invoked an analogy, common amongst mechanical philosophers, with the status of mechanisms contrived to explain the observed motions of the hands of a clock, where mere compatibility with observation is insufficient to establish a conjectured mechanism as the correct one.
27. For an analysis of various uses of the clock analogy by the mechanical philosophers see Laudan (1966).
28. If Boyle's natural minima were to rebound on collision as billiard balls do then they would change direction instantaneously and so be moving in two directions at the same time at the instant of collision. The problem was noted by Leibniz. The problems that collisions posed for the mechanical philosophers have been discussed by Alan Gabbey (1985).
29. Birch (1744, Vol. 1, p. 199, my italics). The corresponding passage in Boyle (2000, Vol. 2, p. 21) is very nearly the same.
30. The quotation is from Boyle's 'Requisite digression concerning those who would exclude the deity from intermeddling with matter' in Boyle (2000, Vol. 3, p. 245).
31. Once again, I acknowledge that I have yet to tackle the details of Boyle's chemistry where Newman, for instance, finds the strongest grounds for the existence of an experimentally-supported corpuscular theory. I focus on that issue in Chapter 8.

Chapter 7
Newton's Atomism and its Fate

Abstract Newton's *Principia* contained a science of mechanics that was able to withstand experimental tests in a demanding way. Newton also articulated and defended an atomic theory of the ultimate structure of matter. His atoms bore the marks of his science insofar as inertia was attributed to them. In other respects they were particles of universal matter with a given shape and size very much like those of Boyle. Newton speculated that there were short-range forces at the atomic level analogous to the force of gravity acting between gross bodies at sensible distances identified in his mechanics. As was the case with the mechanical philosophers who came before him, Newton's atomistic matter theory was accommodated to rather than confirmed by observation and experiment. His atomism did not and could not fruitfully guide his experimenting. Eighteenth-century attempts to develop Newtonian atomism similarly did not bear fruit.

7.1 Introduction

In crucial respects the atomic theory of matter that can be gleaned from Isaac Newton's works is an extension and refinement of Boyle's atomism. As such it suffered from similar shortcomings and tensions. Both in Newton's scientific practice and his own exposition of the methodology involved in it we find a clarification and elaboration of Boyle's conception of experimental science based on matters of fact. But we also find Newton defending a natural philosophy and a matter theory that goes beyond what can plausibly be construed as significantly confirmed by matters of fact. Insofar as Newton defended those broader claims he did so by taking for granted assumptions that were akin to those involved in the mechanical philosophy. The roots of Newton's natural philosophy fed into his theology just as was the case with Boyle.

The comparison of Newton with Boyle needs to be qualified in a major way. A crucial difference was the use Newton made of his new science of mechanics, and especially the notion of force that it involved. As mentioned in Chapter 6, a limitation of Boyle's mechanical philosophy was its failure to identify the general laws of motion presumed to govern the motion of atoms. Not only did Newton correct this deficiency but he showed how his general laws of motion could be

A. Chalmers, *The Scientist's Atom and the Philosopher's Stone*, Boston Studies in the Philosophy of Science 279, DOI 10.1007/978-90-481-2362-9_7,

confirmed empirically. However, the notion of force was decidedly non-mechanical if 'mechanical' is taken in its strict sense and it created tensions within Newton's natural philosophy that his opponents took advantage of.

Newton's atomism, like Boyle's, was accommodated to, rather than confirmed by, experimental phenomena. This was especially the case in chemistry. Newton was able to exploit his notion of force to accommodate phenomena in a more convincing way than Boyle had, but his efforts in this regard in the context of atomism were mere accommodations nevertheless. Newton's new science of mechanics applied to macroscopic phenomena was to progress dramatically, with the identification of measurable forces associated with surface tension, electrical and magnetic attractions and so on, but at the atomic level his speculations, while highly influential, were unproductive. Revelations concerning Newton's extensive experimentation in alchemy, a natural outcome of his atomism, have done nothing to enhance his reputation as a pioneer of experimental science, but nor has it done anything to detract from the magnitude of his achievements in physics. It has not undermined the standing of his gravitational theory as an exemplary paradigm of a science that is extremely general, mathematically formulated yet experimentally confirmed.

7.2 Newton's Science

In referring to Newton's 'science' I refer to what Newton himself called 'experimental philosophy' (Newton, 1979, p. 394 and 1962, p. 547). The science of mechanics as set forth in Newton's *Principia* stands as one of the great scientific achievements of all time, although an adequate grasp of the achievement and its claim to fame requires an appreciation of the detailed way in which Newton brought his highly general, mathematically formulated theory to bear on the world. We need to understand the way and the extent to which Newton's mechanics, and especially his application of it to astronomy, was confirmed by and not merely accommodated to the phenomena.

The mechanics of the *Principia* was based on the three laws of motion. They involved a precise and novel conception of force as a cause of changes in uniform motion, rather than of motion itself. One such force, that of universal gravitation, is identified and given a mathematical formulation in the form of the inverse square law of attraction. Employing a primitive version of the calculus devised by Newton for the purpose, he derived within his theory explanations and predictions of a range of phenomena. At the terrestrial level these included free fall, projectile motion, the motion of pendulums and the laws of collision, while at the astronomical level they included the orbits of the planets and comets and a theory of the tides.

Newton insisted that his astronomy and mechanics were 'deduced from the phenomena' and involved no untestable hypotheses, as opposed to Descartes' system of vortices that Newton clearly saw as hypotheses devised to accommodate rather than genuinely explain or predict the phenomena. The distinctive feature of the *Principia* lies in the way that Newton was able to make good such claims in spite of the difficulties that stemmed from the generality of his claims and the complexity of the real world systems to which he applied his theory. This is most evident in Newton's astronomy, the 'system of the world' referred to in the full-length title of his masterpiece. The key to Newton's success was a method of successive approximation. He applied his theory first to idealised, simplified situations such as the motion of a point planet around a point sun. Guided by the results so achieved he then added corrections allowing, for instance, for the finite size of the bodies in the solar system and their gravitational interaction with each other as well as with the sun. The theory was borne out by the steady improvement of the match between theory and observation as this self-correcting procedure progressed.[1]

Newton's derivation of the law of gravitation 'from the phenomena' can only be adequately understood in terms of the subtleties of Newton's method of approximation, which involved not only refinements of the theory but also of the phenomena. As George Smith (2001b, p. 328) remarks the 'commonplace statement "Newton's theory of gravity explained Kepler's laws" scarcely begins to describe the complex relationship between Newton's theory and Kepler's orbital rules'. Kepler's rules (as they were called before Newton raised their status) were known to Newton to be only approximately borne out by the data and there were competing rules that fitted the data as well as Kepler's rules. What is more, given the attraction of the planets for each other, Newton's theory predicts the falsity of Kepler's laws taken literally so there is no question of the possibility of deriving Newton's theory from laws that are inconsistent with it, a logical point forced on contemporary philosophers of science by Pierre Duhem (1962, pp. 190–195) early in the twentieth century and stressed by Karl Popper. What Newton was able to show was that, in the context of his theory of motion and with appropriate simplifying assumptions, the approximate truth of Kepler's second and third rules (that in their motions the planets sweep out equal areas in equal times with periods the square of which are proportional to the cube of the mean radius of their orbits) implies the approximate truth of the inverse square law.[2] Newton then proceeded to add corrections that increased the match of his theory with observation of planetary positions, but not by accommodating those corrections to the data. Newton took the approximate orbits and used them to calculate corrections to them utilising the inverse square law he assumed to govern attractions between neighbouring planets. The fact that applying them led to an improved match between theory and data was by no means inevitable and the fact that it did ensue constituted evidence for Newton's theory.

Newton's case for his theory by no means stemmed from Kepler's laws and deviations from them alone. In the *Principia* Newton applied his theory to a range of phenomena, including the non-sphericity of the Earth, the orbit of the moon, which was especially complex because of the joint attraction of the sun and Earth, the tides, the precession of the equinoxes and the tracks of comets. In each case approximations were involved and assumptions added to the fundamental laws. But the additional assumptions had independent empirical support and the extent of the ensuing match between the data and the predictions of the modified or augmented theory were by no means guaranteed. Newton's success was not total. The moon's orbit proved to be particularly recalcitrant and in fact required mathematical techniques not available to him. The detailed experimental support for his theory marshalled by Newton in the *Principia* was substantial and already sufficient to show that his theory could not conceivably have been totally on the wrong track. The next century was to see the scope of the successful application of Newton's theory extended, thereby strengthening the case for it even further. Eventually, of course, it was to prove to have its limits, and to stand in need of correction or replacement by the Theory of Relativity. But the fact that contemporary relativity theorists demand that their theories yield Newton's theory as a limiting case for speeds small compared with the velocity of light and as gravity tends to zero implies an acknowledgement of the strength of, rather than deficiencies in, the case made for Newtonian theory constructed by Newton and his followers.

In my introductory chapter to this book I drew a distinction between the confirmation of a theory by data and mere accommodation of a theory to data. I suggested that a theory is confirmed to the extent that it predicts or explains a range of phenomena that follow naturally from it in conjunction with independently testable hypotheses. Newton's mechanics, as I have described it above, can be seen as a detailed instantiation of that claim. Not only can Newton's practice be read as proceeding in accordance to my strictures but he came close to acknowledging as much explicitly. In his earlier writings, and perhaps in his determination to distinguish his *Principia* and its methods from Descartes' *Principles of Philosophy*, Newton stressed his mechanics as being free from hypotheses and derived from the phenomena in so strong a sense that his remarks are difficult to reconcile with his practice if taken literally. But in his later writings he modified and qualified his remarks in a way that brings them into line with the position I have tried to capture with my distinction between confirmation and accommodation. Thus, in the second English addition of the *Opticks* published in 1717, in Query 31, Newton (1979, p. 404) characterised the status of his science in the following way:

> This Analysis consists in making Experiments and Observations, and in drawing general Conclusions from them by Induction, and admitting of no Objections against the Conclusions, but such as are taken from Experiments, or other certain truths. For Hypotheses are not to be regarded in experimental Philosophy. And although the arguing from Experiments and Observations by Induction be no Demonstration of general Conclusions; yet it is the best way of arguing which the Nature of Things admits of, and may be looked upon as so much the stronger, by how much the Induction is more general. And if no Exceptions occur from Phaenomena, the Conclusion may be pronounced generally. But if at any time afterwards any exception shall occur from Experiments, it may then begin to be pronounced with such exceptions as occur.

I presume that Newton's reference to making inductions more general is intended to capture the sense in which the greater diversity of phenomena genuinely supporting a theory the better the theory is. It is also clear that Newton is under no delusion that a theory follows with deductive certainty from the evidence. Rule 4 that Newton added to the third edition of the *Principia* in 1726 re-iterates this latter point.

Newton's views on the experimental testability of science are manifest in his stand on gravity. Although Newton speculated about the cause of gravity in various ways, he was careful to separate such speculations from his science. He posited gravitational attraction, gave it a precise formulation and provided massive empirical support for it as our discussion of the *Principia* makes clear. The law of gravity had strong claims to being supported by experiment in a way that met stringent demands, and, as such, was quite distinct from speculations about the cause of gravity, for which there was no experimental support. Newton (1971, p. 401) appealed to gravity without being able to explain it in much the same way that Boyle appealed to the spring of the air without being able to justify an explanation of it by appeal to experiment. However, Boyle's insistence on sticking to matters of fact, if taken seriously, would have restricted him to low-level empirical claims embodied in his experimental investigations mainly in pneumatics and chemistry. Within a few decades of Boyle's efforts Newton had demonstrated that the experimental method was capable of confirming highly general, mathematically formulated theories. Given the breadth of the phenomena dealt with in quantitative detail by Newton, ranging from the motion of pendulum bobs to that of planets and comets, he had strong grounds for claiming his theory to be true of the mechanics of macroscopic bodies generally.

Optics stands alongside mechanics as the second area in which Newton made significant scientific contributions. He is famous for his experiments on the splitting of white light into colours through refraction and he also discovered colours generated by reflection from and transmission through thin films, the related phenomenon of Newton's rings still bearing his name.[3] However, in this area, Newton was not able to proceed far beyond fairly low-level experimental claims. He clearly favoured a particle theory of light but was aware that he could not support it in the way he had come to demand of his science. Even his formulation of his experimental knowledge in his optics can be challenged for going beyond what his experiments supported, unless his references to 'rays of light' and 'fits of easy reflection and transmission' are interpreted in some vaguer way than is typically implied by the terms 'ray' and 'fit". It is also the case that parts of Newton's *Opticks* presuppose an atomic structure of solids. It is the nature of Newton's atomism and the status of the case he made for it that is the topic of the following two Sections.

7.3 Newton's Atomism[4]

The main published sources for Newton's atomic theory are the Rules of Reasoning and the General Scholium first published in the second edition of the *Principia* in 1713, the *Opticks*, especially Querie 31, first appearing in Latin in 1706 and in English in 1717 and also a short piece 'On the nature of acids' composed in 1692 but not published until 1710. In brief, Newton's atomism was Boyle's mechanical atomism augmented by the addition of inter-particulate attractive and repulsive forces governed by his laws of motion.

Central to Newton's atomism was the homogeneity of matter, the idea, common to the Ancient atomists and the mechanical atomists of the seventeenth century, that there is just one kind of matter. The homogeneity of matter is an assumption not confined to Newton's meta-discourse of the Rules of Reasoning or the more speculative Queries added to the *Opticks*. It appears as an assumption in the body of the text of the *Principia* and is taken for granted in the *Opticks*. In the *Principia*, from the second edition onwards, Newton noted that 'if all the solid particles of all bodies are of the same density, nor can be ramified without pores, a void space or vacuum must be granted'.[5] That is, if the matter composing bodies is of one, homogeneous, kind, then the differing densities of bulk materials must be due to the ratio of full to empty space within them. This explanation is taken for granted in the *Opticks*, where Newton (1979, p. 267) takes the argument further to conclude that the matter of our experience may in fact consist largely of space.

> And hence we may understand that Bodies are much more rare and porous than is commonly believed. Water is nineteen times lighter, and by consequence nineteen times rarer than Gold; and Gold is so rare as very readily and without the least opposition to transmit the magnetic Effluvia, and easily to admit Quicksilver into its Pores, and to let Water pass through it. ... From all of which we may conclude, that Gold has more pores than solid parts, and by consequence that Water has above forty times more Pores than Parts. And he that shall find out an Hypothesis, by which Water may be so rare, and yet not be capable of compression by force, may doubtless by the same Hypothesis make Gold, and Water, and all other Bodies, as much rarer as he pleases; so that light may find a ready passage through transparent Substances.

It is clear that the argument takes the homogeneity of matter for granted. It also, incidentally, takes for granted the material nature of light, magnetic effluvia, gold and quicksilver. The notion that material bodies consist largely of space, implicit in Newton and enthusiastically endorsed by Newtonian atomists of the eighteenth century, has been dubbed the nutshell theory by Arnold Thackray (1968), following Joseph Priestley's remark that on this theory the whole of the matter in the universe might well be collapsed into a nutshell.

Newton's atoms, invoked in the *Principia* in Newton's third Rule of Reasoning, are similar to Boyle's natural minima, although characterised by a slightly different list of properties. For Newton the 'least particles of all bodies' are all 'extended, and hard and impenetrable, and movable, and endowed with their proper inertia'.[6] Inertia is a fruit of Newton's mechanics and is a necessary addition if atoms are to be governed by the laws of motion. Newton (1979, p. 389) gave an argument for the addition of hardness in the *Opticks*. Bodies of our experience are hard or soft to a greater or lesser extent. Softness can be explained by appeal to the alterable space between hard atoms. However, if the atoms themselves are soft it is difficult to see how macroscopic hardness can ensue. Atoms are hard because they lack empty pores. Atoms, the particles at the base of the hierarchy, are incorruptible, no ordinary power being able to overcome their hardness. Were they to be subject to wear, Newton (1979, p. 400) observed, then the properties of the macroscopic bodies made of them would be subject to a corresponding alteration, contrary to what is observed.

> Particles being solids are incomparably harder than any porous Bodies compounded of them; even so very hard, as never to wear out or break in pieces; no ordinary Power being able to divide what God himself made one in the first creation. While the Particles continue entire, they may compose Bodies of one and the same Nature and Texture in all Ages: But should they wear away, or break in pieces, the Nature of Things depending on them, would

> be changed. Water and Earth, composed of old worn Particles and Fragments of Particles, would not be of the same Nature and Texture now, with Water and Earth composed of entire Particles in the Beginning. And therefore, that nature may be lasting, the Changes of corporeal Things are to be placed only in the various Separations and new Associations and Motions of these permanent particles.

The reference in the above passage to God as creator of the atoms is amplified in a way that makes is quite clear that Newton (1979, pp. 402–404) shared Boyle's view that the arrangement of atoms to form the planetary system, the order of the animal kingdom and so on required God as the designer.

Newton's atoms form the lowest level in a hierarchy of particles of increasing complexity as was the case with Boyle's. The idea is made explicit by Newton (1958, pp. 257–258) in 'On the nature of acids' and repeated in Querie 31 of the *Opticks*. Atoms can combine to form 'particles of the first combination' and the latter can combine to form 'particles of the second combination' and so on. In Querie 31 Newton (1979, p. 394) makes it clear that such a series goes through 'divers Successions until the Progression end in the biggest Particles on which the Operations of Chymistry, and the Colours of natural bodies depend and which, by cohering, compose bodies of a sensible magnitude'.

Newton appealed to attractive forces to account for the coherence of particles in complexes, the smaller the particle the larger the force. The heat generated in various chemical reactions is indicative of the strong attractions causing the acceleration of the combining particles. (Newton followed Francis Bacon and Boyle in associating heat with the rapid motion of particles.) The phenomenon of elasticity requires repulsive forces as well, since elastic bodies resist compression as well as extension. The turning of light back on its tracks in the phenomenon of reflection and the great speed at which light travels after reflection was taken by Newton (1979, p. 395) to be evidence of strong repulsive forces. The dispersion of the particles of a solute in a less dense solvent, in defiance of gravity, is attributed by Newton (1979, pp. 387–388) either to weak repulsive forces between particles of solute, or to the attractions between solvent particles being less than the attraction of solute to solvent particle.

Querie 31 of the *Opticks* comes as near to a chemical text as anything Newton published. Many chemical reactions are interpreted in terms of combinations of particles subject to mutual forces of attraction. Here the particles are not atoms but the much more complex clusters of atoms, several steps up the scale of degrees of combination, that constitute the least parts of chemical substances. Precipitation is explained in terms of preferential attraction. So, for instance, the precipitation of (the salt of) a metal from a solution of it in an acid brought about by the addition of salt of tartar (potassium carbonate) is attributed to the greater attraction of the acid particles for the salt of tartar than for the metal. Indeed, successive precipitations of metals from salt solutions by adding further metals suggests that the metals can be arranged in order of their degree of attraction for the particles of acid. For instance, according to Newton (1979, pp. 380–381), the addition of iron to a combination of copper and aqua fortis (copper nitrate) leads to the deposit of copper, the addition of copper to a combination of nitric acid and silver leads to the deposit of silver and so on.

Various physical, as well as chemical, properties are attributed by Newton (1979, pp. 394–395) to atoms, or complexes of atoms, and their attractions.

> If the Body is compact, and bends or yields inwards to Pression without any sliding of its Parts, it is hard and elastick, returning to its Figure with a Force rising from the mutual Attraction of its Parts. If the Parts slide upon one another, the Body is malleable or soft. If they slip easily, and are of a fit Size to be agitated by Heat, and the Heat is big enough to keep them in Agitation, the Body is fluid; and if it be apt to stick to things, it is humid; and the Drops of every fluid affect a round Figure by the mutual Attraction of their Parts, as the Globe of the Earth and Sea affects a round Figure by the mutual Attraction of its Parts by Gravity.

Finally, Newton (1979, pp. 385–386) adopted and adapted the view of the mechanical philosophers that qualities in bodies detected by the senses are caused by interaction between those bodies and the senses. He clearly spells out this view, exploiting his notion of attractive forces, in the case of taste.

> Do not the sharp and pungent Tastes of Acids arise from the strong Attraction whereby the acid Particles rush upon and agitate the Particles of the Tongue? And when Metals are dissolved in acid *Menstruums*, and the Acids in conjunction with the Metal act after a different manner, so that the Compound has a different Taste much milder than before, and sometimes a sweet one; is it not because the Acids adhere to the metallick Particles, and thereby lose much of their Activity? And if the Acid be in too small a Proportion to make the Compound dissolvable in Water, will it not by adhering strongly to the Metal become unactive and lose its Taste, and the Compound be a tasteless Earth? For such things as are not dissolvable by the Moisture of the Tongue, act not upon the Taste.

So much for the content of Newton's atomism. Let us now assess the extent to which Newton was able to make a case for it.

7.4 The Case for Newton's Atomism

Newton laid down stringent conditions that needed to be satisfied if a claim is to be regarded as sufficiently confirmed by observation and experiment to qualify as a part of science. His *Principia* provided a compelling example of how extremely general knowledge claims could be confirmed to an extent that lived up to those conditions and gives ample grounds for making good sense of and defending his claim that hypotheses should not be admitted into science. His optics supplies further evidence of Newton's strictures being put into practice. However, when it comes to Newton's atomism, it not surprisingly fell far short of meeting Newton's demands that ungrounded hypotheses be avoided. Newton's atomistic matter theory is best seen, like Boyle's mechanical philosophy, as a speculative fundamental matter theory supported by accommodating it to, rather than confirming it by, the phenomena.

Newton's recognition of the distinction between claims confirmed by experiment and those transcending such support is evident from his stand on gravity. Here is his exemplary summary of the situation in the General Scholium to the *Principia* (Newton, 1962, pp. 546–547).

> Hitherto we have explained the phenomena of the heavens and our sea by the power of gravity, but we have not yet assigned the cause of this power. This is certain, that it must proceed from a cause that penetrates to the very centres of the sun and the planets, without suffering the least diminution of its force; that operates not according to the quantity of the surfaces of the particles upon which it acts (as mechanical causes used to do), but according to the quantity of the solid matter which they contain, and propagates its virtue on all sides to immense distances, decreasing always as the inverse square of the distance. Gravitation towards the sun is made up out of the gravitation towards the several particles of which the body of the sun is composed; and in receding from the sun decreases accurately as the inverse square of the distances as far as the orbit of Saturn, as evidently appears from the quiescence of the aphelion of the planets; nay, and even to the remotest aphelion of the comets, if those aphelions are also quiescent. But hitherto, I have not been able to discover the cause of those properties of gravity from phenomena, and I frame no hypotheses; for whatever is not deduced from the phenomena is to be called an hypothesis; and hypotheses, whether metaphysical or physical, whether of occult qualities or mechanical, have no place in experimental science. In this philosophy, particular propositions are inferred from the phenomena, and afterwards rendered general by induction. . . . And to us it is enough that gravity does really exist, and act according to the laws which we have explained, and abundantly serves to account for all the motions of the celestial bodies, and of our sea.

Newton's stand on gravity is akin to Boyle's stand on the 'spring' of the air. He cannot explain it but he can accurately characterise it and appeal to it to explain a wealth of phenomena quantitatively and accurately. Newton here is in effect echoing Boyle's point that intermediate explanations are not fundamental, ultimate ones but are highly significant and useful ones nevertheless.

Newton had less dramatic success in optics. Here he clearly favoured a particle theory of light but realised that he could not adequately substantiate it by experiment. True to his own standards, he declined to include the particle theory into his optics, framing his claims in terms of light rays and fits of easy reflection and transmission.[7]

Newton's atomism did not come close to meeting the standards he brought to bear in his mechanics and optics. If one asks what evidence there was, independent of the phenomena to be explained, for the assumption that elasticity arises from attractive and repulsive forces acting between least parts or that chemical combination involves the combination of least parts, then the answer is that there was none. Nothing is added to our knowledge of chemistry by assuming that the measurable affinity between chemical substances is due to affinities between their least parts so long as there is no evidence for those least parts independent of the facility with which those chemicals combine in the laboratory. Needless to say, evidence for atoms and their interaction, a few levels of complexity below the least parts, was even more remote. It should also be remembered that a specification of the force laws involved, a crucial feature of Newton's success in astronomy, is entirely lacking as far as interacting atoms or least parts are concerned. It is not difficult to offer alternatives to Newton's atomism that cannot be ruled out given the evidence available to him. Each portion of a substance, however small, could have the properties attributed to the whole, such as elasticity, density and various chemical affinities, and could have them primitively. Alternatively, the interaction of parts of substances might be mediated by an aether, an assumption that Newton himself flirted with both in his mechanics and optics. He gave some thought to explaining gravitation at a distance in terms of the properties of an all-pervasive ether and he invoked the ether in an attempt to solve a problem in his optics that was in fact a consequence of his commitment to atomism. The regular reflection of light from a plane surface is difficult to reconcile with the fact that, for an atomist like Newton, that plane surface is very irregular on the atomic scale or the scale of least parts. Newton played with the idea that reflection is due to the repulsive force of a film of ether on the reflecting surface. If that isn't a flagrant example of accommodating one's theory to the phenomena then I don't know what is!

The arguments that Newton did mount in favour of his atomism are similar in status to those offered by Boyle to support his mechanical philosophy. Some rest on notions of intelligibility and others are empirical in a weak sense, presuming an analogy between the macroscopic and atomic realms or involving some accommodation of atomism to the phenomena.

Howard Stein (2002) has recently drawn attention to the metaphysics involved in a tract by Newton that he did not publish, now known by the words with which it opens, 'De gravitatione'.[8] In it Newton spells out his general views on the nature of space and of bodies. Just as Boyle made plausible his attribution of some essential properties of a corpuscle by inviting us to imagine what properties a lone particle must have to qualify as matter, so Newton contemplates what God must have to do to create a new material body in the Universe. He concludes that a body must be impenetrable, to distinguish it from space, that it must have a definitive and unchanging shape and size and that it must obey the laws of inertia and collision. The latter fact implies that a body must possess a definite inertia. To the extent that the force of Newton's case draws on common sense notions of what distinguishes body from space it is a dangerous one. That this is the case is a moral that might well be drawn from Newton's own stand on gravity. The action of gravity at a distance, which would appear to involve masses acting where they are not, might well be considered to be unintelligible, just as some of Newton's opponents insisted. Newton's reply, in effect, is that, unintelligible or not, gravity exists and acts just as it is shown to exist and act in the *Principia*. If the successes reported in that book are anything to go by, then science proceeds by explaining the familiar by reference to the unfamiliar and perhaps unintelligible.

Stein (2002) is at least partly right to insist that Newton's case for the nature of atoms is not entirely *a priori*. He does appeal to experience in 'De gravitatione' to defend his notion of body (or atom) just as he does in the *Principia* and the *Opticks*. The mode of argument is most strikingly put in the *Principia*. Rule III of that work is used by Newton (1962, p. 398) to move from properties of observable objects to properties of atoms. Rule III states that 'The qualities of bodies, which admit neither intensification nor remission of degree, and which are found to belong to all bodies within the reach of our experiments, are to be esteemed the universal qualities of all bodies whatsoever'. As it stands there are problems with both the interpretation and the truth of this rule. A reasonably clear idea of how Newton interpreted it is evident from the way he proceeds to employ it to argue for the properties of atoms.

Part of Newton's argument has some force. In his mechanics he has demonstrated that his three laws of motion are scale-invariant, applying alike to large masses like the sun and small ones like pendulum bobs. It is a reasonable assumption, until there is evidence to the contrary, that all bodies, including atoms (if there are such), possess inertia and obey the laws of motion. Further, it might reasonably be concluded that atoms must have some shape and size just as observable bodies do. Such arguments are not logically conclusive but adhering to them make for a sensible research strategy. This is the way Smith (2002b, p. 58) suggests we understand Newton's strictures about accepting well-confirmed theories to be unrestrictedly true until exceptions are discovered.[9] Whatever sense and force these arguments have, they are not shared by the additional moves that Newton (1962, p. 399) makes to ascribe properties to atoms. His argument that they must have (absolute) impenetrability and (absolute) hardness is presented as an empirical one. It goes like this:

> That abundance of bodies are hard, we learn from experience; and because the hardness of the whole arises from the hardness of the parts, we therefore justly infer the hardness of the undivided particles, not only of the bodies we feel but all others. That all bodies are impenetrable, we gather not from reason, but from sensation. The bodies that we handle we find impenetrable, and thence conclude impenetrability to be an universal property of all bodies whatsoever.

There are all sorts of problems with this argument for anyone not already committed to atomism. What experience shows us is that bodies are impenetrable and hard in various degrees depending on what they are made of. Liquids are penetrable in a way that solids are not, glass is penetrable by light but not by tennis balls whilst metals are penetrable by neither. Likewise, the bodies of our experience differ in their degrees of hardness. Why should not atoms show an analogous variability in their degrees of penetrability and hardness? Newton's stand that hardness of macroscopic objects cannot arise from soft atoms is undermined once inter-particulate forces are admitted, and, in any case, the argument presupposes that there are atoms. Even if one concedes the case that atoms must be absolutely impenetrable and hard, there remains the question of what other properties they might have. Newton's assumption that atoms must possess all and only those properties possessed by all observable bodies presupposes that there is just one kind of matter.

It is quite clear that Newton's atomism requires attractive and repulsive forces to act between atoms and the relatively stable complexes they combine to form. Newton's dream, as expressed in the Preface to the first edition of the *Principia* (p. xviii), was that, if these forces were known, then a theory of their action could be developed on a par with his theory of gravity. But of course, as he acknowledged, they are unknown. What is more, given that they must differ from forces between observable bodies, as is implicit in Newton's assumption that the forces between particles are larger the smaller the particle, there is a strong sense in which the micro-world is unlike the macro-world, contrary to the mode of argument that led Newton to attribute impenetrability and hardness to atoms.

The case for Newton's atomistic matter theory in the main rested on the extent to which it could accommodate the phenomena in a way that was superior to its rivals. It must be acknowledged that Newton's theory was an improvement on Boyle's in this respect. By introducing attractive and repulsive forces Newton could accommodate phenomena such as elasticity, gravity and chemical combination and precipitation that Boyle was not able to accommodate or which he was able to accommodate only in a highly contrived way. Given the success of his gravitational theory, Newton's assumptions about other forces acting at the sub-microscopic level were reasonable and appealing however short of experimental confirmation they fell.

But Newton's inclusion of attractive and repulsive forces in his atomism rendered it problematic in a significant respect. As we saw in our discussion of Boyle's mechanical philosophy, a requirement that was placed on matter theory was that it be ultimate, that is, free of entities or claims in need of further explanation. While Boyle aspired to a matter theory that met that demand Newton's atomism fell short of it just because of its inclusion of forces, which Newton acknowledged to be unexplained yet subject to some explanation. In the mechanics of the *Principia* he could defend his appeal to gravity on the grounds that he could clearly specify the law governing it and that he could explain many phenomena by appeal to it. As far as the forces involved in his atomism were concerned, Newton could do neither of those things. It was the fact that Newton's matter theory involved forces that were not ultimate and so in need of explanation that exposed it to criticism from Cartesians and Leibnizians.

The fact that Newton was unable to specify or gain experimental access to the forces he presumed to be operative at the sub-microscopic level, so that he could at best accommodate his atomism to the phenomena, had the consequence that his matter theory was unable to offer useful guidance to experimental science. In this respect the flexibility inherent in Newton's atomism stemming from the freedom to choose force-laws to meet the demands of the phenomena, whatever they might be, was a shortcoming rather than a strength. This assessment is borne out by subsequent developments. Newton's commitment to and elaboration of atomism is best understood in the context of his philosophical disputes, especially with Leibniz, rather than as a component of Newton's endeavour to extend experimental knowledge. His debate with Leibniz, for example, on metaphysical and religious issues, involved the possibility of the existence of space devoid of matter and on the intelligibility of action at a distance. Newton's atomism played a central role in both these issues. (Recall, for example, Newton's use of the assumption that atoms are all composed of the same stuff to argue that bodies of our experience consist largely of space.) I do not pretend to engage with the details of those debates in this book. My focus is on how experimental access to atoms became possible. Newton's speculations certainly did not supply the key.

7.5 The Fate of Newtonian Atomism in the Eighteenth Century

In Chapter 5 we saw how Boyle championed experimental practice by means of which knowledge of intermediate causes can be acquired, his pneumatics standing as an exemplary example of what can be achieved in this regard. In a sense, Newton's theory of gravitation can be seen as a Boylean science offering explanations appealing to an intermediate cause, namely gravity, although that science differed from anything Boyle had to offer because of its great generality and precise mathematical formulation. Much of the progress in eighteenth-century science can be seen as a further development in the quest for knowledge of intermediate causes, involving such notions as surface tension, viscosity, elasticity, electrical and magnetic attraction, temperature and heat and chemical combination. In some areas forces governed by precise laws were identified, ranging from Hooke's law governing elastic deformation in the late seventeenth century to Coulomb's inverse square law governing electrical attractions at the end of the eighteenth century, so that the mathematical apparatus of Newtonian mechanics could be exploited. We should also not forget the spectacular completion of Newton's programme in astronomy, through the efforts of the likes of Euler, D'Alambert and Lagrange and culminating in Laplace's *Mécanique Céleste* early in the nineteenth century. Other areas are best seen as Boylean, rather than Newtonian, sciences.[10] The latter developments include the move towards thermodynamics and a theory of heat with the fashioning of a notion of temperature and, for instance, the discovery of the gas laws, and the elaboration of the notion of chemical combination culminating in Lavoisier's chemistry.

A detailed history of these eighteenth century advances in science is beyond the scope of this book. I do not discuss details here, with the exception of some related to chemistry that are the topic of the next chapter. The general features of these advances which I wish to stress, and which I intend to signal with my labelling of them as Boylean and Newtonian science are as follows. These new fields were both experimental and theoretical. They were experimental insofar as they involved claims that could be pursued and sometimes established experimentally. Such notions as elasticity, temperature and electric charge were all intermediate notions in Boyle's sense insofar as the hidden, ultimate explanation of these notions did not figure in the sciences in question. They were on a par with gravity in Newton's physics and pressure in Boyle's pneumatics. Nevertheless, the new sciences were theoretical in the sense that appropriate conceptions, such as electric charge and temperature, needed to be fashioned and law-like relations governing them established. As we have seen, the physics of the *Principia* could be defended by appeal to observation and experiment yet its contents could hardly be described as eschewing theory!

My construal of the productive aspects of eighteenth century physical science clash with, and is intended to clash with, atomism of the various kinds we have met so far in this book, which, I have argued, sought ultimate explanations, could not be adequately tested by experiment, and is best seen as speculative philosophy or metaphysics rather than science. Newton's atomism was in fact very influential in the eighteenth century. It was also unproductive as far as experimental science was concerned. I elaborate on this theme in the remainder of this section.

Some inadequacies and incoherencies in Newtonian atomism were identified and removed by eighteenth-century figures. Perhaps the best version of the theory was that formulated by the Croatian philosopher, Roger Boscovich, who attempted to reconcile Newton's atomism with the philosophy of Leibniz. One problem he identified in Newton's version had already been picked up by Leibniz.[11] Since Newton's atoms are perfectly hard then the change in velocity they experience on impact must be instantaneous. Boscovich did not allow such lack of continuity to occur in nature any more than Leibniz had, and, in any case, it implies infinite forces acting on colliding atoms. Another problem was already hinted at in my discussion of Newton's argument for hard atoms earlier in the chapter. Once attractive and repulsive forces are granted, and appealed to in order to explain elasticity, then the question of whether atoms themselves are hard or soft becomes incidental. Indeed, given the work that the short-range forces do, or are capable of doing, in Newton's atomism, the shapes and sizes of atoms become a dispensable part of the picture too.[12] Boscovich attempted to remove all these difficulties by construing Newton's atoms as points possessing inertia and as the origin of forces. Here is his own summary of his theory in the Synopsis of his *Theory of natural philosophy* (Boscovich, 1966, p. 10).

> [M]atter is unchangeable, and consists of points that are perfectly simple, indivisible, of no extent and separated from one another; that each of these points has a property of inertia, and in addition a mutual active force depending on the distance in such a way that, if the distance is given, both the magnitude and direction of this force is given; but if the distance is altered, so is the force altered; and if the distance is diminished indefinitely, the force is repulsive and in fact also increases indefinitely; whilst if the distance is increased, the force will be diminished, vanish, be changed to an attractive force that first of all increases, then decreases, vanishes, is again turned into a repulsive force and so on many times over; until at greater distances it finally becomes an attractive force that decreases approximately in the inverse ratio of the square of the distances.

The problem of instantaneous changes in velocity on collision between atoms is removed because the large repulsive force acting close to them prevents them ever touching. At large separation the force becomes the Newtonian gravitation. The various forces in between are to explain forces responsible for cohesion, fluidity, elasticity, electricity, magnetism and chemistry. Boscovich (1966, pp. 76–96) elaborated on the idea that atoms can combine in relatively stable arrangements to form complex particles by analysing the equilibrium conditions for groupings of two, three and four particles, although he did not claim to be able to proceed so far as to be able to derive the more complicated complexes actually existing in nature.

For those eighteenth-century figures who were attracted to Newton's atomism as a natural philosophy and who had learnt to be more relaxed about introducing the notion of force as a primitive than their seventeenth-century precursors had been, Boscovich's theory might well have been seen as having a lot going for it. However, it was not destined to be of much assistance to experimental science. Boscovich applied his atomic theory to a range of physical phenomena in Part III of his *Theory of natural philosophy*. Since he was free to select forces to suit is purpose he was able to meet with some success. He could accommodate his philosophy to the findings of science. But, because he was unable to even formulate, let alone verify, force laws between atoms or complexes of atoms of various kinds, he could not predict any phenomenon and could not give any guidance to experimental research. I will take just one example to illustrate how Boscovitch did relate his natural philosophy to the results of science. I choose an example from chemistry since developments in that subject are to be our concern in the next chapter.

Here is the treatment Boscovich (1966, p. 161) gave of solution and precipitation from solution:

> When certain solids are mixed with certain fluids, we see that the mutual connection which there used to be between the particles of each is dissolved in such a way that the solids are no longer visible; and yet that they are still there, reduced to extremely small particles and dispersed, is shown by precipitation. For, if a certain other body is introduced, there falls to the bottom an extremely fine powder of the original solid, as it rained down. So metals, each in its own solvent, dissolve, and with the help of other substances are precipitated. 'Aqua regia' dissolves gold; and this, on the addition of common salt, is precipitated. It is quite easy to get a clear picture of the matter. Suppose that the particles of the solid have a greater attraction for the particles of the water than for one another; then they will certainly be torn away from their own mass, and each of them will gather around itself fluid particles, which will surround it, in the same manner as iron filings adhere to a magnet; and each would become something in the nature of little spheres similar to what the Earth would resemble, if a sufficiency of water were to be poured over it to submerge it deeply; . . . Hence the solid will be dissolved, and each of the little spheres, so to speak, would represent a little earth with its great abundance of sea surrounding it; and these little earths, on account of their exceedingly small volume will escape our notice; and they cannot fall, sustained as they are by the force that connects them with the sea that surrounds them.
>
> If now another substance is introduced into a fluid of this kind, the particles of which attract the particles of the fluid to themselves with a stronger force, and perhaps too at greater distances, than they are attracted by the particles of the first solid; then his second solid will be dissolved in every case, and its particles will be surrounded by the particles of the fluid, which formerly adhered to the particles of the first solid, being torn away from the latter and seized by the particles of the second solid. The particles of the first solid will then rain down on account of their own weight within the fluid which is specifically lighter, and there will be precipitation.

Boscovich is taking the experimental results of the chemists and adapting his theory to fit them. It could conceivably have been the case that Boscovich employed reference to one phenomenon, precipitation in this case, to formulate a force law that could then be employed to predict other phenomenon, known or unknown. But Boscovich does not do that. Indeed, it is quite clear that the empirical information he has is quite inadequate for leading to any force law. Whatever the status Boscovich's theory had as natural philosophy it did not, and could not, aid experimental science.

Several eighteenth-century figures attempted to construct a chemical theory that would emulate Newton's gravitational theory. They were as unproductive as that of Boscovich as far as aiding chemistry is concerned. A number of chemists, especially G. F. Venel and P. J. Macquer in France, began to distance themselves from such enterprises and recommend an approach more in touch with experiment. Others, such as William Cullen, and Joseph Black, payed lip-service to Newtonian natural philosophy in a way that did not engage with their practice. By the time we reach Lavoisier late in the eighteenth century, we find him explicitly rejecting atomism as an aid to chemistry and defining a chemical element as a substance that cannot be broken down further by chemical means. Newtonian atomism was unproductive as far as eighteenth-century chemistry is concerned, a point made in detail by Thackray (1970). I elaborate on the case of chemistry in the next chapter.

Notes

1. For thorough analyses of Newton's method of successive approximations see Smith (2001a, 2001b, 2002a, 2002b). An analysis of Newton's methodological remarks concerning the status of his experimental philosophy is in Shapiro (2004).
2. As Smith (2002b) has shown the approximate truth of the first law, that planets move in ellipses with the sun as focus, is not sufficient to imply the approximate truth of the inverse square law, which is presumably why Newton did not argue from elliptical orbits to the inverse square law in the *Principia*.
3. For relevant details of Newton's optics see Shapiro (1993).
4. A young Newton flirted with Epicurus's version of atomism, including atomic space and time, but he did not persist with it. These early speculations have been described by McGuire and Tamny (1983) and I will not reproduce their findings here.
5. As quoted in Thackray (1970, pp. 18–19).
6. *Principia*, Vol. 2, p. 399. The same list of properties appears in the *Opticks*, p. 400.
7. For details of Newton's attitude to the particle theory of light see Shapiro (2004) and the more detailed discussion in Shapiro (1993). For a discussion of Newton's 'deductions from the phenomena' in optics see Worrall (2000).
8. Newton's work appears in Hall and Hall (1962, pp. 89–121 (Latin) and pp. 121–156 (English)). Stein warns that the translation is not to be trusted.
9. Smith (2002b, p. 58).
10. My distinction between Newtonian and Boylean science roughly maps onto the distinction between mathematical and experimental science made by Thomas Kuhn (1977) although on my reading Newton's mechanics was experimental as well as mathematical.
11. See Loemker (1969, p. 446).
12. Thackray (1970, p. 15) has stressed this point in the context of the shapes of atoms. He (p. 151) cites expressions of this worry in precursors of Boscovich amongst British Newtonians such as Robert Green and Gowin Knight.

Chapter 8
The Emergence of Modern Chemistry With No Debt to Atomism

Abstract According to William Newman, Boyle set chemistry on its modern course by setting it in the framework of the mechanical philosophy. I challenge this. Talk of corpuscles of impenetrable matter characterised by their shape, size and degree of motion only was too impoverished and too far removed from what could be experimentally tested to be of any help to chemistry. Boyle accommodated to his mechanical philosophy a chemistry acquired by other means, and with less success than either he or Newman imply. The new chemistry focused on substances that could be built up from and broken down into their components, drawing on techniques developed in metallurgy and pharmacy. I draw on the work of Ursula Klein, who has argued this thesis and shown how characteristics of the new chemistry can be seen in what came to be regraded as the first of a series of tables of affinity, published by E. T. Geoffroy in 1718. Chemistry was set on the path to Lavoisier in a way that owed no debt to the atomism of the likes of Boyle and Newton.

8.1 Introduction

My account, in the preceding two chapters, of atomism as developed by the mechanical philosophers and Newton, does not involve a detailed engagement with chemistry. It is time to rectify that deficiency. Newman (2006, p. 3) insists that it is precisely in the area of chemistry that the details of the mechanical philosophy were fleshed out and substantiated. He presents the two as 'indissolubly linked'. In taking this stand Newman is echoing the views of Robert Boyle who described himself as striving to 'make chymical experiments useful to illustrate the notions of the corpuscular philosophy' and who promoted 'the desirableness of a good intelligence between the corpuscularian philosophers and the chymists'.[1]

Robert Boyle has been hailed as 'the father of modern chemistry'.[2] He is reputed to have revived Ancient atomism and applied it to chemistry in a way that led to the banishment of Aristotelian elements and alchemical or Paracelsian principles. This view has had its critics in recent decades. In the early 1990s Sik Yung Kim (1991) and Antonio Clericuzio (1990) challenged the idea that Boyle's chemistry was part of and grew out of his mechanical atomism and I, in (Chalmers, 1993), sought to drive a wedge between Boyle's experimental science, on the one hand,

A. Chalmers, *The Scientist's Atom and the Philosopher's Stone*, Boston Studies in the Philosophy of Science 279, DOI 10.1007/978-90-481-2362-9_8,

and his mechanical philosophy on the other, although I paid scant attention to chemistry. More recently Newman (1996, 2006) has argued that Boyle's chemistry grew, not out of ancient atomism, but out of a corpuscular tradition dating back to Aristotle, incorporated into medieval alchemy and fashioned into an Aristotelian theory of natural minima by Daniel Sennert in the generation prior to Boyle. Whilst downplaying a productive link between Boyle's corpuscular chemistry and ancient atomism, Newman still sees Boyle's corpuscular chemistry as a crucial component of a revolution that put chemistry on its modern course. He retains the thesis that Boyle's chemistry and mechanical philosophy were inextricably linked, his view being facilitated by the broad interpretation he gives to 'mechanical'.[3] Although Boyle drew heavily on the corpuscular tradition, the step from Sennert to Boyle was a major one, according to Newman, insofar as it involved the removal of the remnants of Aristotelian form. In this latter respect Newman endorses an assessment that Boyle himself frequently gave expression to and supports a somewhat attenuated version of the picture of Boyle as the father of modern chemistry.

My characterisation and assessment of the mechanical philosophy in Chapter 6 has already spelt out in general terms the position I will defend in the context of chemistry in the present chapter. In my view, Boyle's mechanical philosophy did not have the resources to productively guide experimental chemistry. That philosophy could at best be accommodated to a chemistry acquired by other means. The flexibility of the mechanical philosophy, stemming from the freedom to adapt the shapes, sizes and motions of corpuscles to the phenomena, extolled by Boyle as a key virtue of it, rendered it empty as far as offering guidance to the experimenter is concerned. In Chapter 6 I described how Boyle noted that experimental knowledge of the phenomena requires the framing of notions capable of grasping 'intermediate' causes capable of experimental investigation rather than ultimate ones that are mechanical in the strict sense. Notions of the spring and pressure of air provided Boyle with just what he needed in pneumatics, as we have seen. From this point of view it is natural to raise the question of what notions were needed to inform chemistry in the way that spring and pressure informed pneumatics. I cast doubt on the extent to which Boyle made helpful contributions in that respect. I regard his chemistry as a premature and unhelpful attempt to reduce chemistry to mechanical atomism.

My articulation and defence of the above view has been much facilitated by the recent work of Ursula Klein, in which she also casts doubt on the debt seventeenth-century chemistry owed to atomism, arriving at this conclusion by a quite different route than myself. I take maximum advantage of her work in what follows.[4] Klein provides an answer to the question of what notions needed to be framed for the guidance of a significant segment of experimental chemistry. What was needed were the notions of chemical substance, chemical compound and chemical combination, notions that are taken for granted in contemporary chemistry but which, according to Klein, were not clearly articulated in a way that was able to inform and make possible a new experimental science until early in the eighteenth century. Klein makes it clear that the formulation and practice of, and historical path to, the new experimental chemistry was quite distinct from philosophical matter theories generally and what I refer to as mechanical atomism in particular.

It suits my purpose to use Klein's construal of the new science of chemical combination as a contrast to Boyle's chemistry. To this end, I summarise Klein's position in the next section before modifying it to my own purpose in Section 8.3. In the next five sections I analyse Boyle's chemistry and its relationship to his mechanical philosophy. I attempt to remove the remnants of the view that Boyle fathered modern chemistry by wedding it to a mechanical version of the corpuscular philosophy. To do so I must engage with and counter Newman's detailed defence of the view I oppose. In the last two sections I return to Newton's chemistry to view it in the light of the considerations of this chapter.

8.2 Klein on Geoffroy and the Concepts of Chemical Substance, Compound and Combination

A feature of seventeenth-century chemistry was the increased use of mineral acids. They were used to an increasing extent in pharmacy, making possible the addition of useful substances derived from minerals to those traditionally extracted from animal and plant materials. As the century progressed the action of acids increasingly became a focus of academic as well as artisanal activity, especially in France in the context of the French Academy and the Jardin des Plantes in Paris. The latter institution had been founded as an institution for training pharmacists. As the century progressed there was a gradual shift in the research done there from a concern with practical applications to more theoretical ones. Etienne François Geoffroy, son of a pharmacist, occupied the chair of chemistry at the Jardin des Plantes from 1712. He was able to take advantage of advances made in chemistry largely through the use of mineral acids to publish, in 1718, his 'Table of different relations observed in chemistry' (Fig. 8.1). Klein construes this table as encapsulating novel notions of chemical substance, compound and combination that had emerged in the chemistry of the time and which served to put chemistry on its modern track. In this section I summarise key aspects of Klein's views on the content and origins of the new conception without any pretence of doing justice to the historical detail she invokes.

Geoffroy's table, the first of a series of what came to be known as affinity tables in the eighteenth century, depicted chemical substances arranged in 16 columns and 9 rows. At the head of each of the 16 columns in the table is the symbol for a chemical substance, or class of substances, which form compounds with the substances appearing below it. The higher a substance is in the column below a reference substance the greater the 'rapport' it has for that reference substance. If a substance high in the list is added to a compound of the reference substance with a substance lower in the list, it will replace that latter substance and itself form a compound with the reference substance. The left half of the table summarises the formation of salts by the action of acids. The right half summarises the combinations of metals with sulphur, mercury and other metals and also the solution of substances in water.

The substances appearing in his table are described by Geoffroy (1996, p. 314) as 'the principal materials with which one usually works in chemistry' and the

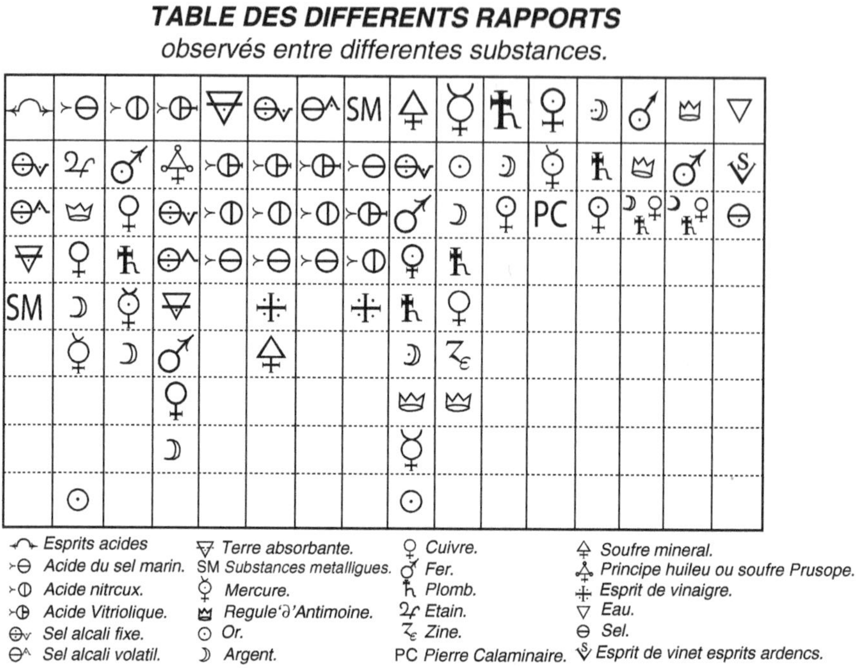

Fig. 8.1 Geoffroy's affinity table

'different substances that are used in chemistry'. Chemical substances are those manipulated in the chemical laboratory. Any such substance is characterised in terms of the substances it can be broken down into, the substances it combines with, the substances it displaces and the substances it is displaced by. Geoffroy's table is a summary of these inter-relations which he refers to as 'laws'. Here is an example of a law for which Geoffroy (1996, p. 313) claimed much evidence and no known exceptions:

> Every time that two substances that have some disposition to join with each other are found united together, if a third arrives that has a greater relationship (*rapport*) with one of the two, then it will unite with them by letting the other go.

Note that this law talks of combining substances, not combining atoms or corpuscles.

In the commentary on his table Geoffroy makes no reference to, and has no need for anything like, atoms. He also refrains from using the term 'affinity' that became associated with some elaborations of Geoffroy's table by later authors. Presumably he wished to dissociate himself from the vitalistic associations of that term. Geoffroy's table and his commentary on it make perfect sense as a reflection on and abstraction from aspects of the chemical laboratory practice of his time without need to invoke atomism or any other deep explanation. It is therefore doubly mistaken to see Geoffroy's table as a version of Newton's views on chemical affinity. Newton's views were couched in terms of atoms and forces of affinity between them, whereas Geoffroy mentions neither atoms nor forces of affinity.

Klein raises the question of the principle of selection employed with respect to the substances included in Geoffroy's table. It is indeed a selection. Geoffroy's description of the substances in the table as 'the principal materials with which one usually works in chemistry' is not accurate since many of those substances, the ones involving the extraction of useful substances from plant and animal materials, are not included. So what is the principle of selection? Klein dismisses various suggestions in the literature as inadequate. One such suggestion is that Geoffroy's table stems from, and is an articulation of, the notion of chemical affinity introduced by Newton in Query 31 of the *Opticks*.[5] We have already noted that Geoffroy did not employ the term 'affinity'. Further, there is no strong evidence that Geoffroy was influenced by the discussion of Query 31 which first appeared in the Latin edition of the *Opticks* in 1706 and not in the 1704 edition which we know Geoffroy was familiar with. It should also be noted that Geoffroy's table involves a wider range of substances than those discussed by Newton. A second suggestion is that Geoffroy's table is simply a convenient summary of empirical knowledge of the time.[6] A major problem with this is that it does not adequately grasp the criteria of selection underlying Geoffroy's table. It is simply not the case that all chemical reactions known at the time are included. As we have already noted, most of the pharmacy of the seventeenth century involved the destructive distillation of animal and vegetable materials to extract their 'essence' but there is no reference to many of the substances involved in these reactions in the table. Larry Holmes suggested that the table summarises the knowledge involved in salt formation developed in the course of the seventeenth century and embodied in the work of Geoffroy's predecessors, especially W. Homberg.[7] But this explains only the left hand side of the table. In any case, interpreting Geoffroy's table merely as a summary of empirical results fails to capture key features of the theoretical conception lying behind Geoffroy's table.

The key to understanding the conceptualisation underlying Geoffroy's table has been identified by Klein. It is what she refers to as the reversibility of the processes involved in the formation of the compounds that can be inferred from the table. The chemical substances that combine to yield compound chemical substances can be recovered from those compounds. Chemical substances can be built up from their components and broken down into those components. Perhaps it is recoverability that should be stressed rather than reversibility, since the recovery of substances from their compounds need not be the exact reverse of the processes involved in their production from their components, and the recovery may involve more than one step. (For instance, silver nitrate can be prepared by adding nitric acid to silver, whilst the recovery of silver involves precipitating silver carbonate by adding potassium carbonate and then heating the silver carbonate.) By an immensely thorough and painstaking piece of historical research Klein has shown that the compounds implied by Geoffroy's table include all and only the chemical substances which could be synthesised from their components and from which the components could be recovered, given the practical knowledge of the time, including the contributions from Geoffroy's own laboratory researches.

Klein rightly stresses the theoretical character of the conceptual scheme at work in Geoffroy's table and his commentary on it. The notion of chemical substance and chemical combination at work involves a theoretical abstraction and goes beyond the deliverances of the senses or of the technological practices from which Geoffroy abstracts. The substances in the table are what Klein calls 'pure substances' characterised in terms of their interactions with other chemical substances. They can be approximated to in the laboratory but many of them are not to be found in nature, and those that are occur in an impure state, mixed up with other substances. At this stage in the history of chemistry, chemical substances are pure to the extent that they exhibit regular and repeatable behaviour as far as their interactions with other substances are concerned.[8] Pure chemical substances interact with others in distinctive ways. In particular they 'combine' to form compounds. The difference between a mixture and a compound is precisely the idea that in a compound the component substances are held together by virtue of a *rapport*. The theoretical notion of combination is distinctive both in respect of what it includes and does not include. It includes the idea that the components of a compound exist in the compound as its components, held together by virtue of a characteristic *rapport*, even though neither the components while in the compound nor their *rapport* are directly observable. As far as what is not included is concerned, the conception of chemical substance and compound does not include a commitment to atoms nor to anything akin to Paracelsian principles, Aristotlelian forms or Newtonian forces of affinity serving to explain *rapport*.

The justification of Geoffroy's conceptualisation lay in the productivity of the experimental practice it informed. How successful it would be and how far it could be taken was not something that could be foreseen. I have already followed Klein and noted that a large part of pharmacy at the time could not be accommodated in the new scheme because it involved the destructive distillation of plant and animal materials, where the term 'destructive' signals the fact that the organic materials from which the essences were extracted could not be recovered. As Klein and Lefèvre (2007, p. 58) note 'the window of opportunity was small for the emergence of the modern concept of chemical compound'. The conception embodied in Geoffroy's table worked only for binary compounds that could be analysed into and synthesised from their components. Nevertheless, this approach to chemistry blossomed over the ensuing decades, and its development set the scene for Lavoisier's chemistry.

Geoffroy's table and commentary on it made no reference to atoms and no such reference was needed. To this observation must be added Klein's claim that the historical path that led to Geoffroy's conceptualisation in fact owed nothing to atomism. Klein has traced this path back to the commercial practices of metallurgy and pharmacy. The notion of recoverability central to the conception implicit in Geoffroy's table and commentary was already present in aspects of those practices, so that the innovation involved isolating it and making it the cornerstone of a novel theoretical conception able to inform a progressive experimental practice. I summarise key elements of Klein's account of the historical path to Geoffroy's table in the next five paragraphs.

Procedures common in the sixteenth century involved the extraction of metals from their ores and the separation of metals from mixtures or alloys by reversible procedures involving notions of recoverability of the kind exploited by Geoffroy. Gold and silver could be extracted from ores by heating the ore with lead, which resulted in alloys of gold or silver with lead. Heating these alloys with carbon reduced the lead to litharge (lead oxide) leaving behind the gold or silver. If necessary, the original lead could be extracted from the litharge. Gold and silver could be separated by a process that we have already come across in connection with Sennert's 'reduction to the pristine state'. Adding nitric acid to a mixture of gold and silver and heating involves the formation of a solution of silver nitrate from which the gold is precipitated and can be filtered off. Adding a metal such as copper, or simply pouring the silver nitrate solution into a copper container, results in the precipitation of silver. Another method involved heating the gold/silver mixture with sulphur, resulting in a solution of silver sulphide from which gold is precipitated. Silver can be recovered by adding a metal such as lead to the sulphide. Those kinds of reactions are incorporated into Geoffroy's table. Those, like myself, who need constantly reminding which metals replace which can refresh their memory with a glance at the table.

As I mentioned above, in sixteenth century pharmacy substances of medicinal value were typically extracted from animal and vegetable matter and only rarely from minerals. This gradually changed with the increased availability and use of mineral acids to form salts. A difference at the practical level was the reversibility of the processes involved in the use of acids to produce salts as opposed to the irreversibility of the distillation of organic materials. In formulating his chemistry of combination, Geoffroy was in part extracting and making explicit what was already implicit in the experimental procedures involved in metallurgy and the transformations associated with the use of mineral acids.

The developments described above took place largely in Continental Europe and the main contributors to the science of salt formation were French. The new developments in pharmacy were published in a series of textbooks appearing in France through the seventeenth century.[9] These books took a typical form. They started with an introductory chapter that included the sketch of a theoretical framework. This derived largely from Paracelsus. Base matter, perhaps the Aristotelian elements or something akin to Aristotle's prime matter, were informed by 'principles', sulphur, mercury and salt to which were sometimes added phlegm and earth. These latter principles were often portrayed, not as material additions to the elements or combinations of them but as non-material 'spirits' that enlivened the elements. Theoretical introductions along such lines typically played a minimal part in the rest of a textbook, which took the form largely of a collection of recipes. As we saw in Chapter 5, Daniel Sennert had conceptualised the action of acids in a way that was both corpuscular and Aristotelian, whilst a few decades later Robert Boyle offered strictly mechanical interpretations. In the latter decades of the seventeenth century leading researchers at the Jardin des Plantes, such as Nicolas Lemery and Wilhelm Homberg employed both mechanical and Paracelsian conceptions.

Apart from these philosophical interpretations, whether Paracelsian, Aristotelian of mechanical, there was more rough and ready talk of the action of acids using metaphors sufficient to engage with artisanal practice. Metallurgists of the mid sixteenth century, such as Georgius Agricola and Lazarus Ercker, understood metals together with impurities as existing in ores as parts needing to be separated. Separation by means of acids was viewed as a process on a par with separation by sieving, washing or heating. Nitric acid, when added to a mixture of gold and silver absorbed the silver, but the silver persisted in the resulting 'dissolution' as a part of it that could be extracted, by adding copper for instance. With the increased use of acids by pharmacists in the seventeenth century this kind of conceptualisation needed to be extended and refined. Attention needed to be given, not simply to the 'dissolutions' resulting from the addition of acids to substances such as metals but to the nature of the new substances so formed. These substances, that came to be classified as salts, needed to be extracted from the dissolution by one of a range of methods depending on the nature of the salt in question. Volatile salts, such as chlorides, were extracted by distilling the dissolutions resulting from the action of hydrochloric acid whilst non-volatile nitrates and sulphates were crystallised from dissolutions made concentrated by boiling. Others salts, such as mercury chloride, could be extracted by sublimation. Precipitation was another key procedure. By the mid-seventeenth century, pharmacists such as Johann Glauber and Christopher Glaser were talking of acids attacking metals and absorbing them. A metal, though persisting in the dissolution in an acid, does not fall to the bottom because the acid fastens on to it. The metal is precipitated through the addition of a substance which weakens the capacity of the acid to fasten on to the metal or which fastens onto the metal more readily.[10]

Geoffroy's predecessors at the Jardin des Plantes went beyond the conceptualisations involved in the talk of artisans, and helped pave the way for Geoffroy's innovations, in a number of ways. They became more concerned with conducting what Francis Bacon had described as experiments of light, designed to understand fundamental chemical processes, rather than experiments of fruit focussed on preparing useful substances. The concept of a salt that they forged was an abstraction that brought together a range of substances that differed widely as far as their observable properties were concerned. Another move implicit in chemical practice in the latter part of the seventeenth century was the erosion of the distinction between natural substances and artefacts. Productions of sixteenth-century artisans, such as alloys, were understood as artefacts; distinct from the natural substances that the metals composing them were presumed to be. The analysis of natural substances into their components and their synthesis from their components undermined this distinction. Nitre prepared in the laboratory came to be recognised as having the same properties as, and as being no less nitre than, the naturally occurring variety.[11]

By 1718, then, the scene was set for Geoffroy to make what Klein construes as his crucial move. His table encapsulated a concept of chemical compound and combination the elements of which were latent in the chemical practice of his time. The historical path to it, as recounted by Klein, owed no particular debt to atomism. Geoffroy, in extracting and focussing on what was implicit in contemporary talk and practice, seems to self-consciously avoid any philosophical notions associated with Aristotelian, Paracelsian or mechanical matter-theories, although he does not explicitly say that that is what he is doing. The merit of his contribution is that he devised a novel conceptualisation that was grounded in and sufficient to productively guide chemical practice whilst setting aside deeper philosophical questions about the ultimate nature of matter that lay beyond the resources at the disposal of chemists. Geoffroy did not explicitly construe his table in this way but, as Klein points out, a commentator on his table published in the 'Histoires' of the Academy did.

> That a body which unites itself to another, e.g., a dissolvent penetrating a metal, which leaves that body to unite with another which is presented to it, is a subject-matter whose possibility could not have been guessed by the most subtle philosophers, and which they cannot easily explain, even today. . . . At first, one imagines that the second metal was better suited to the dissolvent than the first that has been abandoned. But what principle of action can one conceive with regard to this stronger suitability? This is the place where sympathies and attractions would begin to play a role – if they existed. However, leaving as unknown what is unknown, & holding to proven facts, all experience of chemistry proves that a body has more disposition to unite with one body than with another & that this disposition has different degrees . . .. The more chemistry will improve, the more M. Geoffroy's table will improve, as well. Be it through the inclusion of a greater quantity of substances, or through the arrangements and exactitude of the relationships.[12]

One further point about the theoretical conception implicit in Geoffroy's table and commentary is worth stressing. As we have seen, following Aristotle's distinction between 'mixts', such as bronze, and mixtures, such as one of wheat and barley, the distinction was often discussed in terms of homogeneity. Bronze is a mixt because any portion of it, however small, is still bronze, whilst a small sample of a mixture of wheat and barley may well be entirely wheat or entirely barley. This view clashed with Ancient atomism. From the point of view of atomism the distinction between a mixt and a mixture lies in the fact that, in the former, the least parts of a mixt are all alike, whereas in the case of a mixture of two substances, at the level of least parts there will remain a mixture of the least parts of each substance. The distinction between compounds and mixtures in Geoffroy's account differs from both of these. It involves the notion of *rapport*. The components of a compound are indeed *combined* by virtue of the *rapport* that exists between them. Components of a mixture are not combined, there being no such *rapport*.

8.3 Reflections on Klein's Account of Chemical Combination

In this section I reflect on some of the implications of Klein's account of chemical substance, compound and combination as it had emerged by the time Geoffroy published his table in 1718 and mould them to my own purpose.

Klein and Lefèvre (2007, pp. 112 and 301) describe the substances depicted in or implied by Geoffroy's Table as 'pure substances' and they stress their 'artificial' nature. Whilst agreeing with the general tenor of this position I have the odd quibble and wish to put a somewhat different emphasis on it. I wish to stress the extent to which the notion of chemical substances as natural kinds is implicit in Geoffroy's conception. From this point of view there is a sense in which it is misleading to describe chemical substances as 'artificial'.

In the main, chemical substances exist in a pure state only in the laboratory or workshop. Those that are to be found in nature are more or less impure insofar as they are mixed with other substances from which they need to be separated if they are to be 'purified'. There is a sense in which the notion of purity involved here was far from novel in 1718. Archimedes was concerned to develop a method for determining the purity of the gold composing a crown, whilst over half a century before 1718 Boyle discoursed at great length on the purity of chemicals, identifying impurity as one of the main sources of error in chemical experiments. Klein's focus on the 'purity' of the substances in Geoffroy's table and their artificial nature does not serve well to pick out distinctive features of the novel concepts at work. Further, describing pure chemical substances as artificial is inappropriate because it does not accommodate the sense in which many chemical substances are present and at work in nature whether they are extracted in the laboratory or not. There is an important sense in which chemical substances are natural, not artificial.

What is typically involved in the 'purification' of substances serves to support my point. Impure silver can be purified by first adding nitric acid, then adding potassium carbonate, and filtering off the silver carbonate formed as a result and finally recovering silver by heating the silver carbonate. Such a process understood in this way is typical of Geoffroy's conception of chemical substance and compound articulated by Klein. It implies that the silver exists in the original impure mixture in no less real a sense than it does at the end. Purification, and any other transformation that takes place in the chemical laboratory or artisan's workshop, involves chemical substances interacting in the ways that they do because of what they are, independently of us or our experimental practices. Even chemical substances that would not exist were they not produced in the laboratory have objective properties in this sense.

What is distinctive about the substances depicted in or implied by Geoffroy's table is, not that they are 'pure' but that they are substances characterised by their *chemical* properties. Chemical substances are what they are by virtue of the way they combine with other substances to form compounds, can be analysed into component substances, and can displace other substances or be replaced by them. It is precisely these factors that determine the location of substances in the table. Properties other than the ones I have designated as chemical, such as the boiling point of a substance or its smell and colour, whilst undoubtedly used as means of identifying the presence of substances, do not figure, and do not need to figure, in Geoffroy's table and commentary on it. Chemicals are natural kinds. A chemical is of a kind by virtue of the way it combines with other chemicals. Chemicals put themselves into kinds, as it were, by virtue of their mode of interacting with each other. Here I only give increased emphasis to a point that is at least implicit in the writings of Klein, both singly and with her co-author Lefèvre.[13] The latter point out that whereas 'in eighteenth-century plant and animal chemistry chemists drew the boundaries of single substances by referring to observable properties, both chemical and physical, the individuation of pure chemical substances was determined by experimentation and the tracing of substances in experiments'. Klein and Lefèvre (2007, p. 301) stress what I have referred to as the objective character of the distinguishing features of chemical substances by referring to their 'materiality', their 'potential for transformation in series of chemical experiments'. Substances that could be incorporated into Geoffroy's table are chemical substances that are what they are by virtue of the way they interact with other chemical substances. Chemical substances belong to natural kinds.

Geoffroy's table is best seen as a symptom of the new chemistry of combination rather than as constituting it. In the 70 years that separated its publication from the *Tableau de la nomenclature chimique* published by Lavoisier and his collaborators, the number of published affinity tables barely reached double figures.[14] It is also the case that Geofffroy's table was not able to accommodate all that was involved in the chemistry of combination. As critics of it soon pointed out, the position of substances in the table is a variable rather than a given because of the way in which chemical affinities can depend on temperature. These reservations aside, it remains the case that the fact that Geoffroy could abstract from the chemical practice of his day a table displaying a range of chemical compounds that could be analysed into and manufactured from there components is testimony to the fact that there was such a practice to abstract from. Some of the practitioners providing the experimental data drawn on by Geoffroy worked in the context of some matter theory, whether Paracelsian, Aristotelian or mechanical, although many others were artisans subscribing to no articulated matter theory at all. The chemistry of combination at the basis of Geoffroy's table provided a basis for future research that was independent of fundamental matter theory. By the time Torbern Bergman published his table of affinities, in 1783, the number of substances included had been greatly increased and the reactions in which they take part documented in considerable detail.

It will not have escaped the readers notice that Klein's account of the emergence of modern chemistry fits well into my conception, discussed in Chapter 6, of the scientific revolution as involving the emergence of experimental science as distinct from philosophical matter theory rather than as a change from one matter theory to another. The new theory of chemical combination was not a theory of matter in general. It was a theory of chemical combination that did not even cover the totality of substances manipulated by chemists in their laboratories and workshops. In Chapter 6 I construed Boyle as, in effect, recognising the distinction between experimental science and philosophical matter theory. Boyle considered it necessary to invoke intermediate or subsidiary principles and causes in the conduct of experimental science as opposed to ultimate mechanical causes. His pneumatics invoked the spring and weight of the air, and his consequential results stood independently of any assumptions about corpuscular or mechanical atoms. But Boyle included the 'chymical' in his list of subordinate principles and he included the works of 'chymists' amongst claims to knowledge too hastily condemned 'because they cannot be clearly and easily deduc'd from ye doctrines of Atoms, or ye Catholick Laws of motion'.[15] The implication is that chemical knowledge subject to experimental investigation and confirmation needs some 'intermediate principles' on a par with the spring of the air. The 'notions' of chemical substance and compound and the *rapport* responsible for the combination of components in compounds that Geoffroy built into his table were just what was needed to facilitate a line of development that was to culminate in Lavoisier's new system.

In the remainder of this chapter I return, first to Boyle's chemistry and then to Newton's. I hope to show that the atomistic matter theories espoused by Boyle and Newton did not productively inform their chemistry nor were they significantly supported by it. Geoffroy's conceptualisation of chemistry serves as a contrast that helps bring out the significance of my position.

8.4 Boyle's Chemistry: Some Preliminaries

Suppose we accept some version of Klein's view that the notion of chemical substance, compound and combination implicit in Geoffroy's table was just what was needed to productively inform a significant part of experimental chemistry and set it on a track that was to lead to Lavoisier. How does mechanical atomism fit into this picture? More specifically, to what extent did the chemistry developed by Robert Boyle in the context of mechanical atomism contribute to a concept of chemical substance able to fruitfully inform the new experimental chemistry? Whilst acknowledging that Boyle's mechanical atomism did serve a useful negative function, insofar as it provided a case for removing Aristolian elements and substantial forms and Paracelsian principles from chemistry, I maintain that it did not serve a positive function. As a fundamental matter theory, Boyle's mechanical atomism was too far removed from what could fruitfully be experimentally tested to offer useful guidance to the experimenter.

In Chapter 6 I distinguished between Boyle's mechanical atomism and his experimental science, most notably exemplified in his pneumatics. Notions necessary for formulating the claims of an experimental science need to be framed and the claims tested by a range of experiments sufficient to render them strong contenders as 'matters of fact'. Boyle's pneumatics, involving notions of the spring, pressure and weight of air and supported by a range of experiments, many of them involving the air pump, conformed to this pattern. The spring, pressure and weight of air were 'intermediate' notions rather than strict mechanical ones as Boyle openly acknowledged. The support Boyle sought for strict mechanical explanations of phenomena was of a weaker kind than that demanded of experimental knowledge. That support involved the construction of hypothetical mechanisms, mechanical in the strict sense, which would be sufficient to account for known phenomena. That support is especially significant if the phenomena explained poses problems for rival matter theories. Boyle acknowledged that he could not go further and argue that the hypothetical mechanisms were the actual ones operative in nature. He pointed out that testing a fundamental matter theory against the phenomena in the way proposed presupposes knowledge of those phenomena. He complained that many philosophical systems were constructed without paying due heed to the phenomena, and advocated a separation between experimental matters of fact and philosophical systems designed to explain them. He sometimes signalled this distinction in the context of his own exposition of mechanical atomism and experimental matters of fact that he sought to explain by appeal to it. These are the considerations, defended in detail in Chapter 6, that I bring to bear on Boyle's chemistry.

Whatever the merit of my distinction between Boyle's experimental science and his mechanical philosophy, there is no doubt that Newman (2006. p. 3) is absolutely correct in observing that, in Boyle's work, chemistry and his mechanical philosophy were 'indissolubly linked'. There is perhaps not a single chemical text of Boyle in which the chemistry is not related to mechanical explanations of it, and many instances of Boyle invoking chemical phenomena in support of the philosophy. I acknowledge this but maintain, first, that the degree of support for the mechanical philosophy based on Boyle's chemistry was not as strong as is typically supposed and secondly, and more importantly, that the mechanical philosophy did not feed productively into Boyle's chemistry. Insofar as Boyle was able to contrive mechanisms capable of accounting for chemical phenomena, knowledge of the phenomena involved was acquired by other means. As far as furthering the search for knowledge of chemical phenomena is concerned, Boyle did not have much to offer as far as the framing of appropriate notions is concerned, and the limitations of his efforts in this respect were very much tied up with what can reasonably be construed as his premature attempt to reduce chemistry to mechanical atomism.

8.5 Boyle's Mechanical Rather than Chemical Construal of Substances

Boyle did, of course, have some notion of chemical substances and their properties, but that notion did not involve a precision that went beyond the common sense of his time. In 'History of particular qualities' Boyle (2000, Vol. 6, p. 267) singles out chemical qualities as those that have

> bin principally introduc'd and taken notice of by means of Chymical Operations and Experiments; such as are Fumigation, Amalgamation, Cupellation, Volatilization, Precipitation, &c. by which operations among other means, Corporeall things come to appear Volatile or Fixt, Soluble or Insoluble in some Menstruum's, Amalgamable or Unamalgamable, capable or uncapable to precipitate such Bodies, or be precipitated by them, and (in a word) acquire or loose several powers to act on other Bodies or dispositions to be wrought on by them.

This is too imprecise and includes too much to be particularly useful.

There were general features of Boyle's mechanical philosophy, connected with his concern to dispense with notions akin to the substantial forms of the scholastics, which did not lend themselves to a categorisation of chemical substances useful for informing experimental chemistry. One of them was Boyle's view that classifications of substances into kinds are a human imposition rather than one arising naturally from the nature of the substances classified. This view, when taken seriously, barred Boyle from developing a precise notion of chemical substance that went beyond common sense.

In the ‘Origin of forms and qualities’ Boyle takes a stand against the Aristotelian account of essential properties, and of the distinction between generation, corruption and alteration. In taking such a stand he in effect rules out the idea that there are natural divisions between chemical substances. His view makes such distinctions into conventions imposed on the world by us. We can define a globe as a metal sphere, in which case sphericity is an essential property of a globe and any volume of metal that lacks that shape is not a globe. In a similar way, says Boyle (2000, Vol. 5, p. 322), we can define the essential properties of substances by listing the properties essential to that substance. With respect to the classification of substances generally, men ‘did for conveniency, and for the more Expeditious expression of their Conceptions agree to distinguish them into several Sorts’ and have ‘for their Convenience agreed to signifie all the Essentials requisite to constitute such a Body by one Name’.

Given that objects or substances are defined and divided into kinds by way of human conventions, the need for a substantial form as the seat of the essential properties is removed. Also, Boyle can re-interpret and render less significant the Aristotelian distinction between generation, corruption and mere alteration. If liquidity is included in our definition of water, then ice is not water and the freezing of water into ice involves the corruption of water and the generation of ice. If liquidity is not included amongst water’s essential properties, then freezing involves the mere alteration of water from liquid to solid form. Whichever way we go, Boyle implies, there is only one change taking place, and it involves the re-organization of the particles that make up water into a new texture making up ice.[16]

Boyle (2000, Vol. 5, p. 356) drives home the point about the conventional character of kinds a few pages later.

> It was not at random that I spoke when . . . I intimated, That ‘twas very much by a kind of tacit agreement, that Men had distinguished the *Species* of Bodies, and that those Distinctions were more Arbitrary than we are wont to be aware of. For I confess, that I have not yet, either in *Aristotle*, or any other Writer, met with any genuine and sufficient Diagnostic and Boundary, for the Discriminating and limiting the *Species* of Things; or, so to speak more plainly, I have not found any Naturalist has laid down a determinate Number and sort of Qualities or other Attributes, which is *sufficient* and *necessary* to constitute all portions of Matter, endow’d with them, distinct Kinds of Natural Bodies. And therefore I observe that most commonly Men look upon these as Distinct *Species* of Bodies, that have had the luck to have had distinct Names found out for them; though perhaps diverse of them differ much lesse from one another, than other Bodies, which (because they have been hudled up under one Name) have been look’d upon, as one sort of bodies.

In the text following this passage Boyle proceeds to stress that there is no clear-cut distinction between naturally occurring substances and artefacts prepared in the laboratory. He points out, for instance, that there is no good reason to make a categorical distinction between substances formed in nature by the heat of the sun or by a chance fire on a hillside from those formed by artificial heating or burning in the laboratory. His main intent here is to undermine the scholastic notion of substantial form. On a common scholastic interpretation of that distinction, naturally occurring substances differ from artefacts precisely by virtue of the substantial forms possessed by the former. Rather than focussing on the artefacts as opposed to natural substances, as the site for developing a notion of chemical substances understood in terms of their interactions, Boyle is denying any useful distinction.[17] Boyle adds to his point by appealing to his own mechanical viewpoint. Natural substances do not differ from artefacts because in both cases the properties of the substances arise alike from the arrangements and motions of the corpuscles of which they are composed. No substantial forms are necessary. Boyle makes this point in the case of the likeness of naturally occurring and 'factitious' vitriol, but the fact that his point is a general one, about the inadequacy of the Aristotelian notion of form and the superiority of the mechanical philosophy rather than some specific point about chemistry, is illustrated by the fact that he proceeds to give, as a second example, the similarity between a normal pear and one grown from a tree grafted onto a thorn.

According to Boyle (2000, Vol. 5, pp. 359–360), what he refers to as 'chemical concretes', whether natural or artificial, are to be characterised in terms of a 'concourse of accidents'.

> Since, then, these Productions of the Fire, being of Nature's own making, cannot be deny'd to be Natural Bodies, I see not why the like Productions of the Fire should be thought unworthy of that Name onely because the Fire, that made the former, was made by chance in a Hill, and that which produc'd the latter was kindled by a Man in a Furnace. And if flower of Sulphur, Lime, Glass, and colliquated mixtures of Metals and Minerals, are to be reckon'd among Natural Bodies, it seems to me but reasonable that, upon the same grounds, we should admit flower of Antimony, Lime, and Glass, and Pewter, and Brass, and many other Chymical Concretes (if I may so call them), to be taken into the same number; and then 'twill be evident that, to distinguish the *species* of Natural Bodies, a Concourse of Accidents will, without considering any Substantial Form, be sufficient.

Boyle (2000, Vol. 5, pp. 322–323) does specify the essential properties of substances more easily recognised by us as *chemical* substances than glass. Gold, for instance, is designated as a body 'that is extremely ponderous, very malleable and ductile, fusible and yet fixed in the fire, and of a yellowish colour'. In the ensuing discussion the ability to 'resist aqua fortis' is added. Boyle's intent is to undermine the need to characterise substances in terms of substantial forms. Natural and artificial substances are what they are by virtue of the characteristic set of properties they possess, and those properties are presumed to arise from various arrangements and motions of component corpuscles. This treatment of properties in general does not point in the direction of the notion of a chemical substance understood in terms of what it does and does not combine with and to what degree.

Boyle's view that distinctions between kinds of substances are not natural but imposed on nature by us for our convenience fitted in with another view of his that did not help with the elaboration of a notion of chemical substances as 'intermediate' causes in chemistry. This was the idea that any substance could in principle be changed into any other by bringing about the appropriate change in the underlying mechanical structure.

> So that though I would not say, that Any thing can immediately be made of Every thing – as a Gold Ring of a *wedge* of Gold, or Oyl, or Fire of Water; yet, since Bodies, having but one common Matter, can be differenc'd but by Accidents, which seem all of them to be

> the Effects and Consequents of Local Motion, I see not, why it should be absurd to think, that (at least among Inanimate Bodies), by the Intervention of some very small *Addition* or *Subtraction* of Matter (which yet in most cases will scarce be needed), and of an orderly *Series of Alterations*, disposing by degrees the Matter to be transmuted, almost of any thing may at length be made of Any thing . . . (Boyle, 2000, Vol. 5, p. 332)

Boyle's views on the possibility of changing anything into anything else, and on the distinctions between substances being a matter of convention, both of them fitting naturally into his mechanical atomism, turned attention away from the task of understanding chemicals as kinds that combined in distinctive ways with chemicals of other kinds. They suited Boyle's purpose of undermining the need to appeal to substantial forms but they did not provide useful guidance to chemistry.

Another factor that stood in the way of Boyle framing notions that were able to inform an experimental program that was distinctly chemical was the extreme generality of his mechanical matter theory. It was designed to give an account of how qualities in general arise from the 'primitive affections' of pieces of universal matter. Boyle's use of chemistry to support his mechanical matter theory involved him in offering possible mechanisms for explaining the whole range of changes in qualities accompanying chemical change. His focus was as much on the changes of colour accompanying chemical change, for example, as on the changes of substances bearing the colours.

The treatment of chemistry by Boyle in the service of his general mechanical matter theory had the consequence that a focus on chemical combination of chemical substances was far from central, if not conspicuously absent. This point is well illustrated by Boyle's essay 'On the mechanical causes of chemical precipitation'. Boyle (2000, Vol. 8, p. 484) rejects appeal to antipathies and sympathies as the cause of precipitation and proposes, in its place, 'a greater congruity as to bigness, shape, motion and pores of the minute parts between the *Mestruum* and the *Precipitant*, than between the same Solvent and the body it kept before dissolved'. The reasons he rejected sympathies and antipathies can be gleaned from the article in question and elsewhere. First, Boyle (2000, Vol. 8, pp. 415 and 484) construed them as unacceptably mysterious and anthropomorphic. Secondly, he argued that the facility of substances to combine with or precipitate others need to be understood as relations between substances rather than as properties of single substances, which Boyle assumed to be the practice of those resorting to sympathies and antipathies. So, for instance, substances with a great antipathy towards each other can nevertheless react in a similar way with some third substance.[18] The notion of *rapport* to be found in Geoffroy is sufficient to indicate that sympathies and antipathies between chemical substances can be put to work in chemistry in a way that construes them as relational properties that are not anthropomorphic. Boyle did not take that line, even though it would have been in accord with what he had to say elsewhere about the importance of appeal to 'intermediate causes and explanations'. He explicitly declined to offer a 'History of Precipitations' that would co-ordinate knowledge of them at the experimental level. Rather, Boyle (2000, Vol. 8, p. 481) proceeded directly to contrive 'the *Mechanical Causes* of Precipitation'. In like manner, Boyle devises mechanical explanations of the action of acids and alkalis, volatility and fixedness and so on. In all these instances there is an emphasis on the mechanical breaking down or coherence of corpuscles by reference to mechanisms that are frequently highly contrived and which are not of a kind that can usefully guide the experimenter. An emphasis on chemical combination is, in the main, conspicuously absent.[19]

8.6 Boyle on the Properties of Chemical Corpuscles

The fact that Boyle proposed a mechanical or corpuscular chemistry suggests that his position can easily be accommodated to, and can even be read as an anticipation of, the view of chemical combination identified by Klein. Chemical substances are what they are by virtue of the nature of the corpuscles that compose them, and chemical combination comes about as the result of the association or dissociation of corpuscles. To the extent that such a position represents Boyle's position, Klein's view that the idea of chemical combination emerged as a significant novelty in Geoffroy's paper of 1718 would appear to be undermined. In this section I explore the precise character of, and role played by, Boyle's corpuscles and their relation to the chemical substances they are presumed to compose. I conclude that the details of Boyle's position do not undermine Klein's position. Boyle's corpuscular theory, as he construed it, did not provide him with a notion of a chemical substance adequate for chemistry.

If chemistry is to be explained by reference to corpuscles then there is a basic question that needs to be answered. What properties do the corpuscles possess that enable them to fulfil their role as explainers of chemical phenomena? Did the corpuscles figuring in Boyle's chemistry possess only strict mechanical properties, or did he need to attribute some further 'chemical' properties to them?

We have seen that Boyle's corpuscular chemistry grew out of a medieval corpuscular tradition via the work of Daniel Sennert. In that tradition, the minima of a substance possessed properties characteristic of the substance they were minima of. As Newman (1996) and Antonio Clericuzio (2000) have observed, the first version of atomism to appear in Boyle's writings shared this feature. The surviving pages of a manuscript on atomism written by the young Boyle (2000, Vol. 13, p. 228) contain the view that atoms are particles that nature cannot divide and which possess the properties of the homogeneous substances they are the least parts of. The main argument given by Boyle for the existence of atoms is a reproduction of Sennert's reduction to the pristine state, the recovery of silver after its dissolution in nitric acid.[20] By the late 1650s, when Boyle was composing the *Sceptical chemist*, he had adopted the mechanical philosophy. In that work, Boyle (2000, Vol. 2, p. 230) introduced his hierarchy of particles with mechanical atoms at the base and corpuscles of various degrees of complexity composed of them. (He also added further examples of reductions to the pristine state to strengthen the case for the existence of corpuscles.) Boyle (2000, Vol. 2, p. 229) made it clear that his 'natural minima', corresponding to what I call mechanical atoms, are composed of universal matter 'actually divided into Particles of several sizes and shapes variously mov'd'. There remains the question of the properties that Boyle attributed to the corpuscles made from these atoms. Did they possess some 'chemical' properties over and above strictly mechanical ones?

Textual evidence apart, there are conceptual difficulties associated with the claim that corpuscles have chemical properties. Such a position is implied by Boyle's youthful assertion that atoms possess the properties of the wholes they are atoms of. But such a view cannot be coherently sustained. Insofar as atoms are invoked to explain various chemical properties they cannot also possess them. Take, for example, the dissolution of silver in nitric acid and its subsequent recovery, the reaction so central to the arguments of the atomists. The property, possessed by bulk silver, of being dissolved in nitric acid cannot also be a property of the corpuscles of silver for atomists like Sennert and Boyle. The whole point of their argument for the existence of corpuscles or natural minima of silver is that these particles persist as such in the solution thus accounting for the fact that they can be recovered. Again, silver melts at high temperatures but an atomist could not afford to conclude from this that corpuscles of silver do the same. Whilst it is true that corpuscles need to possess properties sufficient for them to play the chemical roles required of them, it cannot coherently be claimed that they possessed chemical properties in an unqualified way.

It is undoubtedly the case that the corpuscles figuring in Boyle's chemistry are not mechanical atoms but structures built up from those atoms. In spite of their compound character, strict mechanical properties can be attributed to corpuscles in a straightforward and non-mysterious way. A corpuscle will possess a shape, size and motion that is the resultant of the shapes, sizes and motions of the mechanical atoms that compose it. The shapes, sizes and motions of corpuscles are derivative, but they are strict mechanical qualities nevertheless. I maintain that the textual evidence strongly points to the fact that, from the *Skeptical chemist* onwards, Boyle attributed only strict mechanical properties to chemical corpuscles.

According to Boyle, semi-permanent corpuscle composed of mechanical atoms will have a shape, size and degree of motion that is determined by and is the resultant of the shapes, sizes and motions of the mechanical atoms that compose it. Boyle referred to the structured arrangement of mechanical atoms comprising a particle as its texture. The shapes, sizes and motions of corpuscles that result from their texture are not primitive mechanical properties but they are strict mechanical properties nevertheless. Because corpuscles are complex structures of mechanical atoms, their shapes, sizes and motions typically change when those structures are modified by adding or removing mechanical atoms or by rearranging or changing the motions of the mechanical atoms composing a corpuscle. This is made quite explicit by Boyle (2000, Vol. 5, p. 326) in the following passage from the 'Origin of forms and qualities':

> That as well each of the *Minima Naturalia* as each of the Primary Clusters above mention'd having its own Determinate Bulk & Shape, when these come to adhere to one another, it must *alwaies* happen that the Size, and *often*, that the Figure, of the Corpuscles compos'd by their Juxta-position and Cohesion, will be chang'd; and *not seldome*, too, the Motion either

> of the one or the other, or both, will receive a new Tendency, or be alter'd as to its Velocity or otherwise. And the like will happen, when the Corpuscles, that compose a Cluster of Particles, are disjoin'd, or anything of the little Mass is broken off. And whether anything of Matter is added to a Corpuscle or taken away from it, in either case, (as we just now intimated,) the Size of it must necessarily be alter'd, and for the most part the Figure will be so too, whereby it will both acquire a congruity to the Pores of some Bodies (and perhaps some of our Sensories), and become Incongruous to those of others, and consequently be qualify'd, as I shall more fully show you hereafter, to operate on divers occasions, much otherwise than it was fitted to do before.

Qualities of substances are due to the sizes, shapes, motions and arrangements of the corpuscles that compose them and will change if those shapes, sizes, motions and arrangements are changed in the kinds of ways listed in the above quotation.[21] Boyle embraced chemical qualities and chemical change in this scheme. He aspired to reduce chemistry to mechanism in the strict sense.

Both Newman and Clericuzio find it necessary to depart from the strict mechanical interpetation of Boyle and construe him as attributing chemical properties to corpuscles. The text they use to this end involves some experiments by Boyle on colour changes accompanying chemical reactions. White mercury sublimate (mercuric chloride) is dissolved in water to form a colourless solution. This turns orange when salt of tartar (potassium carbonate) is added. Addition of oil of vitriol (sulphuric acid) results in a colourless solution once again.[22] Boyle employs this experiment to aid him defend his 'mechanical' account of colours generally. On this account, colours do not inhere in coloured objects, as forms or principles. Rather, they are modifications in light, itself an (unspecified) mechanism, brought about by its interaction with the object appearing coloured. Because colour results from an interaction of light with a structured object, a change in that structure can result in a change in colour, the removal of colour, or the introduction of a colour where there was none before. On the one hand, Boyle regards these considerations to be the 'fittest to recommend the Doctrine propos'd in this Treatise', that is, the Treatise defending a mechanical account of colour. One the other hand, he acknowledges that his account of the colour changes involved in his experiments with mercury sublimate employs 'Chymical' reasoning and 'Chymical Notions'. He acknowledges that there is a difference between a 'chymical Explication of a *Phenomenon*' and a 'truly Philosophical or Mechanical' one and admits that his account of colour change falls short of the latter, such truly mechanical explanations being 'more than I dare as yet pretend to'.

Newman (2006, p. 185) takes Boyle to be appealing to 'chymical properties of corpuscles in order to explain the source of their mutual association and dissociation'. For Newman, this does not undermine the status of Boyle's corpuscular chemistry as 'mechanical'. It remains mechanical in the sense that clocks and watches are. The behaviour of chemical substances are explained by structures of corpuscles with relevant properties just as clocks are explained by structured arrangements of rigid gear-wheels, heavy pendulum bobs and so on. That is, he construes Boyle's chemistry as mechanical in a common, rather than a strict, sense. I have already aired my dissatisfaction with such a move. It runs counter to Boyle' persistent characterisation and defence of a strict, not a common, version of the mechanical philosophy.

Clericuzio (1990, p. 145) takes Boyle's discussion of colour changes in his experiments with mercury sublimate as evidence for his claim that Boyle's corpuscular chemistry was not mechanical because the explanation of colour change offered by Boyle 'is based on the substitution of compound corpuscles having chemical properties'. In the elaboration of his position Clericuzio invokes Boyle's distinction between intermediate and ultimate, mechanical causes that I have described in Chapter 6. It is true that if corpuscles are to be situated in Boyle's scale of causes then they will not be at the top of the scale since they are compound structures composed of mechanical atoms. In this sense they will be intermediate causes. But the whole point of Boyle's distinction is to demarcate the ultimate mechanical causes that are remote from what is empirically accessible and causes, such as the spring of the air in his pneumatics, that are empirically accessible. The corpuscles Boyle invoked in his chemistry were no more accessible than the mechanical atoms of which they were composed.

My own view retains the idea that Boyle aspired to a corpuscular chemistry that was mechanical in the strict sense. His chemical corpuscles were to bear only the strict mechanical properties of shape, size and motion, notwithstanding the fact that they were compound particles. Boyle hoped, at best, to contrive, but not vindicate, corpuscular mechanism that would serve to explain chemical phenomena. Knowledge of the chemical phenomena was itself to be established by experiment. The fact that a colourless solution of mercury sublimate turns orange on the addition of salt of tartar and can be rendered colourless again by adding oil of vitriol constitutes chemical knowledge. At the level of experiment we have 'chymical reasoning' invoking 'chymical notions'. The 'chymical explication of a phenomenon' established at this level is distinct from a 'truly philosophical or mechanical' one.[23] This interpretation avoids the problematic ascription of 'chemical' properties to corpuscles, and spares us the task of identifying in Boyle just what those chemical properties were and how he sought to identify them.

8.7 Chemical Properties and Essential Properties

There is a view on chemical substances to which a corpuscular theory would seem to readily lend itself. On this view, chemical substances are divided into kinds insofar as the corpuscles that constitute them are divided into kinds. Newman (2006, p. 198) has recently attributed a position akin to this to Boyle. He interprets Boyle's mechanical chemistry involving semipermanent corpuscles as providing Boyle with a means to identify the essential properties of substances by way of their chemical as opposed to physical properties.

> It was chymistry that allowed him to distinguish the essential differences of bodies in a relatively certain fashion, and without such stable essences Boyle could not argue that the qualitative mutability of the phenomenal world was mostly a matter of alterations in texture imposed on fundamentally unchanged corpuscles by mechanical means.

The essential properties of substances stem from the character of the semi-permanent corpuscles that compose them. Chemical substances fall naturally into kinds because corpuscles fall naturally into kinds, distinguished from each other by the essential character of those corpuscles. Non-essential properties arise from arrangements or motions of the corpuscles or perhaps their interaction with other particles such as those presumed to constitute light. On this view, the essential properties of gold, such as its resistance to nitric acid, stem from the nature of the corpuscles that comprise it. Inessential properties such as temperature are to be attributed to the rapid motions of those corpuscles that retain their identity, so that gold is gold whether it is hot or cold.

The main text that Newman appeals to in order to support his attribution of a position such as this to Boyle is the latter's 'History of particular qualities', published in 1670. In that essay Boyle (2000, Vol. 6, p. 280) does make a distinction between essential and what he terms 'extra-essential' qualities.

> For here it is to be considered, that besides that peculiar and Essential Modification which constitutes a Body, and distinguishes it from others that are not of the same Species, there may be certain other Attributes that we call *Extra-essential*; which may be common to that Body with many others, and upon which may depend those more external Affections of the Matter which may suffice to give it this or that Relation to other bodies, divers of which relations we style Qualities.

Boyle's position is illustrated by examples. Degree of hotness is an extra-essential property of iron, an iron rod being as much iron after it is made red hot by beating as it was before. Pieces of iron, silver and wood retain their essential properties whether they are rough or rendered smooth enough to regularly reflect light.

In 'History of particular qualities' Boyle writes of the essential properties of 'bodies', where he clearly intends by that term samples of substances, such as iron, gold and so forth. But Boyle (2000, Vol. 6, p. 281) also refers to the essential properties of corpuscles. Given Boyle's mechanical matter theory it can be said, in general terms, that corresponding to the essential properties of a body, such as a lump of iron, there must be some permanent underlying corpuscular structure responsible for those essential properties.[24] Any change in the structures responsible for the essential properties of iron will result in the iron changing into some other substance or substances. By contrast, the structures responsible for the extra-essential properties of the iron, such as its temperature, can change without the iron losing its identity as such. This much is clearly implied by Boyle's text. But Newman (2006, p. 197) assumes more. He construes Boyle as identifying the essential properties of a substance with its chemical properties. 'The heat that one feels upon rubbing iron, brass, wood, or stone can be induced and allowed to depart without altering the chemical properties of the material being rubbed, which are taken as a measure of its essential character.'

I do not find grounds for identifying essential qualities with chymical qualities in Boyle's text. There is the further difficulty of identifying which qualities are chymical ones in Boyle's schema. When Boyle gives examples of the essential properties of substances he lists qualities other than what might reasonably be regarded as chemical ones. For instance, yellowness and ductility are included in the list of the essential properties of gold given by Boyle (2000, Vol. 5, pp. 322–323). If there are semi-permanent particles in nature distinguished by their essential structure and responsible for the chemical properties of the substances they are least parts of, then there is a sense in which corpuscles, and hence chemical substances fall into natural kinds. This clashes with Boyle's view, that I have described at length, that the division of substances into kinds by specification of the essential properties of those kinds is a result of human convention.

There is a further difficulty posed by Boyle's text for Newman's position. Whilst it is clear that in Boyle's view the essential properties of bodies must correspond to some underlying mechanical structure,that structure need not consist in some semi-permanent particle. According to Boyle (2000, Vol. 6, pp. 281–282,in the very article drawn on by Newman as his source for the views he attributes to Boyle, explains that only part of the structure of corpuscles can be regarded as essential to the nature of the bodies they form, other parts being 'extraessential.'

> For if Corpuscles without looseing that Texture which is Essential to them, may (as we have show'd they may) have their Shape, or their Surfaces, or their Scituation changed; may also admit of Alterations, (especially as these Corpuscles make up an Aggregate or *Congeries*,) as to Motion or Rest; as to these or those degrees or other circumstances of Motion; as to Laxity and Density of parts, and divers other Affections; why should we not think it possible, that a single (though not indivisible) Corpuscle, & much more an Aggregate of Corpuscles, may by some of these, or the like changes, which, as I was saying destroy not the Essential texture, be fittd to produce divers other Qualities, besides those that necessarily flow from it.

Newman (2006, p. 197) raises the question of how Boyle distinguished essential from non-essential properties. 'How did Boyle know which properties of a body were essential?' Given Boyle's views on the conventional character of essential properties it would seem that fixing a list of essential properties is something to be decided rather than discovered. But this issue aside, I find the answer to the question that Newman (2006, p. 198) attributes to Boyle problematic.

> [I]t was above all the classification into chemical species that allowed Boyle to determine the essential differences of the aggregate corpuscles. *Colours* contains large sections devoted to indicator tests for deciding whether a particular substance belongs to the class of 'acid salts'. 'alkalizate salts', or 'urinous salts'. In other contexts, he employs such time-honoured tests as cupellation, dissolution in different mineral acids, and colour of flame to detect a metal or other substance when its presence is not obvious to the senses.
>
> Many of these tests are based on the assumption that the aggregate corpuscles being tested for are not destroyed by the test itself – instead they remain undivided during the procedure and hence retain their identity, At the same time, the reagent employed to reveal the hidden substance is assumed to react selectively on the latter's aggregate corpuscles (by causing them mutually to disperse or coalesce, for example) and hence to circumvent the sort of generalised mechanical effect that Boyle describes as 'extraessential'.

A problem I have here is that I am not convinced that tests of a substance that leave it unchanged are more revealing and significant than those that do not. I can distinguish between a box of genuine matches and a box of fakes by striking a few of each. The effectiveness of the test is not impaired by the fact that the genuine matches are destroyed in the process. For Boyle, some colour changes arise from essential changes (changing mercury into its red oxide by heating) and others do not (raising steel to a high temperature). Further, the tests for identifying and classifying salts mentioned by Newman do involve chemical transformation of them. It is true that the metallic bases of the salts remain unchanged by the indicator tests but it is salts that are divided into families by the tests, not metals.

If Newman's interpretation of Boyle's texts were correct, then we would have a neat account of what a chemical substance amounts to in Boyle's chemistry. Substances are characterised by their essential, chemical, properties as opposed to their inessential, physical, ones. What is more, chemical substances retain their identity so long as the semi-permanent particles that compose them remain intact. I reject this view because I cannot find in Boyle the identification of essential with chemical properties nor do I find a distinction in Boyle between chemical and other kinds of properties. Boyle's mechanical contrivances aimed at explaining qualities in general and chemical properties in particular are too diverse and ad hoc to lend themselves to an interpretation that would identify semi-permanent particles as the seat of chemical properties, whatever those latter might be. We have seen that Boyle divided 'chemical concretes' into kinds by means of conventional definitions specifying a 'concourse of accidents' that involved properties, such as colour and degree of malleability, not obviously or unproblematically classified as chemical.

8.8 The Mechanical Philosophy Versus the Experimental Philosophy

Like his fellow advocates of the mechanical philosophy, Boyle used the term 'mechanical' not only in the context of a mechanical matter theory but also in a more common and less strict sense that fitted well with the emphasis the mechanical philosophers placed on the centrality and importance of experiment. When Boyle's contemporary Henry Powers published a book summarising an approach very similar to that of Boyle he chose to call it 'Experimental philosophy' although 'The mechanical philosophy' would have served as well, given the contents. I have argued that seventeenth-century chemistry owed little to mechanical matter theory in the strict sense. But what of the more general senses of mechanical that had natural links with experiment? To what extent were advances in seventeenth-century chemistry the fruits of the mechanical philosophy interpreted in some of the common senses described in 6.8 rather than in the strict sense?

There is no doubting that seventeenth-century chemistry was advanced and substantiated by way of experiment. Chemists, whether they were Aristotelians, Paracelsians, mechanical philosophers, or apothecaries and metallurgists subscribing to no explicit philosophy at all, conducted their search for chemical knowledge by way of experiments conducted in the contrived situations of the workshop or laboratory. Their practice clashed with a hardline Aristotelian view to the effect that it is misguided to attempt to understand the natural world by upsetting the course of nature by way of artificial interventions. Seventeenth-century chemistry was mechanical in the sense of artisanal.

We have noted that another common sense of mechanical was that whereby the behaviour of a clock is understood in terms of the relations between its component parts. Some of chemistry was mechanical in this sense insofar as the nature of complex substances was understood in terms of their components. Understanding chemicals by reference to their chemical structure is like understanding clocks and watches by reference to their mechanical structure, both involve breaking down wholes into their 'parts' and building them up from their 'parts'.

Having pinpointed some common senses in which chemistry might be said to be mechanical, it is important to be clear about what this means for the nature of seventeenth-century chemistry. Chemical substances can be learnt about and even brought into existence in the artificial conditions of a laboratory experiment but this does not mean that chemicals are mechanical artefacts. Boyle makes part of the point I am getting at. He makes it about the products of tradesman but it applies equally to the productions of the experimental chemist. Many artificial productions, Boyle (2000, Vol. 6, pp. 467–468) writes,

> do differ from those that are confessedly natural, not in essence, but in efficients; there are very many things made by tradesmen, wherein nature appears manifestly to do the main parts of the work: as in malting, brewing, baking, making of raisins, currants, and other dried fruits, as also hydromel, vinegar, lime etc. and the tradesman does but bring visible bodies together after a gross manner, and then leaves them to act one upon another, according to their respective natures.

In the *Sceptical chemist* Boyle makes this point in the context of chemical productions of the laboratory. The instruments used in the laboratory such as acids or heat are 'Agents of Nature's own providing and whose chief Powers of Operation they receive from their own Nature or Texture, not the Artificer' so that their effects are produced 'whether the Artificer intended it or no'. Boyle (2000, Vol. 2, p. 300) proceeds to draw an analogy between chemistry and gardening.

> And, indeed, the Fire is as well a Natural Agent as Seed; and the Chymist that employs it, does but apply Natural Agents and Patients, who being thus brought together, and acting according to their respective Natures, perform the worke themselves; as Apples, Plums, or other fruit, are natural Productions, though the Gardiner bring and fasten together the Sciens of the Stock, and both Water, and do perhaps divers other ways Contribute to its bearing fruit.

Chemists can bring about chemical changes by mixing, dissolving, grinding, heating and so on. They can create the situation in which chemicals combine, but which chemicals combine with which and to what degree is not something the experimenter can determine. Chemical substances, whether produced in nature or in the laboratory, 'act upon one another according to their respective natures'.

The above points do not rest upon knowledge of some ultimate matter theory that explains the origins of the 'natures' of chemical substances. Boyle, of course, considered those natures to arise from the shapes, sizes and motions of corpuscles according to the mechanical philosophy in the strict sense. A problem here lies in dis-analogies between chemical substances and mechanical artefacts such as clocks and watches (disregarding the point that the latter are not mechanisms in the strict sense). Parts of a watch or clock do not combine spontaneously to form the whole in the way that chemicals combine to form compounds. Further, watches do not combine with other watches in ways characteristic of watches, whilst chemical compounds do combine with each other in ways characteristic of their kind. Admitting that chemistry is mechanical in the common senses I have identified gives a defender of the mechanical philosophy in the strict sense much less than is required.

These considerations make it possible to appreciate that Newman (2006, pp. 198–215) has given too much credence to some key arguments of Boyle to the effect that chemical qualities, along with all other qualities, are basically mechanical.[25] As Newman documents, Boyle gave many examples of how properties of substances can be changed 'mechanically'. He identifies experiments that Boyle described as the 'fittest' to support his position. It involves (mechanically) adding one colourless liquid to another colourless liquid with the result that a third liquid is produced that is permanently deeply coloured. The mechanical addition of a third colourless body results in this deep colour disappearing leaving a colourless liquid once again.[26] Experiments such as this do help to establish that colour is an emergent property of a body depending on underlying features of that body which can change when those underlying features are changed. It also undermines the idea that colours stem from immutable substantial forms or Principles. But does it establish that colours are 'mechanical'? An analogy will help illustrate the point that the argument does not take us as far as Boyle or Newman imply. An acorn can be induced to grow into an oak tree by 'mechanically' placing it in the ground and 'mechanically' adding water and nutrients. This is hardly sufficient to establish that the process is 'mechanical' any more than the fact that I can be rendered unconscious by a mechanical blow to the head establishes a materialist theory of the mind. As Boyle put it when discussing changes brought about by the interventions of tradesmen, a chemist who adds one colourless liquid to another 'leaves them to act on one another according to their respective natures'. In the experiment highlighted by Boyle the natures of the combining colourless liquids are such that they combine to yield a deeply coloured resultant liquid. The nature of that coloured liquid is such that it can be transformed into a colourless liquid by the addition of a small particle of an appropriate solid. The additional assumption, that the natures of the substances involved and the processes involved in their transformation are 'mechanical' in a sense sufficiently strong to support Boyle's mechanical philosophy is gratuitous. Of course, if a reliance on contrived experiments is sufficient to qualify a practice as 'mechanical', then Boyle's arguments do support the mechanical philosophy. But this move makes a host of Aristotelians and Paracelsians into mechanical philosophers and renders the appellation relatively innocuous.

Boyle did attempt to take his case for the mechanical philosophy further, as we have seen. He attempted to contrive possible mechanisms, in the strict sense, that can serve as possible explanations of qualities and their transformation. Some of his attempts were more contrived and less plausible than others. His suggestion that the temperature rise of a piece of lead is due to an increase in the rapidity of the motions of the corpuscles that comprise it enabled Boyle to accommodate the fact that lead is lead as much when it is hot as when it is cold. But lead changes colour when its temperature is sufficiently high, as Boyle noted. Here Boyle was less able to contrive a possible mechanism. He believed colour of a body to result from the interaction of the corpuscular structure of that body with light corpuscles but was unable to supply details let alone substantiate them. There was a further complication here. Whilst lead remains lead through the colour changes brought about by heating, the colour changes involved in his mixing of liquids involved chemical changes. Maybe I am employing a distinction here not available to Boyle. His characterisations of substances by reference to a 'concourse of accidents' did not lend itself to a precise identification of chemical substance as we have seen. But this simply reinforces one of my central points with respect to Boyle's chemistry. So intent was he on invoking chemistry in support of a strict sense of the mechanical philosophy that he speculated on possible mechanical causes of the full range of changes accompanying chemical change without having framed appropriate chemical notions able to guide and be substantiated by experiment. This is the point I have been emphasising by contrasting Boyle's chemistry with the chemistry of chemical combination encapsulated in Geoffroy's table of *rapports* highlighted by Klein.

Like many of his contemporaries, Boyle was an experimental philosopher. He attempted to understand nature by intervening in it. He recognised that concepts, that he referred to as notions, needed to be constructed that would make possible the formulation of hypotheses capable of guiding research and liable to be converted into 'matters of fact' once the degree of independent experimental support warranted it. This aspect of his work is best exemplified in his pneumatics. Boyle was also a mechanical philosopher in a strict sense. He aspired to make a case for the mechanical philosophy by contriving mechanisms capable of explaining the phenomena. Chemistry was the main area in which he sought to make his case. He declined to do so in pneumatics, freely admitting that he was unable to contrive mechanisms for the weight and spring of the air. The moral I draw from this is that experimental science called for appropriate concepts and hypotheses capable of guiding, and liable to confirmation by, experiment rather than the mechanical philosophy. Boyle made major contributions to pneumatics by way of concepts that were not mechanical in the sense of the strict mechanical philosophy. He did not make comparable contributions in chemistry because he was over-concerned to contrive mechanical explanations and less concerned with framing notions capable of productively informing chemical research.

8.9 Newtonian Affinities

I have argued in the foregoing sections that Boyle's mechanical philosophy did not help him make experimental progress in chemistry and that the mechanisms he contrived to explain known chemical phenomena were *post hoc* and highly artificial.

Newton's atomic chemistry, made possible by his transformation of the mechanical philosophy by introducing forces, is subject to a similar critique.

As we have seen, Newton, like Boyle, assumed a hierarchy of particles, with mechanical atoms as the foundation, combining in various ways to yield particles of higher degrees of composition. A difference was that Newton assumed attractive forces to be responsible for holding composite particles together. Such a stance did not point in the direction of a notion of chemical kinds of the sort implied in Geoffroy's table. Given chemical substances and their mode of interacting with other substances, Newton could attribute this behaviour to attractive forces acting between the composite particles he presumed to be the least parts of those substances. But this presumes knowledge of the substances and their properties. Newton's atomic chemistry did not have the resources to non-arbitrarily postulate kinds of corpuscles, nor any access to the force laws that might govern the attractions acting between them.

As we saw in the previous chapter, the most detailed treatment of chemistry that Newton published was Querie 31 of the *Opticks*. Several features of his treatment of chemistry in that text supports my contention that his atomic chemistry involved adapting his atomism to chemical, and other knowledge, acquired by other means. His matter theory was not capable of guiding experimental chemistry any more than Boyle's was. In Querie 31 Newton (1979, pp. 380–381) does mention series of precipitations of the kind central to Geoffroy's classification and he does refer to the important regularity that salts result from the combination of an acid and a 'dry earth'. But to interpret these passages as important advances in chemistry emerging from Newton's atomism is to ignore several features of the contents of Querie 31. Firstly, the experimental facts that Newton accommodates to, and takes as evidence for, his atomism are not novel fruits of Newton's theory but experimental knowledge common to all chemists of the time. Secondly, Newton's remarks about series of precipitations and salt formation are not singled out as important features of experimental chemistry. Rather they are interspersed with many other accommodations of Newton's atomism to experimental facts, many of them not 'chemical' at all. I proceed to illustrate and support these claims.

An interpretation of Newton's paragraphs on precipitation as the source of the affinity tables that helped to shape eighteenth-century chemistry is historically false and attributes to Newton's atomism a facility to guide chemical experimentation that it did not possess.[27] As Klein (1994, 1995, 1996) has insisted, by the early eighteenth century preferential precipitations had a history of two hundred years or more and had become a theoretical and experimental focus of chemists working in association with the Botanic Gardens in Paris in the last few decades of the seventeenth century. One of them, Christopher Glaser, described a series of precipitations in much the same way that Newton came to do, in his text, *The Compleat Chemist*, published in English in 1677,

> The Silver dissolv'd in the Aqua-fortis, and poured into the Vessel of water, precipitates and separates itself from its Dissolvent, by putting a plate of copper into it . . . The Silver is found in the bottom. It must be wash'd, dry'd, and kept (if you please) in form of Calx, or else reduc'd into an Ingot in a Crucible, with a little Salt of Tartar. But if into this second

> water, which is properly a Solution of Copper, you put a body more earthy and porous than Copper, as Iron is, the Copper precipitates, and the Corrosive Spirits of the Aqua-fortis fasten to the substance of Iron; which may likewise be precipitated by some Mineral more earthy and porous than Iron, as Lapis Calmonaris and Zink.[28]

Far from being a product of atomism these series of precipitations posed a problem for it, as is clear from the text of 1975 written by Nicolas Lemery, Glaser's successor as Professor at the Botanic Gardens in Paris. If the absorption of silver, copper and iron by nitric acid is explained in terms of some congruity between the shapes of acid particles and pores in the metal particles then how can one explain why the silver particles, once absorbed by the acid, are displaced by added copper particles which are in turn replaced by iron particles? To be sure, Newton was able to counter the problem by appealing to attractions of varying degrees between particles. He was able to accommodate phenomena of precipitation to his atomism in a superior way to previous atomists, but it was a mere accommodation nevertheless, and an accommodation to phenomena that had been known for decades.

Any temptation to read Newton's remark that salts are the product of the combination of an acid with a dry earth as a reference to an experimental law of the kind identified by Geoffroy should be dissolved once Newton's remark is placed in its context. I reproduce the whole paragraph, which was quoted in part in 7.3, to show the extent to which Newton's remarks about salt formation occur alongside a range of other phenomena, including tastes of acids, which Newton (1979, pp. 385–386) accommodates to his atomism.

> When Mercury sublimate is re-sublimed with fresh Mercury, and becomes *Mercurius Dulcis*, which is a white tasteless Earth scarce dissolvable in Water, and *Mercurius Dulcis* re-sublimed with Spirit of Salt returns into Mercury sublimate; and when Metals corroded with a little acid turn into rust, which is an earth tasteless and indissolvable in Water, and this Earth imbibed with more acid becomes a metallick Salt; and when some Stones, as Spar of Lead, dissolved in proper *Mentruums* become Salts; do not these things shew that salts are dry earth and watery Acid united by Attraction, and that the earth will not become a salt without so much acid as makes it dissolvable in Water? Do not the sharp and pungent Tastes of Acids arise from the strong Attraction whereby the acid Particles rush upon and agitate the Particlees of the Tongue? And when Metals are dissolved in acid *Mentruums*, and the Acids in conjunction with the Metals act after a different manner, so that the Compound has a different Taste much milder than before, and sometimes a sweet one; is it not because the Acids adhere to the metallick Particles, and thereby lose much of their Activity? And if the Acid be in too small a Proportion to make the Compound dissolvable in Water, will it not by adhering strongly to the Metal become unactive and lose its Taste, and the Compound be a tasteless earth? For such things are not dissolvable by the Moisture of the Tongue, act not upon the Taste.

Elsewhere in Querie 31 Newton gives an atomic interpretation of a range of other phenomena, including the absorption of water by deliquescent substances, the solution of salts in water and the generation of heat accompanying the mixing of acids and alkalis. Because Newton's discussion involves an accommodation of chemical, along with other, phenomena to his atomism involving attractions, it does not involve a conceptualisation of chemistry that can feed productively into its experimental practice.

8.10 Chemistry from Newton to Lavoisier

I have followed Klein and highlighted the significance of the general notion of chemical combination implicit in Geoffroy's table, that was abstracted from metallurgy and pharmacy and from the studies of salt formation carried out by his predecessors at the Botanical Gardens in Paris, Lemery and Homberg. The centrality of the idea that chemical compounds can be synthesised from as well as analysed into their components was made explicit by Geoffroy, as Klein has noted.[29] In none of his published works on chemistry did Geoffroy invoke, nor did he need to invoke, atoms or corpuscles.[30] The path to the empirical regularities implicit in Geoffroy's table, which he referred to as 'laws', owed no debt to atomic or corpuscular theories and their formulation required no reference to them.

Having said this, it must be acknowledged that Geoffroy's achievement did not usher in a clear separation of an experimental chemistry and philosophical matter theories. Those in the business of articulating matter theories were very quick to take advantage of the new chemistry and accommodate their matter theory to it. Newton's atomism could readily be adapted to it by interpreting Geoffroy's *rapports* as representing attractions between atoms or corpuscles. So natural was this step that, as we have seen, Geoffroy's table was read by contemporaries and by some subsequent historians as a rendering of the affinities invoked in Querie 31 of Newton's *Opticks*. However, whilst atomism could be readily accommodated to 'laws' of the kind referred to by Geoffroy, and this fact may have contributed to their ready acceptance, atomic theories were not capable of predicting them or guiding experimental chemists towards them. Atomic theories were totally open and unspecific on the question of where the corpuscles presumed to take part in chemical combinations were to be placed in the hierarchy of structured particles, mechanical theories of the kind championed by Boyle were completely open and unspecific on the question of the shapes, sizes and motions of atoms or corpuscles whilst Newtonian atomism was correspondingly open and unspecific on the question of the specification of inter-atomic or inter-corpuscular forces. Articulations of atomic matter theories and their accommodation to the phenomena owed a debt to advances in chemistry of the kind embodied in Geoffroy's table but the reverse is not true.

As a fundamental matter theory, Newton's atomism was an improvement on its competitors provided one could learn to live with the unexplained (and unspecified) forces that it involved. It was an improvement because of the extent to which it could be accommodated to the phenomena and because the mechanics of the *Principia* served as a model of how complex systems could be explained by appeal to the forces governing their components. Because of this, Newton's matter theory was, in a sense, highly influential in many quarters. Thackray (1970) has traced the development of Newtonian matter theory in the eighteenth-century. One striking feature of it is its lack of productiveness as far as experimental chemistry is concerned, a point already implicit in my discussion of Boscovich's elaboration of Newtonian atomism described in the previous chapter. Robert Siegfried has recently published a book, *From elements to atoms* (2002), in which he traces the history of chemical composition from the seventeenth century to Dalton's atomism early in the nineteenth.

The fact that he invokes pre-Daltonian atoms only in the context of the failure of the mechanical philosophers to develop an atomic chemistry, and that there are only three minor references to Newton's matter theory in his study, lends support to my view that eighteenth-century developments in chemistry owed little to atomic matter theory.

Eighteenth-century developments in the chemistry of combination can reasonably be seen as extensions and elaborations of the notions of chemical compound and combination implied in Geoffroy's table of 1718. Subsequent affinity tables, as they became known, expanded the range of substances that were included. By the second half of the century the gases became recognised as chemical substances, beginning with 'fixed air' (carbon dioxide) discovered by Joseph Black in the 1850s.[31] This move enabled consistent sense to be made of the weight relations involved in chemical combination and helped to distinguish between reactions that involved building up and those that involved breaking down of component substances. The culmination of these developments was Lavoisier's chemical 'revolution'. It involved the recognition that certain chemical substances, namely the elements (that Lavoisier referred to as 'simple substances'), are components of chemical compounds but do not themselves have chemical components. A basic principle of the new chemistry was that the weight of elements is conserved in chemical reactions.[32] The new chemistry made it possible to recognise that the combustion of metals involves the combination of metals with oxygen rather than the expulsion of a substance (phlogiston) from them. It is significant for my story that Lavoisier explicitly separated his chemistry from speculations about atoms. He understood elements as those substances which cannot be broken down further by chemical means. As for the atoms that might be supposed to compose the elements Lavoisier (1965, p. xxiv) judged that 'it is extremely possible but we know nothing at all about them'.

Besides the chemistry of combination that had resulted in Lavoisier's chemistry of elements and compounds by the end of the century there was the chemistry of the materials comprising plants and animals. As Klein and Lefèvre (2007, Part III) describe in detail, this branch of chemistry could not be accommodated by the chemistry of combination highlighted by Geoffroy. What came to be described as 'organic' substances could be analysed into components but they could not subsequently be synthesised from those components as 'inorganic' substances could. The way in which organic chemistry was eventually subsumed into the chemistry of combination is a highly significant moment in our investigation of the history of atomism as we will see in the next chapter.

Notes

1. From the opening of 'Some specimens of an attempt to make Chymical Experiments Useful to illustrate the notions of the Corpuscular Philosophy' (Boyle, 2000, Vol. 2, pp. 85–92).
2. See, for example, Partington (1961, Vol. 2, p. 496).
3. See the 'Concise conclusion' of Newman (2006) for his characterisation of Boyle's chemistry as revolutionary.
4. Key references are Klein (1994, 1995 and 1996) and, more recently, Klein and Lefèvre (2007).

5. See, for instance, Thackray (1970, pp. 85–88).
6. See, for instance, A. M. Duncan (1964, pp. 177–194).
7. See F. L. Holmes (1989) for a detailed account of the emergence of the chemistry of salt formation.
8. As is noted by Klein and Lefèvre (2007, p. 111) this notion of purity of a substance falls short of the notion that involves composition from elements in the right proportion, which comes later.
9. Notable examples are Jean Beguin (1624), Nicaise Le Febvre (1664), Christopher Glaser (1677), and Nicolas Lemery (1677).
10. For details and documentation of these modes of construing the action of acids by metallurgists and pharmacists see Klein and Lefèvre (2007, pp. 136–147).
11. In their account of the Paracelsian philosophy employed by the French chemists, Klein and Lefèvre (2007, pp. 39–44) stress the non-material nature of the principles informing chemical substances, rendering them akin to Aristotelian forms in this respect. They insist that those principles did not exist in compounds as material parts of them. As a consequence, the 'essences' generated by the distillation of plant and animal materials could be seen as resulting from transmutations rather than extractions of already existing components. Stressing this aspect of the Paracelsian view helps Klein and Lefèvre portray the view of combination involved in Geoffroy's table as a novelty. Newman (2008, pp. 176–182) objects, finding plenty of examples in Paracelsian texts of talk of analysis and synthesis that seems to imply the continued existence of constituents, and even of 'principles', in the substances they are components of. Newman is correct to point out that the Paracelsian texts are more complex than the generalisations of Klein and Lefèvre allow for. Further, it is not surprising that some talk of analysis and synthesis of chemical substances can be found in Paracelsian texts since reversible reactions (in the sense of Klein and Lefèvre) were as much a reality for the authors of those texts as for any other chemist of the time. I am not surprised that chemical philosophers of a Paraselsian or Aristotelian persuasion were able to adapt their systems to fit reversible as well as irreversible reactions. Those philosophies were sufficiently imprecise and flexible to make such accommodations possible. (I cannot resist reminding the reader that Boyle's mechanical atomism was also highly flexible and adaptable, a point I will be putting to use later in this chapter.)
12. As cited by Klein (1996, p. 276). I have omitted the italics added by Klein.
13. See, for instance, Klein and Lefèvre (2007, p, 111).
14. For a discussion of eighteenth-century affinity tables see Duncan (1964 and 2001) and Klein and Lefèvre (2007, chapter 9)
15. Boyle (1990, Vol. viii, f184, reel 5, frame 189).
16. Boyle's views are spelt out in 'Origin of forms' (Boyle, 2000, Vol. 5, pp. 322–330).
17. For other references to this distinction in Boyle's 'Origin of forms' see Boyle (2000, Vol. 5, pp. 322, 324–325, 330 and 351–352).
18. 'I observe also, that a Dissolution may be made of the same by *Menstruums*, to which the Chymists attribute (as just now I observed they did to some bodies) a mutual Antipathy, and which therefore are not like to have a Sympathy with the same third body, as I found by trial, that both *Aqua Fortis*, and Spirit of Urine, upon whose mixture there ensues a conflict with a great efforvescence, will each of them apart readily dissolve crude Zince, and so each of them will, with Filings of Copper.", from 'Experiments and Notes about the Mechanical Origine or Production of Corrosivenesss and Corrosibility' (Boyle, 2000, Vol. 8, p. 467).
19. The various essays in Boyle's 'Experiments, Notes &c. About the Mechanical Origine or Production of divers particular Qualities' (2000, Vol, 8, pp. 315–523) well illustrate the feature of Boyle's mechanical theory that I am referring to. As an example of the contrived nature of some of Boyle's mechanical explanations consider the following passage in which Boyle (2000, Vol. 8, p. 470) offers possible explanations of how Mercury sublimate (mercuric chloride) can lose its corrosiveness when converted to *Mercurius dulcis* (mercurous chloride) by grinding it with mercury. ' It may perhaps somewhat help us to conceive, how this change

may be made, if we imagine, *that* a company of mere Knife-blades be first fitted with Hafts, which will in some regard inhibit their wounding power by covering or casing them at that end which is design'd for the handle; (though their insertion into those Hafts, turning them into knives, makes them otherwise the fitter to cut and pierce and *that* each of them be *afterwards* sheathed, (which is, as it were, a hafting of the Blades too;) for then they become unfit to stab and cut, as before, though the blades be not destroyed: Or else, we may conceive these Blades without Hafts or Sheathes to be tied up in bundles, or as it were in little faggots with pieces of wood, somewhat longer than themselves, opportunely placed between them. For neither in this new Constitution would they be fit to cut and stab as before. And by conceiving the edges of more or fewer of the Blades to be turn'd inwards, and those that are not, to have more or less of their points and edges to be sheath'd, or otherwise cover'd by interpos'd bodies, one may be help'd to imagine, how the genuine effects of the Blades may be variously lesson'd or diversifi'd. But, whether these or any like changes of Disposition be fancy'd, it may by Mechanical Illustrations become intelligible, how the Corrosive Salts of common Sublimate may lose their efficacy, when they are united with a sufficient quantity of quicksilver in *Mercurius dulcis*.'

20. See Newman (1996) for a detailed account of Boyle's borrowings from Sennert..
21. Boyle's position provides an answer to a longstanding question, that of the way in which components of a compound exist in the compound. They exist in the compound in some strong sense since they can be recovered from it. But they cannot exist in too literal a sense because the compound does not have properties that are the sum or average of those of its constituents. The answer implicit in Boyle's theory is that the corpuscles of the ingredient substances are present as parts of the corpuscles of the compound, but the shapes, sizes and motions of the corpuscles of the compound, on which its properties depend, differ from the shapes, sizes and motions of the corpuscles of the ingredients. Boyle has an answer to the puzzle. Whether it was a satisfactory or the right answer is another question. (The contemporary answer to the question of the sense in which elements exist in their compounds is quantum mechanical.)
22. Boyle's account is in his 'Experiments touching colours' (Boyle, 2000, Vol. 4, pp. 150–153). Clericuzio's discussion is in Clericuzio (1990, pp. 578ff.) and Newman's account, and his critique of Clericuzio, is in Newman (2006, pp. 181ff.).
23. My interpretation is not completely borne out by Boyle's words. When Boyle spells out the 'chymical reason' for the happenings in his experiments with mercury sublimate most of his discussion is couched in terms of experimental manipulations perfectly in line with my interpretation, but corpuscular talk slips into the discussion. Boyle talks, for instance, of the 'Coalition of the Mercurial particles with the Saline ones'. So convinced is Boyle of the corpuscular nature of chemical substances that he can substitute talk of the combination of mercury sublimate with salt of tartar in terms of combining particles without realising the shift in levels.
24. Boyle (2000, Vol. 5, p. 324) talks of the mechanical structure responsible for the essential qualities of a body in 'Origin of forms' referring to that mechanical structure as the 'stamp' of the body.
25. Newman (2004, pp. 271–283) makes many of the points I have made about Boyle on the relation between the natural and the artificial. He explicitly makes the point that Boyle's arguments 'would have been effective only against the most rigid proponents of an absolute distinction between art and nature' clearly implying that a range of Aristotlean alchemists from Geber to Sennert would have had no problem about endorsing Boyle's views. This does not sit well with Newman's insistence in *Atoms and Alchemy*, that Boyle's chemistry supported the mechanical philosophy in a way that distinguished his position from that of his Aristotelian and chymical opponents.
26. Newman (2006, pp. 182–185) utilises Boyle's discussion of colour changes in Boyle (2000, Vol. 4, pp. 150ff.)
27. For an example of the identification of Newton as the source of affinity tables see Maurice Crossland, (1963, pp. 369–441), especially p. 382.
28. As cited by Klein (1995, p. 89).

29. In a 1704 article Geoffroy wrote: 'What completely assures us that we have succeeded in investigating the composition of bodies is, having reduced mixta into their simplest substances that chemistry can provide, we can recompose them by reuniting these same substances', as cited by Klein (1996, p. 272).
30. Geoffroy did invoke a corpuscular theory in a posthumously published pharmaceutical work as Klein (1995, p. 93) has noted.
31. The work of Black serves to illustrate the point that, on the one hand, many eighteenth-century chemists did not separate their chemistry from matter theory and that, on the other hand, their experimental progress owed little debt to matter theory. In his lectures at the University of Edinburgh Black articulated a Newtonian matter theory and he also construed affinity tables in terms of Newtonian attractions. See Thackray (1970, pp. 223ff.). But the experiments involved in his preparation and identification of fixed air (carbon dioxide) involved noting the weight loss accompanying the conversion of limestone into quicklime by strong heating and the recovery of limestone by heating an aqueous solution of quicklime with potassium carbonate. The experimental argument was quite independent of his atomism. The same can be said for Black's experimental investigations of heat that led to his identification of latent heat.
32. Attention to weight relations in chemistry had a long history that goes back at least as far as Geber. Lavoisier himself drew on the seventeenth-century researches of Von Helmont. On the latter point see Newman and Principe (2002, pp. 296–309).

Chapter 9
Dalton's Atomism and its Creative Modification via Chemical Formulae

Abstract Dalton proposed that each element is made up of identical atoms distinctive of it and that least parts of compounds consist of characteristic combinations of small numbers of them. The laws of proportion predicted by his theory were borne out by experiment. For the first time, tentative experimental contact was made with a property of atoms, their relative weight. However, not much chemistry can be done armed only with such weights. It took a creative shift in Dalton's programme for significant progress to be made. This occurred in organic chemistry with the use of formulae and the representation of properties other than weight by suitable arrangements of symbols in them. The formulae could serve their function without interpreting the symbols in them as representing atoms. Most of the chemists involved did interpret them as representing atoms, but they differed in a crucial respect from the atoms in the philosophical tradition dating back to Democritus. The properties of atoms were to be discovered by chemical research rather than set down at the outset. The property of valency was one such property, the necessity of which became evident from the 1860s. Progress in nineteenth-century chemistry was a precondition for rather than result of the introduction of atomism into chemistry, Dalton notwithstanding.

9.1 Introduction

Our story so far has not yet reached the stage where it can be said that experimental contact with atomism has been accomplished. A plausible candidate for the first version of atomism that made such a thing possible is John Dalton's chemical atomism, formulated in the early nineteenth century. By assuming that chemical combination takes place via combining atoms Dalton drew connections between the combining weights of chemical substances in the laboratory and the weights of combining atoms. For the first time, it would appear, a line was opened to gaining some knowledge of a property of atoms, their relative combining weights.

My own view, to be developed in this chapter, is that there is much that is mistaken and misleading about seeing Dalton's atomic theory as the beginnings of an experimentally testable version of atomism. As far as chemistry is concerned, significant progress in the nineteenth-century was made, but in a way that is better

A. Chalmers, *The Scientist's Atom and the Philosopher's Stone*, Boston Studies in the Philosophy of Science 279, DOI 10.1007/978-90-481-2362-9_9,

construed as paving the way for atomism rather than resulting from it. Dalton's theory had no testable content that went beyond the laws of proportion that it entailed and so could not productively guide chemistry in a way that could not be achieved by way of the laws of proportion alone.

The situation was to change with the deployment of chemical formulae in organic chemistry. That practice was able to go beyond the consequences of the laws of proportion in a dramatically successful way. Chemical properties other than combining weight ratios were depicted by appropriate ordering of the symbols in the formulae for chemical substances. By about 1860 the demands placed on chemical formulae had resulted in a unique set of them. The arrangement of symbols in the formulae were able to capture isomerism, stereochemistry and valency, and made possible the classification and prediction of chemical reactions to the extent that it spawned a massive synthetic chemical industry. It became increasingly plausible to interpret the symbols in formulae as representing atoms, and such an interpretation was clinched by experiment around the close of the century. Dalton's own version of atomism which grew out of, and to some extent remained anchored in, physics was soon forgotten and was unproductive. The reformulation of his theory utilising formulae that made progress possible was at best a chemical atomism differing markedly from Dalton's physical atomism. There is even a case for doubting that the success of formulae in nineteenth-century chemistry represented an experimentally testable version of atomism at all. It is reasonable to argue that progress in nineteenth-century chemistry was a precondition for, rather than the result of, the incorporation of atoms into experimental science. I devote this chapter to an articulation and defence of these claims.

9.2 Dalton's Atomism

Newtonian atomists of the eighteenth century aimed to explain properties of bulk matter, such as coagulation and chemical combination, by appeal to inter-atomic forces. In the case of chemical combination the forces were referred to as affinities. A major problem with this approach was the gulf between the speculations about inter-atomic forces on the one hand and what could be investigated experimentally on the other. Near the end of the eighteenth century, Claude-Louis Berthollet, himself a Newtonian atomist, spelt out the futility of trying to derive inter-atomic affinities from experiments on chemical combination in the laboratory because an atomist must recognize that any affinities measurable at the macroscopic level are a function of the state, temperature and masses of the combining substances and arises from unknown arrangements of large numbers of atoms.[1] These problems could be supposed to be at a minimum in the case of the physical properties of gases, where an atomist could assume the atoms to be sufficiently far apart for forces of coagulation and chemical affinities to be ignored. (As it happened, the first atomic theory to gain significant empirical support, the kinetic theory of gases, made headway by ignoring inter-atomic forces altogether and admitting only the impulsive forces experienced by colliding atoms.)

Dalton's atomism emerged out of what was in key respects a Newtonian atomistic theory of gases. The details of the path that led Dalton to his theory have been much studied and debated.[2] Here I extract some of the uncontroversial features.

An atomistic theory of gases took shape in the context of one of Dalton's early research preoccupations, namely, meteorology. In 1793 we find him insisting that the absorption and precipitation of water vapour by the atmosphere is a physical rather than a chemical process. The fact that the amount of water vapour that can be absorbed by a given volume of air at a fixed temperature is independent of the pressure of the air in that volume told against the prevailing idea that the absorption was due to some chemical affinity between air and water. Dalton's understanding, atomistic from the start, was of atoms of water interspersed amongst other atoms composing air and acting independently of them. The idea that each gas in a volume makes its contribution to the total pressure independent of the other gases in the mixture was soon confirmed experimentally and has survived as 'Dalton's law of partial pressures'. Dalton developed his atomistic understanding by adapting Newton's observation, in the *Principia* (Book II, Proposition 23) that a gas made up of a static array of atoms repelling each other with a force inversely proportional to their separation will obey Boyle's law, notwithstanding the problems with this conjecture that Newton himself had already discerned.[3] Dalton speculated that the atoms of each gas repel atoms of like kind with a force inversely proportional to their separation whilst exerting no force on atoms of other gases. This explained both the law of partial pressures and also why the gases in the atmosphere remain a homogeneous mixture rather than separating out with the more dense gases settling in layers below the less dense.

Dalton soon extended his research to consider the solubility of gases in liquids, and here he was able to join forces with William Henry, who had done experimental work in the area. In particular, the latter had shown that the amount of gas absorbed in an adjacent liquid is proportional to the pressure of the gas at the liquid surface, once again suggesting a physical process rather than one involving affinities. Dalton, faced with the question of why some gases dissolve more readily in a given liquid than others, invoked the weight of the atoms of the respective gases as the likely cause of the difference. In the 1805 paper where these ideas were developed, Dalton (1805) observed:

> An enquiry into the relative weights of the ultimate particles of bodies is a subject, as far as I know, entirely new: I have lately been prosecuting this enquiry with remarkable success. The principle cannot be entered upon in this paper; but I shall just subjoin the results, as far as they appear to be ascertained by my experiments.

Entries in Dalton's notebooks of 1803, now lost but referred to by Henry E. Roscoe and Arthur Harden (1896, pp. 26–29), imply that the atomic weights 'subjoined' to the 1805 paper were arrived at via Dalton's chemical atomism.

Alongside these studies of the behaviour of gases and integrated into them were Dalton's views on the nature of heat. His acceptance of the caloric theory of heat put him in the majority in the first decade of the nineteenth century, but what was peculiar to Dalton's treatment was the extent to which he took a strong view on the materiality of caloric and the way in which he integrated that fluid into his atomic theory. An early version of these views appeared in a five page note 'On heat', an entry in Dalton's notebook dated May 23, 1806 (Roscoe and Harden, 1896, p. 71, italics in original).

> According to this view of the subject, every atom has an atmosphere of heat around it, in the same manner as the earth or any other planet has an atmosphere of air surrounding it, which cannot certainly be said to be held by chemical affinity, but by a species of attraction of a very different kind. Every species of atoms or ultimate particles of bodies will be found to have their peculiar powers of attraction for heat, by which a greater or less quantity of heat will be conglomerated around them in like circumstances: this gives rise to what has been called the *different capacities* of bodies for heat or their *specific heat*.

Whatever its initial attraction, Dalton's efforts to pursue his physical atomism soon ran into serious trouble, and his attempts to square it with threatening experimental results had the effect of it losing whatever coherence it had. For example, Dalton's attempts to give an account of the specific heats of gases, which, as can be inferred from the quotation from Dalton reproduced above, focused on the atmospheres of caloric surrounding atoms of a gas, met with no significant support from experiment. The same can be said of Dalton's attempt to link the solubility of gases in liquids to atomic weights. There was a fundamental tension in Dalton's theory concerning the cause of the expansion of gases. The caloric theory attributed expansion of a substance to an addition of caloric, which, for an atomist, insinuated itself between atoms and pushed them further apart. Yet Dalton explained vapours in terms of 1/r repulsions between like atoms. Dalton (1810, p. 548, italics in original) did respond to this difficulty by attempting to explain the even distribution that a mixture of gases settles into by appeal to caloric. His account involved the assumption that '*every species of pure elastic fluid has its particles globular and all of a size; but that no two species agree in the size of their particles*, the pressure and temperature being the same'. The rough idea seems to be that where unlike atoms meet, there is a discontinuity in the state of caloric, because of the difference in size, and that the forces arising from this discontinuity give rise to the motions that result in the uniform mixing of the gases. It is difficult to disagree with Roscoe and Harden (1896, p. 23) when they remark that this idea 'does not appear to have been very carefully thought out, and although the conditions of equilibrium would certainly be disturbed, it is doubtful whether the intestine motion of which Dalton speaks would have been set up in a vessel filled with atoms'. Dalton expanded on his idea and illustrated it with diagrams in Part 2 of his *New system of chemical philosophy* published in 1810. Whilst these diagrams do exhibit the discontinuities arising from differing particle size, they did not help Dalton to show what he needed to show, namely, that the result of the discontinuities is a homogeneous mixture of gases and a force varying as 1/r between like atoms. It should also be noted that Dalton's assumption that the atoms of unlike substances differ in size had no independent support and Dalton himself violated the condition on a number of occasions.

Dalton's attempts to build on the early successes of his atomic theory of gases were unsuccessful. This is not surprising from a modern point of view given that Dalton worked with a static model of the arrangement of atoms in a gas and utilized a quite specific and detailed version of the caloric theory. According to an authority on the caloric theory, it does not require a modern vantage point to appreciate the shortcomings of Dalton's theory. Robert Fox (1968, p. 197) concludes an analysis and appraisal of Dalton's caloric theory with the observation that 'to his contemporaries, whether in 1800 or 1842, Dalton's work on the theory of heat must have seemed almost as wrong-headed and irrelevant to current problems as it does to us now'.

9.3 Dalton's Atomic Chemistry

Dalton's chemical atomism emerged out of his theory of gases because he saw in it the possibility of opening up an avenue for gaining experimental access to the relative weights of atoms. The bare bones of his theory appeared in the closing pages of his *New system of chemical philosophy* (1808) and can be separated from his physical atomic theory as his contemporary chemists soon learnt to do but which Dalton himself did not do.

Dalton was able to take for granted and exploit Lavoisier's chemistry and also the law of constant proportion of elements in compounds. Dalton took for granted the notion of chemical element, and the fact that the weight of each element is preserved in chemical reactions. He also took for granted the outcome of the debate between Proust and Berthollet concerning the law of constant proportions. Berthollet had likened compounds to saturated solutions, using the comparison to cast doubt on the law of constant proportions. Proust countered by establishing that the proportion of solute in a saturated solution varies with temperature whereas the proportion of elements in a compound does not, and also stressed the independence of those proportions of the method of preparation and of the physical state of the compound.[4] A third crucial component of the background, that Dalton could take for granted in a way that could not have been done half a century earlier, was the notion of gases as distinct chemicals that could be isolated, identified and weighed.

Against this background, Dalton assumed chemical elements to be composed of 'ultimate particles' or atoms that cannot be changed by 'chemical agency'.[5] The least parts of a chemical compound are assumed to be made up of atoms of the combining elements. Dalton referred to these as 'compound atoms'. According to Dalton (1808, p. 113), 'all atoms of the same substance, whether simple or compound, must necessarily be conceived to be alike in shape, weight and every other particular'. Figs. 9.1 and 9.2 show Dalton's representations of atoms and his explanation of them. The position they illustrate straightforwardly entails the law of constant proportions. The latter law already had empirical support as we have noted. But Dalton's atomic theory predicted two other laws, the law of multiple proportions and the law of equivalent proportions. The former law states that, if two

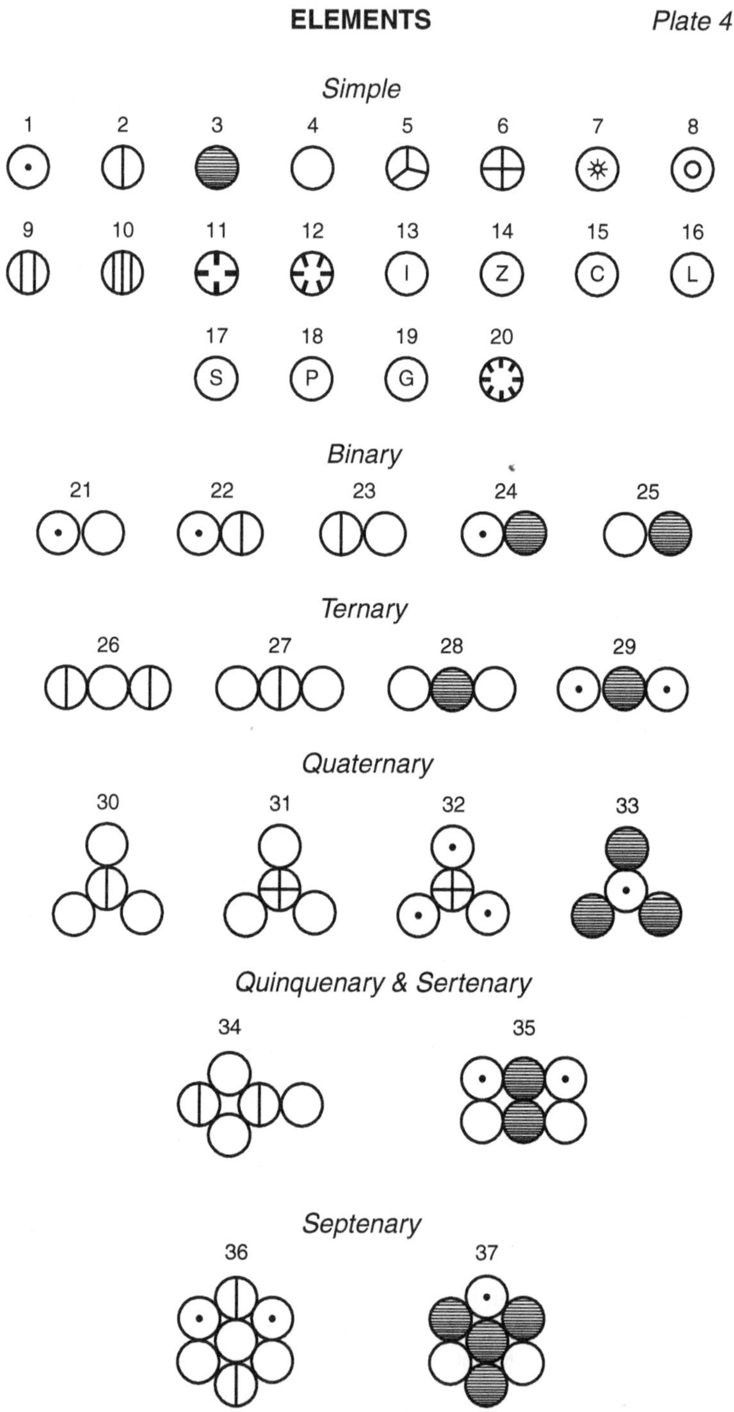

Fig. 9.1 Dalton's representations of atoms

PLATE IV. This plate contains the arbitrary marks or signs chosen to represent the several chemical elements or ultimate particles.

Fig.			Fig.		
1	Hydrog.; its rel. weight	1	11	Strontites	46
2	Azote	5	12	Barytes	68
3	Carbon or charcoal	5	13	Iron	38
4	Oxygen	7	14	Zinc	56
5	Phosphorus	9	15	Copper	56
6	Sulphur	13	16	Lead	95
7	Magnesia	20	17	Silver	100
8	Lime	23	18	Platina	100
9	Soda	28	19	Gold	140
10	Potash	42	20	Mercury	167

21. An atom of water or steam, composed of 1 of oxygen and 1 of hydrogen, retained in physical contact by a strong affinity, and supposed to be surrounded by a common atmosphere of heat; its relative weight =	8
22. An atom of ammonia, composed of 1 of azote and 1 of hydrogen	6
23. An atom of nitrous gas, composed of 1 of azote and 1 of oxygen	12
24. An atom of olefiant gas, composed of 1 of carbone and 1 of hydrogen	6
25. An atom of carbonic oxide composed of 1 of carbone and 1 of oxygen	12
26. An atom of nitrous oxide, 2 azote + 1 oxygen	17
27. An atom of nitric acid, 1 azote + 2 oxygen	19
28. An atom of carbonic acid, 1 carbone + 2 oxygen	19
29. An atom of carburetted hydrogen, 1 carbone + 2 hydrogen	7
30. An atom of oxynitric acid, 1 azote + 3 oxygen	26
31. An atom of sulphuric acid, 1 sulphur + 3 oxygen	34
32. An atom of sulphuretted hydrogen, 1 sulphur + 3 hydrogen	16
33. An atom of alcohol, 3 carbone + 1 hydrogen	16
34. An atom of nitrous acid, 1 nitric acid + 1 nitrous gas	31
35. An atom of acetous acid, 2 carbone + 2 water	26
36. An atom of nitrate of ammonia, 1 nitric acid + 1 ammonia + 1 water	33
37. An atom of sugar, 1 alcohol + 1 carbonic acid	35

Enough has been given to show the method; it will be quite unnecessary to devise characters and combinations of them to exhibit to view in this way all the subjects that come under investigation; nor is it necessary to insist upon the accuracy of all these compounds, both in number and weight; the principle will be entered into more particularly hereafter, as far as respects the individual results.

Fig. 9.2 Dalton's explanation of his representations of atoms

chemical substances combine in more than one way to form compounds, then the ratios of the weights of one of them that combine with a fixed weight of the other are small integral numbers. (For example, the weights of nitrogen relative to a fixed weight of oxygen in the three oxides of nitrogen, nitrous oxide, nitric oxide and nitrogen peroxide, are in the ratios 4:2:1.) The law of equivalent proportions states that if weight x of substance A combines with weight y of substance B to form a compound and with weight z of substance C to form another compound, then, if B and C combine to form a compound they will do so in weights that are in the ration ny:mz where n and m are small integers. These laws were soon confirmed by a range of experimental evidence, some of it supplied by Dalton himself.[6]

Given the situation as I have described it, then, Dalton's chemical atomism was confirmed by experiments on combining proportions insofar as they supported the three laws of proportion. Three independent laws that were a natural consequence of the theory were borne out by experiment. Here I differ from Paul Needham (2004) who denies that Dalton's theory explains why chemical substances combine in constant proportions. The relationship between our respective views has some subtleties an appreciation of which illuminates the subsequent discussion of this chapter. I see no reason to doubt that Dalton's theory explains why chemical substances combine in *constant proportions*. The three laws of proportion are, after all, a straightforward and natural consequence of his theory. On the other hand, Dalton's theory does not explain why chemical substances *combine* in constant proportions. It does not explain why substances combine at all. It takes it as given. The views of Needham and myself come much closer together when it comes to the question of the prospects of developing Dalton's atomism in a direction which would take it beyond merely explaining constant proportions. Both Needham and I agree that there were no such prospects. Dalton's atomism did make contact with experiment in the arena of combining proportions. By contrast, as we have seen, the search to explain chemical combination by appeal to inter-atomic force laws had failed to engage with experiment. Was there a way of going beyond combining proportions without losing contact with experiment? As we shall see later in this chapter, there was such a way. It involved a creative transformation of Dalton's theory through the deployment of chemical formulae. It was a move that Dalton himself vehemently resisted.

As Dalton (1808, p. 163, italics in original) stressed, the 'one great object' of his new chemical philosophy was 'to show the importance and advantage of ascertaining *the relative weights of the ultimate particles, both of simple and compound bodies, the number of simple elementary particles which constitute one compound particle, and the number of less compound particles which enter into the formation of one more compound particle*'. What exactly was the 'importance and advantage' of determining such numbers? Given the form of Dalton's own version of his theory, the importance for chemistry lay in the consequences for combining proportions that followed from the numbers. However, the knowledge of combining proportions could be handled, via the laws of combining proportions and the experimental measurement of those proportions, without invoking atoms at all. Those of Dalton's contemporaries, like Humphry Davy, who identified the laws of proportion as the core of Dalton's theory and separated that from the speculative atomism had a point so long as the theory offered no prospect of moving beyond combining weight ratios.[7]

Even within the domain of combining weights, Daltonian atomism confronted a basic problem. The atomic constitution of compounds is underdetermined by measurements of combining weights of chemical substances in the laboratory. For instance, Dalton's measurement of 7 as the weight of oxygen relative to hydrogen in water is compatible with water consisting of one atom each of hydrogen and oxygen, with an atomic weight of 7 for oxygen relative to hydrogen. But it is also compatible with water consisting of two atoms of hydrogen with one of oxygen, and an atomic weight of 14 for oxygen, or with water consisting of two atoms of oxygen and one of hydrogen and an atomic weight of $3\frac{1}{2}$ for oxygen and so on.[8] Dalton had a solution to the problem that referred back to the physics underlying his theory. Since like atoms are presumed to repel each other, the most stable arrangements are those that minimise their proximity.[9] Dalton settled for one atom each of hydrogen and oxygen in water for this reason. In cases where there are multiple compounds of the same elements, Dalton suggested arrangements of atoms that maximised the separation of like atoms. So, for instance, the diagrams for nitrous oxide in the *New System* shows one oxygen atom separating the two nitrogens on either side, whilst a compound atom of alcohol (which at the time Dalton believed to be what we would write as CH_3) is depicted as a carbon atom with three hydrogen atoms arranged at 120 degree intervals around it.[10]

One problem with the simplicity rule used by Dalton to help get around the under-determination of atomic structure and atomic weights by measurements of combining weights was the question of its truth. It was not destined to be of great help to organic chemistry and, after all, it did lead Dalton to settle on one atom each of hydrogen and oxygen for water. A second problem was its insufficiency. In those cases where the same elements combined to yield more than one compound, so that no more than one of the compounds could involve the simplest arrangement of one atom of each element, the rule could not discriminate between equally simple alternatives. It could not decide, for example, which of the two common oxides of carbon is binary and which tertiary. C_2O and CO is just as simple and just as compatible with the experimental weight measurements as CO and CO_2. There were other strategies for solving the under-determination problem in Dalton's lifetime. One of them was to assume the hypothesis put forward by Avogadro in 1811 that equal volumes of gases contain equal numbers of molecules and to compare weights of molecules by comparing vapour densities. Another was to assume the law of Dulong and Petit, that the product of atomic weight and specific heat is a constant. A third assumed a correlation between atomic arrangements and crystal structure according to the 'law of isomorphism' suggested by E. Mitscherlich in 1821. The details of these attempts to determine atomic weights are sufficiently instructive to warrant a special chapter. For the moment it is sufficient for our purposes to appreciate that all of the methods involved difficulties and limitations and none of them produced definitive results. In any case, Dalton did not take advantage of them. As we shall see in 8.6, the problem of under-determination was solved, by about 1860, mainly through advances in organic chemistry made possible by use of formulae, the move that Dalton resisted.

The development of Dalton's atomic chemistry in his own hands adds weight to the charge that the content of his theory did not go beyond what can be captured by laws of proportion alone. After the publication of Volume 1, Part 1, of *A new system of chemical philosophy* in 1808, Dalton published Part 2, Volume 1 in 1810 and Volume 2 in 1827. Part 1 of a second edition of Volume 1 was published in 1842, two years before his death. This was a word for word reproduction of the 1808 edition. The chapters in Volume 1, Part 2 and in Volume 2 follow a common pattern. Beginning with simple compounds and proceeding to more complex ones, Dalton deals with their key chemical properties and their mode of preparation and then proceeds to detail the results of analysis that give the proportions of the elements in each compound. At the end of such considerations Dalton suggests an atomic constitution for the compound in question. That constitution is a result, coming at the end of the investigation. In no case does an atomic constitution guide or inform the research. This inability of Dalton's atomism to inform research in chemistry is a feature of it that will stand out more starkly when we are able, in a the following sections, to compare it with more fruitful approaches involving the deployment of chemical formulae.

9.4 The Introduction of Chemical Formulae by Berzelius

It was the Swedish chemist Jacob Berzelius who first introduced into chemistry formulae of the kind now commonplace for representing the composition of compounds. By the time he did so, in 1813, he was able to take advantage of the addition of a further experimental law that had been added to the three laws governing combining weights in 1809. That was Gay Lussac's law specifying that when gases combine at some definite temperature and pressure they do so in volumes that bear a simple ratio to each other and to the volume of the product if gaseous. Berzelius (1813, 1815) argued that using formulae was preferable to using Daltonian diagrams because the latter, in conjunction with a table of 'atomic weights', could capture all that was warranted by experiments on combining weights and volumes without commitment to the atomic hypothesis. This point is central to what follows and needs to be clarified in a way that will explain the use of italics around 'atomic weights' in the previous sentence.

An atomist will typically take the hydrogen atom as a standard so that the atomic weight of any other element will be the weight of an atom of it compared to the weight of an atom of hydrogen. From the atomic point of view, a formula of H_2O for water indicates that a compound atom of water consists of two atoms of hydrogen combined with one of oxygen. The measured equivalent weight of 8 for the amount of oxygen in water relative to hydrogen yields a relative atomic weight of 16 for oxygen. But there is no compulsion to take the weight of a hydrogen *atom* as the standard. More in keeping with what is actually done in the laboratory, any portion of hydrogen whatsoever can be taken as the standard and the 'atomic weight' of a second element can be defined relative to it. The formula H_2O will then indicate two portions of hydrogen of a given weight for a portion of oxygen that is 16 times that weight. Of course, if HO is taken as the formula for water then the atomic weight of oxygen will be 8 rather than 16. Some decision needs to be made to remove the under-determination of formulae and 'atomic' weights' by weight and volume measurements in the laboratory. But that is the case whether one is an atomist or not.

Berzelius (1815, 125–126) explained that he differed from Dalton insofar as he considered 'the atomic theory as imperfect, and as clogged with difficulties'. He promoted his introduction of formulae as an alternative to Dalton's diagrams precisely because they could be interpreted as representing combining weights and volumes without a commitment to atoms. Berzelius (1813, 359) described the view embodied in his formulae as 'founded on something very analogous to the corpuscular hypothesis of Dalton' but considered himself to have the advantage over the later 'of not founding my numbers on an hypothesis, but upon a fact well known and proved'. Two years later Berzelius (1815, 127) re-iterated this point.

> I placed beside the corpuscular theory, a theory of volumes; because that theory is in some measure connected with facts that may be verified. To those who think that the theory of volumes may be fatal to the corpuscular theory, I would observe, that both are absolutely the same thing; but that the theory of volumes has this immediate advantage over the other that it may be more easily verified. ... The only difference between the two theories consists in the words *atom* and *volume*, that is to say, in the state of aggregation of the elements.

Berzelius's claims are problematic as they stand for it cannot at the same time be the case that his theory amounts to the same thing as Dalton's whilst being less hypothetical. What Berzelius clearly intends is that his theory is equivalent to Dalton's as far as the experimental evidence available at the time is concerned. That circumstance draws into question the extent to which chemical atomism can be said to be supported by that evidence.

Berzelian formulae, in conjunction with a table of relative 'atomic weights', can be used to represent chemical constitution without a commitment to atomism. Berzelius himself did not use this as a reason for denying atomism. Rather, he attempted to develop Dalton's atomism further so that it would go beyond the prediction of combining weights and volumes to explain a mechanism for chemical combination. Inspired by the phenomenon of electrolysis he presumed that atoms were held together in compounds by electrostatic forces. It is doubtful whether Berzelius's theory did have testable content in excess of the evidence for the laws of chemistry and of electrolysis that he was attempting to explain. In any case, at least in 1815, he clearly separated this hypothetical part of his theory from the account of combining weights, claiming not to attach too much significance to the former.

> I do not consider the conjectures which I hazarded on the electro-chemical polarity of the atoms as of much importance. I scarcely consider them in any other light than as an ideal speculation deriving some little probability from what we know of the chemical effects of electricity.
>
> (Berzelius,1815, p. 123).

As we have seen, Berzelius argued for his formulae as an alternative to Dalton's diagrams on the grounds that the former can be used to express weight and volume relations involved in chemical composition without a commitment to atomism. Klein (2003, p. 20) has pointed out that those chemists who were inclined to take this path talked of combining equivalents (Wollaston), proportion (Davy), combining weight (Young), portion (Thomson) and parcel (Whewell) rather than atoms. To her list can be added doses (Donovan), combining quantities (Brande) and stoichiometrical numbers (Gmelin) as noted by Goodman (1969, p. 45). Berzelius acknowledged the hypothetical status of the atomic theory that he favoured. However, it was an hypothesis that he took seriously enough to incorporate it into his deployment of chemical formulae in organic chemistry. That deployment was to have dramatic consequences that he had not intended or anticipated, as we discuss in the next section.

9.5 The Binary Theory of Berzelius

Berzelian formulae were not much used in chemistry before the late 1820s, not even by Berzelius himself (Klein, 2003, p. 250, n. 2). This is understandable in light of the fact that, in inorganic chemistry where they were first introduced, they express little more than combining weights and volumes that can be just as well expressed in other ways. As Klein has argued in detail, this was to change when formulae were used in the much more complicated area of what is now referred to as organic chemistry. A large number of elements figure in the composition of inorganic compounds, with each compound consisting of fixed proportions of just a small number of those elements. By contrast, organic compounds are made up of a small number of elements, mainly carbon, hydrogen and oxygen and to a lesser extent nitrogen. As a consequence, knowledge of the proportions of elements in a compound is by itself an inadequate indication of its properties. In addition, a reaction involving the production of some organic substance of interest is, in organic chemistry, unavoidably accompanied by parallel reactions involving the production of by-products. In this section and the next, drawing heavily on the work of Klein (2001 and 2003) and Rocke (1984), I outline some of the ways in which order was brought to organic chemistry through the use of chemical formulae to such a degree that, by around 1860, a fairly unique set of formulae adequately characterising the properties and composition of organic compounds had emerged.

There are three features of Berzelius's chemistry that are basic to an understanding of his application of them to organic compounds. His 'binary' theory, which understands complex compounds as a combination of two less complex compounds, his electrochemical interpretation of those combinations as involving electropositive and electronegative components, and the central role that oxygen (Lavoisier's 'acidifying principle') played in his system. The idea that complex compounds have components that are themselves compounds goes back as far as the chemistry of salts, summarised in Geoffroy's table as discussed in the previous chapter, where a salt is considered as the combination of an acid and a base. Berzelius distinguished between the 'immediate constituents' of a compound into which they could readily be separated by experiment and their 'elemental' constitution, into which a compound could be divided only by a more complex series of experiments. Berzelius typically construed the immediate constituents as oxides. So, for him, copper sulphate was a combination of oxides of copper and of sulphur ($CuO + SO_3$) and sulphuric acid was a combination of water and sulphur oxide ($H_2O + SO_3$). These representations could readily be interpreted in terms of Berzelius's electrochemical theory. The pairs of immediate constituents of a compound consisted of an electropositive and electronegative component.

Berzelius was aware of the difficulties of transferring formulae to organic chemistry. One key problem that he well appreciated was the status of the law of multiple proportions in that field. That law holds that the various weights of one element that combine with some fixed weight of a second are in simple numerical ratios to each other. If those ratios are allowed to become indefinitely large, then the law loses its empirical content, because numbers can always be chosen in such a way that the law is compatible with the data.[11] The problem of reconciling the law of multiple proportions with the measured weights of carbon, hydrogen and oxygen in complex organic compounds, and of deciding on some definite integral numbers, proved to be a major one.

9.6 Chemical Formulae and the Rise of Organic Chemistry

In the 1820's a number of European chemists were less wary of applying formulae to organic chemistry than Berzelius himself had been. They employed them to display what they proposed to be the immediate constituents of organic compounds rather than simply recording their elemental constitution. Liebig, for example, represented alcohol as $C_4H_{10}O$. H_2O, which made it easy for him to regard the formation of ether (which he represented as $C_4H_{10}O$) from alcohol as involving the extraction of water. As we shall see, other chemists disagreed with Liebig and offered different formulae. In the remainder of this section I review some of the ways in which problems associated with identifying correct formulae were overcome. But one key point should be stressed. In using formulae to represent 'immediate constituents' of compounds, clearly with the intent of conveying something of chemical significance thereby, chemists were already going beyond using formulae simply to summarise what is contained in laws of combining weights and volumes.

I will not attempt to give a detailed account of, nor even summarise, the historical path that led to chemists arriving at a fairly definitive set of chemical formulae for organic compounds by the 1860s. Rocke (1984) and Klein (2003) can be consulted for those details. Here I give a schematic account of some of the demands placed on chemical formulae that were eventually to lead to a unique set of them that were able to embody chemical knowledge going way beyond laws of proportion.

One device that proved productive, exemplified in the display of 'immediate constituents' in formulae by Berzelius, involved introducing some order into the symbols representing the elements in an organic compound so that they represented properties other than mere combining weights and volumes. So-called radicals were understood as groupings of elements that remained intact through a chemical reaction and played a role similar to that of elements in inorganic chemistry. So, for instance, series of compounds could be understood as resulting from various additions to the methyl radical, CH_3, so that we have methyl alcohol, CH_3OH, methyl chloride, CH_3Cl and so on, using modern atomic weights. A fruitful idea was that of homologous series, an example of which is that involving the successive addition of CH_2 to the methyl radical to form ethyl, butyl, propyl and higher order compounds. Using this device, the properties, and even the existence and method of preparation, of higher order substances could be predicted on the basis of knowledge of the lower order ones. Berzelius introduced the terminology that distinguished 'empirical formulae', which simply indicated the proportion by weight of elements in compounds, from 'rational formulae' which included some ordering of symbols to reflect chemical properties other than combining weights. So CH_2O is the empirical formula for acetic acid whereas $C_2H_3O_2H$ is a rational formula for that acid.

The complexity of organic reactions, due to the many by-products invariably accompanying the preparation of some product, was confronted by using chemical equations to track the formation of each product. The formation of ether from alcohol by the action of sulphuric acid can be represented, using modern atomic weights, by the equation $2C_2H_6O = C_4H_{10}O + H_2O$. The numbers of occurrences of C, H and O on each side of the equation must balance, so that the weight of each element remains unchanged. Equations representing the formation of the various by-products can be represented by other balanced equations. In this way the messy process involving several parallel reactions and the formation of a mixture of products is comprehended by representing it as a superposition of identified reactions independent of each other and each represented by a balanced equation. Klein (2003, 118–129) has shown how, in the late 1820s, Jean Dumas and Polydore Boullay first used this technique to understand the formation of ether and its by-products from alcohol, thereby bringing order to a reaction that had caused confusion for decades. Thereafter, the use of chemical equations became commonplace and indispensable.

To illustrate the way in which the demands placed on formulae to adequately represent the properties of compounds and trace chemical reactions eventually led to a fairly unique set of formulae up to the task, I give examples abstracted from the historical detail.

The simplest empirical formula for acetic acid is CH_2O as pointed out above. This formula cannot be used to reflect the experimental fact that the hydrogen in acetic acid can be replaced by an equal volume of chlorine in the laboratory in four different ways yielding four distinct chemical compounds. Three of those compounds are acids similar to acetic acid and in which the relative amounts of chlorine vary as 1:2:3. The fourth compound has the properties of a salt rather than an acid. These experimental facts can be captured in a formula by doubling the numbers and rearranging the symbols in the empirical formula so that we have $C_2H_4O_2$, rearranged to read $C_2H_3O_2H$. The experimental facts can now be readily understood in terms of the substitution of chlorine for one or more of the hydrogens, with the three chloro-acetic acids represented as $C_2H_2ClO_2H$, $C_2HCl_2O_2H$ and $C_2Cl_3O_2H$ and the salt, acetyl chloride, as $C_2H_3O_2Cl$. This notion of 'substitution', which first emerged in the work of Dumas in the 1830's, where the replacement of one element or radical in a compound in the laboratory is represented by the replacement of one symbol or set of symbols by another in a formula, was to become a commonplace and powerful technique. The manipulation of formulae on paper, substituting symbols by others, could be highly suggestive of experiments to be conducted in the laboratory.[12]

The notion of substitution had a ready application in the understanding of the action of acids in terms of the substitution of hydrogen. This is already exemplified in the previous section, where the hydrogen, at the end of the formulae, responsible for the characteristically acidic behaviour of acetic acid, is separated from the other hydrogens and the formation of the salt, acetyl chloride, is represented as the substitution of chlorine for the hydrogen. It became recognised that some acids are polybasic, with two or more replaceable hydrogens and capable of forming two or more series of salts.

My third example involves Alexander Williamson's experiments with ethers published in !850. I draw heavily on Rocke (1984, pp. 215–223) but abstract from some of the chemical detail.[13] The experiments showed formulae for alcohol and ether championed both by Dumas, on the one hand, and Justus von Leibig on the other, were inadequate to accommodate his experimental results, whereas Williamson's own formulae could readily do so.

Dumas considered alcohol to be composed of the etherin radical (our ethylene) and two portions of water, and conceived of the formation of ether as the removal of one of the waters. So we have the equation

$$C_4H_8.2H_2O = C_4H_8.H_2O + H_2O$$

A corresponding formula for the formation of methyl ether would be

$$C_2H_4.2H_2O = C_2H_4.H_2O + H_2O$$

Liebig, by contrast, saw alcohol as the hydrated oxide of the ethyl radical, and once again understood the formation of ether as the extraction of water. On this view, the equations for the formation of ethyl and methyl alcohol are

$$C_4H_{10}O.H_2O = C_4H_{10}O + H_2O$$

and

$$C_2H_6O.H_2O = C_2H_6O + H_2O$$

Williamson's version of the formation of ether is represented by the equation

$$2C_2H_5OH = (C_2H_5)_2O + H_2O$$

With the corresponding equation for the formation of methyl ether

$$2CH_3OH = (CH_3)_2O + H_2O$$

Williamson's representation strongly suggests that the addition of concentrated sulphuric acid to a mixture of ethyl and methyl alcohol could yield, not only ethyl and methyl ether but a mixed ether, methyl-ethyl ether, the formation of which is represented as follows:

$$CH_3OH + C_2H_5OH = CH_{3.}C_2H_5O + H_2O$$

This mixed ether cannot be reconciled with the formulae of Dumas and Liebig, so the experimental preparation of that ether by Williamson told in favour of his formulae.

By the 1860s, then, formulae in organic chemistry had made possible major advances sufficient to spawn a synthetic chemical industry, and were such that formulae, and hence relative 'atomic weights', could be established. By the time this had been done it was possible to ascribe a novel property to chemical elements, namely, their valency. Armed with this notion, chemists were able to devise structural formulae for chemical compounds, as illustrated most dramatically in the work of Kekulé. Phenomena such as isomerism and optical rotation could readily be comprehended by the three-dimensional formulae built up from elements with a specific valency. The question I now raise is the extent to which this development can be considered a triumph for, and confirmation of, atomism.

9.7 Chemical Formulae a Victory for Atomism?

I have already hinted that I think there is a case for the view that nineteenth-century chemistry paved the way, rather than constituted a case, for an experimentally based and testable version of atomism. In describing the steps that led to unique formulae in organic chemistry I carefully avoided reference to atoms. The symbols in chemical formulae can of course be taken to refer to atoms. But, as we saw when discussing Berzelius's first introduction of formulae, those symbols can also be taken as referring to combining portions or volumes. On this interpretation, the structural formulae of organic chemistry depict some structure possessed by chemical compounds that is related to their chemical properties. But that structure need not be an atomic structure.

To do justice to the points I wish to make in this section I introduce a distinction between three positions that I refer to as physical atomism, chemical atomism and agnostic anti-atomism. My distinctions are related, but not identical, to those made by others, Rocke (1984) and Klein (2003) in particular.

Physical atomism involves atoms that are embedded in some physical theory such as those of the mechanical philosophers or Newton and possess physical properties such as mass, shape, size and the propensity to attract or repel other atoms. The kinds of properties possessed by physical atoms are determined in advance of chemical research by the physical theory that governs them.

Chemical atoms are the least parts of chemical elements. As well as mass, a property shared by chemical and physical atoms, chemical atoms are presumed to possess chemical properties characteristic of the elements they are atoms of. The kinds of properties it is necessary to attribute to chemical atoms is to be determined by chemical research. An example is valency, interpreted as a property of atoms by chemical atomists, which emerged in the course of advances in organic chemistry as we have seen.

Agnostic anti-atomism involves a refusal to commit to atomism. It is not a denial of atomism, which is a claim of similar strength to its affirmation. An agnostic anti-atomist would claim that the practise and success of the chemistry with which we are concerned does not require a commitment to atoms and is compatible with the idea that chemicals retain their properties however much they are divided. According to this view, the dramatic successes of the enterprise cannot straightforwardly be taken as evidence for atomism.[14]

It is clear that both a physical and a chemical atomist is free to use chemical formulae, interpreting the symbols in those formulae as representing physical and chemical atoms respectively. But an agnostic anti-atomist is free to use them too. The discussion in the previous sections of the path that led to unique rational formulae for compounds makes perfect sense if formulae are taken simply as describing chemical properties as well as combining weights and volumes. As already indicated, I deliberately omitted any reference to atoms in that section. The appearance of OH at the end of the formulae for a compound indicates that it has the properties of an alcohol, whilst CO_2H is indicative of an organic acid and so on. Further, the substitution of one element for another in a compound in the laboratory is mapped by the replacement of one symbol for another in a chemical formula. The formula of a compound represents some structure of that compound, but it does not have to be an atomic structure. The compound could possess the structure all the way down, as it were.

An analogy will help illustrate the coherence and intelligibility of agnostic-anti-atomism and its assumption that the indefinite divisibility of chemical substances is compatible with their characterisation using formulae. The electric field, $\underline{E}$, has the symmetry of an arrow whilst the magnetic field, $\underline{H}$, has the symmetry of a spinning disc. These facts led Maxwell and his followers to assume that $\underline{E}$ represents a strain in the aether whilst $\underline{H}$ represents a vortex in that aether. But there is no aether. The fields of classical electromagnetism are continuous and possess what structure they have all the way down. Agnostic anti-atomism was viable up until 1860 and beyond because no chemical evidence told against the possibility that chemical compounds possess their structure all the way down. Given the state of affairs in 1860 there was no guarantee that physical and chemical atoms would not be banished from science in the way that the aether came to be.[15]

A number of chemists involved in the developments of concern in this paper can be classified as chemical atomists. Kekulé (1867, 303–304), for example, made the distinction between physical and chemical atomism, and his commitment only to the latter, quite explicit. The historian Christoph Meinel (2004, 257) confidently invokes 'the usual distinction between chemical and physical atoms' which 'provided a common denominator for those who did not want to engage in metaphysical debates about the existence of atoms, but sought to pursue chemistry pragmatically'. Chemical atomists were certainly judicious in distancing themselves from physical atomism. The gap between the abstract claims of physical atomic theories and chemical experimentation could not be bridged prior to the 1860s at least, and physical atomic theories gave no useful guidance to organic chemistry. If the rise of the latter owed a debt to Dalton at all, it was to the chemical atomism that others creatively extracted from his work, rather than to the physical atomism that he espoused.

The fact that the use of chemical formulae is compatible with agnostic anti-atomism raises the possibility that the rise of organic chemistry did not constitute a strong case for atomism at all. The productive enterprise of arranging symbols in chemical formulae to capture chemical properties other than combining weights or volumes, and the representation of the replacement of one element or group of elements in a compound in the laboratory by the replacement of one symbol or group of symbols by others in a formulae made perfect sense without a commitment to atomism as we have seen. Pierre Duhem (2002) spelt out a detailed defence of this position at the turn of the century. Many of the relevant nineteenth-century research papers invoke formulae with no mention of atoms, whilst use of the term 'atom' is dispensable in those that did invoke the term. Further, many of the chemists that did refer to atoms interpreted them as useful fictions when pressed on the matter.[16] As David Knight (1992, 120) puts it, 'chemists were almost all atomists, but recognised atomism as an optional extra' when pressed. The scepticism of many chemists concerning the existence of atoms is borne out by the 'atomic debates' that took place in Britain in the 1860s and 1870s, documented by Brock (1967). There is little doubt that, as the century progressed and as links were forged between chemistry and physical processes such as the behaviour of gases, electrolysis, optical rotation, the osmotic pressure of electrolytes and non-electrolytes, spectroscopy and so on, the case for interpreting the symbols in formulae as representing atoms became increasingly powerful. An experimental path to knowledge of atoms was eventually forged. I have raised doubts about the extent to which nineteenth-century successes in chemistry constituted such a path. To the extent that structural formulae in chemistry do signal a commitment to atomism, it was a chemical atomism that emerged out of chemical practice and owed little or nothing to the physical atomism that had a long history from Democritus to Dalton.

9.8 Dalton's Resistance to Chemical Formulae

Now that we have traced the path that led to major advances in chemistry via the use of chemical formulae, it is instructive to reconsider Dalton's chemistry in contrast to it.

Dalton's diagrams did not play the suggestive and constructive role that Berzelian formulae came to play in organic chemistry, as Klein (2003, pp. 23–40) has argued. Dalton's diagrams appear in only a few places in his work. They appear in Appendices to both parts of Volume 1 of *The New System*, they were made public in a lithograph shown at the 1835 meeting of the British Association for the Advancement of Science in Dublin, and a few of them appear in the piece 'On the phosphates and arseniates' printed in 1840. They are never used by Dalton as a heuristic aid or in any other way in the body of any of the texts. This is in stark contrast to the constructive use of formulae evident in the writings of organic chemists such as Dumas and Liebig on the Continent, and by Frankland and Williamson in Britain soon after Dalton's death. The productive use of formulae as 'paper tools' for guiding and describing the work of organic chemists in the nineteenth century is stressed by Klein (2003, p. 33) who refers to the 'graphic suggestiveness and maneuverability' of formulae as opposed to diagrams. The constructive use of formulae is as striking in the work of Dumas, for example, as is the absence of any such use of diagrams in Dalton's texts.

To some extent Dalton did employ the notion of the 'immediate constituents' as opposed to the elemental constituents of a compound, although he did not use that terminology. This was already the case in 1808 when Dalton (1808, p. 163) stressed the importance, not only of the number of 'ultimate particles of simple bodies' in a compound, but also the number of 'less compound particles which enter into the formation of more compound particles'.[17] It is also evident in Dalton's commentary on his diagrams of some atoms. For example, as can be seen from Figs. 9.1 and 9.2, diagram 34, representing nitrous acid (N_2O_3), is described by Dalton (1808, p. 164) as composed of one atom of nitric acid (NO_2) plus one of nitrous gas (NO). However, Dalton did not put the notion of immediate constituents to productive use in the way that organic chemists using formulae learnt to do. Diagram 34, in Fig. 9.1, representing an atom of nitrous acid does not of itself show two component parts, one consisting of nitrous gas and the other of nitric acid. It is only the verbal commentary on the diagram, shown in Fig. 9.2, that does that. Exactly the same diagram appears in Volume 1, Part 2 of the *New system*, but this time the text describes nitrous gas as made up of one atom of oxygen combined with two atoms of nitrous gas.[18] Plus signs, subscripts and brackets could be used to introduce structure into 'rational formulae' in a way that could not readily be done with diagrams.

My case concerning the unproductiveness of Dalton's own version of atomic chemistry stands in contrast to what a number of Dalton scholars have claimed. Arnold Thackray, for instance, talks of Dalton's chemistry being conditioned by 'a conviction of the importance of structural chemistry' and asserts that 'his views on structure possessed a power far beyond his critics' perceptions'. 'Precisely the sort of three-dimensional thinking he [Dalton] pioneered', writes Thackray (1972, pp. 117 and 124) 'was to prove vital not only to the spectacular progress of organic chemistry in the nineteenth century, but also to the triumphs of molecular biology in our own time'. W. V. Farrar (1968, p. 297), in an article on 'Dalton and Structural Chemistry' goes so far as to claim that 'if more chemists had been playing with balls and sticks in the same way as Dalton, then we would not have had to wait so long for the theory of structure', although the substance of his paper suggests an interpretation more in keeping with my own.

One of the advantages of the use of formulae in chemistry that became of increasing importance from the late 1820's on was the need to distinguish between isomers. So, for instance, methyl ether could be written as $(CH_3)_2O$ as distinct from its isomer ethyl alcohol, C_2H_5OH, with the form of these formulae immediately conveying that the first has properties typical of ethers and the latter those typical of alcohols. It is frequently claimed that the ability to deal with isomers by invoking atomic structure was one of the early triumphs of Daltonian atomism. I was certainly told so at school. The historian C. A. Russell (1968, p. 263) makes such a point quite unambiguously when he refers to isomerism as 'the first successful prediction of the atomic theory' whilst the philosopher Alexander Bird (1998, p. 152) is equally forthright in claiming that 'the phenomenon that eventually clinched the debate in its [Dalton's atomic hypothesis] favour was the ability of the theory to account for the phenomenon of isomerism, which was discovered some considerable time after Dalton proposed his view'. Dalton never used his diagrams to distinguish between isomers so far as I am aware. Nor could they be used to do so with anything like the facility of chemical formulae. But, more to the point, atomism did not and could not predict isomerism because no assumptions about the relationship between structure of atoms and chemical properties were included in its premises. Isomerism was discovered experimentally by chemical means. The results of analysis showed the existence of compounds containing the same elements in the same proportions which nevertheless had differing chemical properties. Formulae could be and were readily deployed to reflect this experimental fact in a way that could be emulated with Daltonian diagrams only with difficulty and with the aid of accompanying text. Isomers were accommodated by utilising the conventions underlying the use of formulae to represent chemical properties. This could be done in a way that did not even commit one to atomism.

The nature and status of Dalton's atomic chemistry compared with that on the Continent, the former wedded to physical atoms as represented by Daltonian diagrams and the latter making use of Berzelian formulae, was a question that became a matter of concern for the members of the British Association for the Advancement of Science. It is clear from the records of those deliberations that they sensed the fact that British chemistry was being left behind by developments on the Continent, which was certainly the case as far as organic chemistry is concerned. The *Report of the Committee on Chemical Notation*, published in the official *Report* of the meeting of the British Association in Dublin in 1835 cites a number of observations made by William Whewell, one of which expressed the recognition that most chemists in other nations had adopted Berzelian formulae and 'the only effect of our keeping back would be, to throw us behind science'.[19] Whewell's remarks also show that he was aware that use of Berzelian formulae involved less of an ontological commitment than use of Daltonian diagrams.

> Dr Dalton's method supposes a theory, Berzelius only states a fact. The notation of the Swedish chemist shows that such and so many atoms are present. Dr Dalton's, on the other hand, attempts to show their method of molecular arrangements, of which we have no positive knowledge whatsoever.[20]

Whewell was right to infer that there was a strong sense in which the use of formulae initiated by Berzelius involved less theoretical commitment than Dalton's atomism, the latter involving a fairly literal interpretation of diagrams of compound atoms. But he was wrong to imply that use of formulae involved no theory. The deployment of, and disputes concerning, rational formulae by the Continental chemists assumed low-level chemical theories, involved, for example, in the understanding of acids in terms of substitutable hydrogen, in the claims that various radicals persisted as such through a wide range of chemical change, that atoms or portions represented by the algebraic symbols could be substituted one for another and so on. The key point, not quite made by Whewell, is that the theory involved in the use of formulae was of a kind that could be tested by chemical experiments, whereas, at the time, the strong claims about atoms involved in Dalton's theory could not be.

The records of the Dublin Meeting show that Dalton strongly resisted replacement of his diagrams by formulae. He is reported as insisting that 'regard must be had to the arrangement and equilibrium of the atoms (especially elastic atoms) in every compound atom, as well as to their number and weight'.[21] Dalton had prepared a lithograph showing his illustrations of elements and compounds. He regarded his method as 'the only one representing nature'. These remarks reveal the extent of Dalton's commitment to the physical reality of his spherical atoms and the extent to which his thoughts on them were still tied up with the physical theories from which his atomism had emerged. The lack of chemical utility of Dalton's approach compared to the one involving rational formulae is starkly apparent once we focus on the organic compounds represented in Dalton's lithograph. If we replace Dalton's spheres by symbols then we get CHO for acetic acid. This leaves absolutely no scope for the quite different compounds arrived at by replacing hydrogen by chlorine either inside or outside the radical. Whether diagrams could be put to suggestive use or not, the fact is that Dalton himself put them to no such use. Two years later, in 1837, Dalton reaffirmed his opposition to Berzelian formulae.

> Berzelius's symbols are horrifying: a young student in chemistry might as soon learn Hebrew as make himself aquainted with them. They appear like a chaos of atoms. Why not put them together in some sort of order? [They] equally perplex the adepts of science, discourage the learner, as well as to cloud the beauty and simplicity of the Atomic Theory.[22]

Dalton's persistence on this point is presumably a reflection of his commitment to a physical theory of atoms, a theory which I have argued was ill-supported and unproductive during Dalton's lifetime and beyond.

One more point gives some historical support to my somewhat negative view of the productiveness of Dalton's atomism. Edward Frankland was an English chemist who had became familiar with the use of formulae in organic chemistry during a period in Germany working with Hermann Kolbe and Robert Bunsen. He did pioneering work on organo-metallic compounds. In 1851 he gave his inaugural lecture as the first professor of chemistry at Owens College, later to become the University of Manchester.[23] One might have expected him to ingratiate himself with the dignitaries of Manchester that were present by making much of John Dalton, the figure they would have been keen to claim as one of their own. In fact, Frankland made just one passing reference to Dalton, mentioning him in the same breath as William Henry. The lack of significance of Dalton's chemistry that is implied here, made public by Frankland in 1851 in Manchester, is totally explicable on the assumption that Frankland took for granted the views on Dalton I have canvassed in this chapter.

9.9 Is My Critique of Nineteenth-century Atomism Positivist?

Am I not making too much of my critique of atomism in nineteenth-century chemistry? After all, from a historical point of view most of the chemists who contributed to its advance, especially in organic chemistry, were atomists of some kind or other, the atomic interpretation of formulae gained more credibility as the century advanced and it was eventually fully vindicated. If the reader is inclined to respond to my insistence on the viability of agnostic anti-atomism in the nineteenth-century as the stance of a positivist who does not appreciate the role of theory in science, then he or she would certainly not be the first to do so. I devote this final section of the chapter to a clarification of what I see as the substance and the point of my critique.

In my analysis of seventeenth-century developments I distinguished between philosophical matter-theories, which were theories of matter in general accommodated to the phenomena, and scientific theories that were designed to explain a limited domain of phenomena and were confirmed by experiment to some significant degree. I argued that the distinction was present in practice in the work of Boyle and Newton and made explicit by them on occasions. By the nineteenth century the distinction between philosophy and science had become more pronounced and was becoming institutionalised. The term 'science' came to have its modern connotations and the term 'scientist' was introduced. In the first half of the nineteenth century Auguste Comte attempted to capture this distinction with his 'postivism'. He distinguished between mythical or theological explanation, metaphysical explanation and scientific explanation. The former predominated in the era before the philosophers of Ancient Greece replaced it by their critical metaphysics that in turn was eventually challenged by scientific knowledge based on observation and experiment. I do not endorse the details of the way in which Comte and later positivists such as Ernst Mach explicitly characterised science. It was much too narrow and unable to adequately capture the theoretical dimension of science and the ways in which it can be borne out by experiment (as opposed to mere observation). But it should be clear that I do endorse the importance of the distinction between science and philosophy or metaphysics. If this makes me a positivist then so be it.

The way in which I characterised the distinction between experimental science and philosophical matter theories in the seventeenth century needs to be modified to characterise the situation that had emerged by the early nineteenth-century. When, for instance, Laplace attempted to explain elasticity of solids by speculatively specifying inter-atomic forces he was not offering a theory of matter in general but rather of a quite specific set of phenomena. In that respect it was analogous to experimental science rather than philosophical matter theories. On the other hand, to the extent that there was no support for Laplace's atomic assumptions independent of the phenomena involving elasticity that they were designed to explain, his theory was accommodated to, rather than confirmed by experiment just like philosophical matter theories. Those chemists who objected to atomism can sometimes be read as doing so because it lacked the experimental support appropriate for a science and at other times because it constituted an inappropriate reduction of chemistry to physics. Chemists had come to see their enterprise as autonomous from physics. As we have seen, those chemists who did endorse atomism endorsed chemical atomism. The properties of atoms were to be discovered by chemical research rather than imported from some matter theory of the physicists. The property of valency, for example, posed problems for physicists' conceptions rather than being a consequence of them

Nineteenth-century chemists had every reason to be wary of importing physical atomism into their science. The time was not yet ripe for any such importation to be of any guidance to their experimental work. But I have conceded that once chemical formulae were showing their worth, most chemists were chemical atomists of some kind or other. The chemists who took a stand against atomism are best interpreted as objecting to physical, rather than chemical, atomism. The objections to atomism in chemistry generally that were articulated by Pierre Duhem were very much the exception rather than the rule. Am I, then, giving to much credence to agnostic anti-atomism as opposed to chemical atomism?

In my construal of agnostic anti-atomism I suggested that the structures exhibited by chemical formulae need not be interpreted as atomic structures. Substances could conceivably be continuous, possessing their structures all the way down. I claim that chemistry at the time of Kekulé did not involve evidence that required a chemical atomistic rather than an agnostic interpretation of formulae. There is an instructive analogous situation. By the mid-nineteenth century evidence for a wave-theory of light was considerable and the Newtonian particle theory was no longer a serious contender. But a distinction can be made between an undulatory theory of light, on the one hand, that remains agnostic on the question of what the undulations are undulations of, and a theory that attributes light to transverse waves in an elastic aether, on the other. Since the weaker assumption could stand tests against the available evidence just as well as the aether theory the question arises of what warrants the stronger assumption. Ernst Mach was one of the few nineteenth-century figures who saw things this way. In his optics he endorsed an undulatory theory but declined to attribute the undulations to oscillating states of an aether. The line of questioning with respect to the aether was stressed by Einstein around the turn of the century, at a time when the aether theory had been extended to electromagnetism and light waves were recognised as electromagnetic waves. The fact that the aether could be removed from optics and electromagnetic theory without any loss of empirical content became an argument for dispensing with the aether. A point I wish to stress is that, for most of the nineteenth century, there was no guarantee that the atom would not be banished just like the aether.

My suggestion that chemical substances might possess the structures exhibited by chemical formulae all the way down is strange and counter-intuitive. But so is the claim that electric and magnetic fields are states that have a structure but are not states of any underlying medium. Twentieth-century science was to undermine many of the intuitions that lay behind philosophical assumptions informing metaphysics. Any adequate understanding of the nature of science needs to include an appreciation of how this came about. However 'obvious' the atomic interpretation of chemical formulae may be or have been, it was not a necessary consequence of nineteenth century chemistry. In this respect I am in agreement with Pierre Duhem. In another respect I totally disagree with the French philosopher. Duhem expressed the view that knowledge of atoms was impossible in principle whereas I maintain that in the nineteenth century the case had yet to be made. Subsequent developments were to show just how wrong Duhem was. The remainder of this book shows how (after a digression in the following chapter).

Notes

1. This aspect of Berthollet's work is summarised in Thackray (1970), pp. 230–233.
2. For details see Roscoe and Harden (1896), Nash (1956), Thackray (1972), Cole (1978) and Rocke (1984, Chapter 2).
3. The assumption that atoms of a gas repel each other with forces proportional to 1/r has the consequence that the pressure of a large sample of a gas will be greater than that of a smaller sample when the density of the two samples are the same, contrary to what experiment straightforwardly shows to be the case.
4. The outcome of the debate was not as decisive at the time as it became in retrospect and it may well be that the emergence of atomism increased the attractiveness of the law of constant proportions.
5. Dalton (1808, p. 163). All quotations from this work refer to the 1964 Citadel edition.
6. The issue of whether or not Dalton was aware of the simple ratios borne out by the composition of the oxides of nitrogen and used it in the construction of his theory is a tricky one. See, for example, the discussion in Rocke (1984, pp. 29–33). Further support was soon coming, in 1808, from experimental work on the oxalates and carbonates of strontium and potassium by Thomas Thomson (1808) and William Wollaston (1808).
7. Davy made this distinction when commenting on the award of a Royal Medal to Dalton in 1826. See Knight (1967, pp. 18–20).
8. Dalton stuck to a relative atomic weight of 7 for oxygen long after experiment was well able to single out 8 as a much more accurate figure. On page 6 of a short paper 'Quantity of acids' that an ageing Dalton had printed along with sundry other pieces he wrote: 'In 1807 I first published in my system of Chemistry Part1, the atom of water; it was 1 for hydrogen and 7 for oxygen = 8, the relative weight of an atom of water. I have seen no reason for alteration from that time to this, in 1840'. In a talk delivered in 1830 he did use 8 for oxygen, 'not because I think it is the most correct, but because it is frequently met with in the books of chemistry' (Thackray, 1972, p. 98).
9. Dalton (1811) made the basis of his simplicity rule explicit in response to John Bostock's charge that it was arbitrary.
10. As a matter of fact, Dalton was not entirely consistent on this point in the *New System* as observed by Klein (2003, p. 39). The last three of Dalton's diagrams (Figure 9.1) show like atoms in contact. This inconsistency was removed in subsequent versions of the diagrams.
11. Berzelius's awareness of this problem is discussed by Klein (2003, pp. 20–23). The problem is also discussed by Maureen Christie (1994).
12. For details of the emergence of the notion of substitution, as an unintended and unanticipated consequence of Dumas's research, and for a discussion of its powerful suggestiveness see Klein (2003, 188–206).
13. Williamson did not prepare his ethers directly from the alcohols, as in my simplified scheme, but from 'ethalate of potash' (potassium ethoxide) obtained from alcohol my dissolving potassium in it.
14. My distinctions between the various kinds of atomism overlap with those made by Rocke and Klein but are not identical. The main difference lies in my distinguishing agnostic anti-atomism from chemical atomism. Rocke (1984, pp. 10–15) distinguishes between physical and chemical atomism. For him, it is the latter, rather than the former, that productively informed the rise of organic chemistry in the nineteenth century. In Rocke's view, chemical atomism

'was universally (if implicitly and often unknowingly) accepted throughout the course of the nineteenth century' whereas physical atomism 'was controversial and far from universally accepted' (p. 10). He characterises chemical atomism as affirming the existence for each element of 'a unique 'atomic weight', a chemically indivisible unit, that enters into combination with similar units of other elements in small integral multiples' (p. 12). He insists that this chemical atomism 'has greater content than stoichiometry' (p. 13). Most of this is consistent with chemical atomism and agnostic anti-atomism as I have defined them The exception is Rocke's inclusion, in his chemical atomism, of the notion of chemical atoms assumed to be 'chemically indivisible units'. My claim is that the story of the rise of organic chemistry, as Rocke himself has told it, did not require a commitment to chemical atoms and could be accommodated by an agnostic anti-atomist. Organic chemistry in the third to the sixth decades of the nineteenth owed its dramatic success to assumptions that went beyond what could be sensibly construed as generalisations from observations made in the laboratory, and certainly went beyond what is contained in the laws of proportion. The idea that the properties of organic compounds are related to an invisible structure that goes beyond weight relations and which can be mapped by rational formulae, the representation of the replacement of one element by another in a compound by the substitution of on symbol by another in a rational formula, the construal of acids in terms of replaceable hydrogen and so on were all theoretical assumptions. In light of this, I can agree with Rocke (1984, pp. 12–13, 84–87, 177–180) that those chemists, notably William Wollaston and Leopold Gmelin, who expressed themselves in terms of a definite set of equivalents rather than atomic weights, did not thereby avoid a commitment to theory. Such a commitment was necessarily involved in use of formulae to express more than combining weights and volumes. However, my position differs from Rocke's because my distinction between chemical atomism and agnostic ant-atomism allows me to deny that this theoretical commitment was necessarily a commitment to chemical atomism. An anonymous referee of a paper of mine objected to my claim that Ursula Klein's portrayal of the emergence of organic chemistry is compatible with agnostic anti-atomism. That referee interpreted Klein as defending chemical atomism in this context and invited me to read her book more closely. I have done so, and find no need to alter my original claim. Klein agrees with Rocke in claiming that the use of formulae by organic chemists went beyond what is implied by laws of proportion insofar as they were used 'to model the invisible constitution of organic substances' (2003, p. 11). The symbols in chemical formulae indicated 'scale-independent bits or portions of elements, which overlapped but [were] not identical with the concept of "atom" in the philosophical and physical tradition' (2003, p. 12) so that 'Berzelian formulas had a theoretical meaning that differed from "atomic" composition" ' (p. 14–15). The notion of a 'scale-independent bit or portion' is ambiguous. According to one interpretation, the bits or portions are discrete ontological entities which have a definite weight and any other property that they possess whatever scale one might choose to measure them by. According to a second interpretation, the bits or portions are any sample of a chemical element, however small. These bits or portions will all alike posses the chemical properties of the substances they are bits or portions of, whilst the 'atomic weights' involved in specifications of weight relations in chemical combinations and substitutions will be scale invariant because they are *relative* weights. It is the second of these interpretations that makes most sense of Klein's work. The distinction made by Klein (2003, p. 252, n. 24) between 'scale independent chemical portions' and 'particles in the micro-world' would seem to require this as does her insistence that the object of the work of the organic chemists in deploying formulae 'was not the behaviour of sub-microscopic atoms but rather, in a traditional intellectual framework, the macroscopic level of substances or substance components and their recombinations' (Klein, 2003, pp. 265–266, n. 24). In short, Klein's work on the introduction of Berzelian formulae into organic chemistry fits well with what I have termed agnostic anti-atomism.

15. This comparison is discussed in more detail in Chalmers (2008).
16. For instance, Edward Frankland, whose work helped in the formation of the concept of valency, referred to talk of atoms as 'a kind of ladder to assist the chemist' (as cited by Brock, 1967, p. 21).

17. Dalton (1808, p. 163).
18. Dalton (1810). The text is on page 366 and the diagram on p. 561.
19. Thackray (1972, p. 118).
20. *Ibid.*, p. 118.
21. *Ibid.*, p. 117. The diagrams in Dalton's lithograph are reproduced on p. 119.
22. This is from a referees report on a paper on salts by Thomas Graham cited in. Brock (1993, p. 139). Graham's paper was published in spite of Dalton's negative appraisal. It would appear that the editor of the *Philosophical Transactions* sought a second opinion, that of W. H. Pepys. The latter's report, held in the library of the Royal Society (Referees Reports, 1832–1849), recommended publication and remarked 'In respect to the chemical symbols used by the author, I have not studied them but I see they were printed in his former paper'. It would appear that the editor had specifically asked for Pepys' view of Graham's use of formulae in the light of Dalton's negative remark.
23. See Frankland (1852) for the text of the lecture.

Chapter 10
From Avogadro to Cannizzaro: The Old Story

Abstract A common story is that in 1858 Cannizzaro put an end to confusion over values to be attributed to atomic weights by showing how they could be calculated from equivalent weights and densities using Avogadro's hypothesis and checking against molecular weights calculated from specific heats using the law of Dulong and Petit. There is a sense in which it was possible for Cannizzaro to do this in 1858 where others before him had failed. However, the significance of this needs to be downplayed for a number of reasons. Firstly, chemists had arrived at definitive molecular weights by way of formulae in organic chemistry without need of Avogadro's hypothesis. Secondly, Cannizzaro's method yielded only raw formulae not the structured formulae that organic chemists needed. Cannizzaro's paper of 1858 was parasitic on the work of the organic chemists in this respect. Thirdly, not much chemistry can be done via knowledge of atomic and molecular weights alone, so the importance of their determination for chemistry is over-rated by the common story.

10.1 Introduction

In the previous chapter we saw how proponents of some form of Daltonian atomism were faced with the problem of how to overcome the under-determination of chemical formulae and atomic weights and we saw how developments in organic chemistry guided by formulae led to a solution of the problem by the 1860s. This is not the way the story is usually told. A more common one recognises that, in the light of Avogadro's hypothesis (that equal volumes of gases at the same temperature and pressure contain the same numbers of molecules), molecular weights can be compared by comparing gas densities.[1] Molecular weights can then be divided into weights of the atoms composing them using equivalent weights in the way clearly spelt out by Cannizzaro (1961) in 1858. The atomic weights, and hence formulae, arrived at in this way can be augmented, for non-volatile substances, and checked for volatile ones, against the atomic weights arrived at via the law of Dulong and Petit, which holds that the product of atomic weight and specific heat is a constant. Cannizzaro was at pains to indicate the extent to which the two methods yielded consistent results.

A. Chalmers, *The Scientist's Atom and the Philosopher's Stone*, Boston Studies in the Philosophy of Science 279, DOI 10.1007/978-90-481-2362-9_10,

Avogadro published his hypothesis in 1811. For many historians and chemists, this raises the problem of why it took so long to apply it in the modern, post-Cannizzaro, fashion to establish definitive atomic and molecular weights and chemical formulae. As we shall see, and as recent historians such as John Hedley Brooke (1981) and Nicholas Fisher (1982) have argued in detail, to say that Avogadro's hypothesis could have been employed at the time to establish atomic and molecular weights is to seriously misread the historical situation. There are problems, stemming from the questions of what Avogadro actually claimed in his hypothesis, whether it could be known to be true and what could be done with it at the time. We shall see that what was possible for Cannizzaro in 1858 was by no means possible in earlier decades. We shall also see that, as a matter of historical fact, Avogadro's hypothesis was not neglected and that those who attempted to utilise its consequences to determine atomic and molecular weights ran into trouble. As Fisher (1982, p. 84) puts it, 'chemistry was not ready for Avogadro in 1811 or 1821 as it was beginning to be for Cannizzaro in 1858'.

In the previous chapter I indicated how the problem of atomic and molecular weight determination was solved by 1860 by way of establishing a unique set of formulae capable of meeting the chemical demands put on them. Rocke (1984) has argued that case in great detail and I have availed myself of the results of his work. In the light of it, the significance of Cannizzaro's paper takes on a different complexion. His method, while it does yield atomic and molecular weights, did not give the organic chemists the structural formulae that they needed, and which they had fashioned for themselves without its help. Cannizzaro's method yields the formula $C_2H_4O_2$, for acetic acid, for instance, and not one that can discriminate between the different kinds of hydrogen substitution that we discussed in the previous chapter. Cannizzaro's work was parasitic on the work of the organic chemists in a way that is not captured in the usual story of his 'innovation'.

In the final section of this chapter I shall query the extent to which focussing on the determination of atomic and molecular weights was the key problem for chemists that the common story of the path to Cannizzaro presumes it to be.

10.2 Avogadro's Hypothesis According to Avogadro

Avogadro's own version of his hypothesis as published in 1811 cannot be interpreted as equivalent to the modern version of that hypothesis. Avogadro's notion of a molecule does not readily map onto the modern molecule or the modern atom. Avogadro worked in the context of Berthollet's affinity theory. His molecules were particles of a gas that exerted forces of affinity on other molecules, with no ready and identified distinction between physical and chemical forces. Avogadro recognised that Gay Lussac's law for the combining of gases in simple ratios by volume had implications, for an atomist, for the relation between the numbers of molecules of various gases in unit volume. If a volume of one gas chemically combines with a volume of a second that bears a simple ratio to it, then the atomist's idea that the chemical combination involves the combining of a small number of atoms (or 'molecules') of each substance implies that the number of atoms ('molecules') of each gas in a unit volume bear a simple ratio to each other. Avogadro *assumed* that the numbers were equal.

Avogadro's assumption that the numbers of molecules per unit volume are equal had the immediate consequence that some molecules are divided in chemical reactions. Since two volumes of hydrogen combine with *one* volume of oxygen to yield *two* volumes of steam the oxygen molecules need to be split in half to be spread over the twice as numerous steam molecules. Avogadro had no trouble with the idea that molecules forming parts of a gas be chemically divisible into smaller units, but he had no concept of a chemical atom. In that sense, his oxygen molecules were not diatomic; they were just chemically divisible.

As Martin Frické (1976) has argued, the divisibility of molecules in Avogadro's theory was introduced in a totally ad hoc way. Avogadro had no independent evidence for the divisibility of oxygen molecules into two. It was simply assumed in order to reconcile the experimental facts with his hypothesis. The same was the case for other molecules that needed to be split if they were to be accommodated to measured combining volumes.

Another problem stemmed from the fact that the application of his hypothesis was limited, or should have been limited, by the small number of chemical substances available in gaseous form. This, in fact, did not stop Avogadro. He extended his attribution of molecular weights to non-gaseous elements by taking their chemical equivalent weights, postulating what volumes those weights represented and then specifying the corresponding vapour density. By 1821 he had proposed molecular weights for all the elements known to him. An indication of the degree of speculation involved is given by the fact that his value for the vapour density of phosphorous is 3/5 the correct value and that for sodium eight times too large.

Contemporaries of Avogadro who saw potential in his hypothesis were free to interpret it in their own way. One natural attempt for a chemist was to interpret Avogadro's 'molecule' as chemical 'atom'. This could not be done in an unqualified way because, as we have seen, such an interpretation would involve the division of atoms of oxygen in the formation of steam. Berzelius avoided this problem by interpreting Avogadro's hypothesis as claiming that equal volumes of elementary gases contain equal numbers of atoms. Steam, being a compound, did not need to conform, and could be assumed to have half as many 'compound atoms' (H_2O in Berzlius's case) per unit volume as hydrogen and oxygen. When even this assumption clashed with Berzelius's chemical ideas he restricted the hypothesis still further to apply to the common gases only. The changing views of Berzelius highlight the problems with the precise interpretation, range of applicability and truth of Avogadro's hypothesis in the decades following its first publication.

One consideration that loomed large in Berzelius's reaction to Avogadro's hypothesis stemmed from his view that chemical combination involved the joining of atoms with opposite polarity. For instance, the hydrogen and oxygen atoms combining to form water were understood to be positively and negatively charged, respectively. From this point of view, diatomic molecules, which would have been pressed upon Berzelius had he applied Avogadro's hypothesis to all gases, were impossible. Most chemists of the time followed Berzelius in this view. It was only from the 1840s onwards, when the notion of substitution took hold, that this particular perspective was undermined. The fact that what Berzelius regarded as electropositive hydrogen could be replaced by electronegative chlorine in acetic acid, for instance, undermined Berzelius's theory.

10.3 Ampère's Version of Avogadro's Hypothesis and Geometrical Atomism

Another version of Avogadro's hypothesis occurs in the work of André Marie Ampère (1814). He developed a version of the hypothesis, independently of Avogadro but with an acknowledgement to that scientist. Ampère's terminology shows that a clear conception of the modern distinction between atom and molecule had yet to be put in place. He devised what is in effect a geometrical atomic theory, with the least parts of chemical substances taking the form of regular geometrical solids, the regular tetrahedron being the most simple. So, for instance, both hydrogen and oxygen were made up of tetrahedrons, which Ampère referred to as 'particles' with four points, one at each vertex, referred to as 'molecules'. Knowing that two volumes of hydrogen combine with one of oxygen to give two of steam, Ampère presumed that one hydrogen particle, with its four point molecules, combined with half of an oxygen particle (two point oxygen 'molecules') to yield a 'particle' of steam involving four molecules of hydrogen and two of oxygen. This was consistent with Ampère's version of Avogadro's hypothesis, which, on his formulation, read 'equal volumes of gases contain equal numbers of particles'.[2]

The speculative character of Ampère's theory is evident. Guided by the equal numbers hypothesis, which he assumed, and the experimental findings concerning the chemical combination of gases by volume, Ampère constructed geometrical models of 'particles' of chemical substances that would fit the data. (There are plenty of simple geometrical assumptions for the geometry of particles that would be incompatible with Avogadro's hypothesis.) In the case of Ampère, no less than of Avogadro, a speculative atomic theory is accommodated to, rather than tested against or borne out by, experimental evidence. It was natural for an atomist like Ampère to assume some relationship between crystal structure and the shapes and arrangements of particles and molecules (or atoms and molecules), but the time was far from ripe for such hypotheses to make contact with experiment. (It is worth recalling that the crystal structure of sodium chloride, the simplest molecular structure for which is an asymmetric dumbbell, is cubic! In general there is no simple correlation between molecular and crystal structure.)

The geometrical theory found little support, although it was taken up in earnest by Marc Antoine Gaudin, culminating in his book *L'Architecture du monde des atomes* (1873). The fate of the theory in his hands supports my negative judgement concerning the productiveness of the approach. Seymour Mauskopf (1969, p. 70), in an otherwise sympathetic account of Gaudin's theory, sums up the situation thus:

> His [Gaudin's] assumptions, particularly his chemical assumptions, became more untenable as time progressed. Obsessed as he was with symmetry, Gaudin's contention that there was a complete reorganization of all the atoms in a chemical combination was in contradiction with the principal line of development in organic chemistry – the development of the radical theory and then the type theory. Moreover, Gaudin refused to accept the concept of valency, and he seemed to ignore the developments of stereochemistry in the 1860's. His assumptions remained basically unchanged over forty years; yet what might have been novel and suggestive in 1831 had become obsolete by 1873.

A more modest stand on the relation between crystal structure and atomic arrangements was proposed by Eilhard Mitscherlich around 1819. His idea was that substances with similar formulae, which for an atomist means similar atomic arrangements, should have similar crystal structure. This was borne out for a range of substances for which both formulae and crystal structure were known and, as Berzelius was quick to appreciate, it offered the prospect of inferring an otherwise unknown formulae for a substance from the identity of its crystal structure with that of a substance with a known formula. This principle was able to assist in a limited way with the determination of atomic weights via formulae, although there were known exceptions, glaringly exhibited by the phenomenon of dimorphism, where the same substance possesses more than one crystalline form.

10.4 Vapour Densities and Specific Heats as a Path to Atomic Weights

If Avogadro's hypothesis is true, then relative molecular weights can be measured by comparing vapour densities. However, at the time Avogadro and Ampère proposed the equal numbers hypothesis few elements were readily accessible in the gaseous state. Dumas set out to rectify this in the 1820s. He developed techniques for measuring vapour densities at high temperatures. Far from helping to solve the problem of atomic weight determination, his experiments exacerbated it. The densities of mercury, phosphorous and arsenic differed from what his chemistry led him to expect, leading Dumas to despair of any simple path from molecular weights determined from vapour density measurements and atomic weights needed for chemical formulae. The problem was soon to be made worse once it was realised that some vapour densities, relative to that of some standard gas such as hydrogen, were in fact a function of temperature. (The explanation lies in the phenomenon of thermal dissociation. Molecules of sulphur consist of six atoms at temperatures around 200 degrees centigrade and of two atoms at around 900 degrees centigrade. At intermediate temperatures sulphur vapour contains a mix of the two in thermal equilibrium, in proportions depending on the temperature! As we shall see, it was thermodynamics rather than atomism that first led to an understanding of thermal dissociation.)

While vapour densities offered a path to relative molecular weights only for gases, the law of specific heats proposed by Dulong and Petit in 1819 applied only to solids. That law claimed that the product of atomic weight and specific heat for a solid is a constant. As its authors noted, for an atomist the law implies that each atom, of whatever substance, makes the same contribution to specific heat. The status of the law as an empirical truth was problematic. Specific heats could be measured readily enough, but testing the law required knowledge of atomic weights also. Dulong and Petit needed to alter many of those weights accepted at the time. For instance, of the atomic weights accepted by Berzelius in 1819 they needed to halve eight, take two thirds of the value for bismuth, one third for cobalt and one fourth for silver.[3] Such results could either be taken as casting doubt on the validity of the law or that of the atomic weights prior to their modification. The latter move meant changing formulae that were proving their worth in chemistry, and it is not surprising that a chemist like Berzelius made use of the specific heats law when it suited but resisted its implications when it threatened to play havoc with his chemistry. The law of Dulong and Petit could be extended to apply to molecules, as suggested by Gaudin in 1831.[4] This extension involves the assumption that each atom in the compound makes the same contribution to specific heat.

As we shall see, in 1858 Cannizzaro was able to show that atomic weights derived from specific heats and from vapour densities could be rendered consistent. The main obstacles to such a move in earlier decades stemmed from the reluctance to admit diatomic or higher orders of atomicity to gaseous elements and from a range of chemical formulae that had shown their chemical usefulness but which implied atomic weights differing from those defended by Cannizzaro (which happen to be the correct ones). As we have seen, that latter problem was in effect removed by 1858 because developments in organic chemistry had resulted in unique formula and atomic weights, with the latter agreeing with Cannizzaro's estimates.

It is interesting to note, D. Ladenburg (1900. p. 107), a chemist himself involved in advances in organic chemistry in the second half of the century, and writing in 1869, much less removed from the historical situation than we are, saw the problem of atomic weights as solved by developments in organic chemistry rather than in a convergence of measurements of vapour densities and specific heats. Having described the attempts to fix atomic weights in the ways described above, he concluded:

> It had come to this then: Inorganic chemistry, in connection with physics, had not been able to maintain the conception of the atom. It is my business to show, in the next lectures, how it was reintroduced into the science by means of organic chemistry.[5]

10.5 Cannizzaro Reappraised

According to Frické (1976, p. 304), it was Cannizzaro who transformed chemistry for the better by showing how to arrive at definitive formulae and atomic weights. Contemporary historians of chemistry, not to mention most chemists, typically share this view. David Knight (1967, p. 97), for instance, writes:

> Until after Cannizzaro had successfully resurrected Avogadro's hypothesis, chemists did not agree on atomic weights and it was therefore not possible in practice to discriminate to the satisfaction of all between the two kinds of isomerism [involving isomers and polymers] which Berzelius had distinguished. Until general agreement could be reached on the formulae of the various isomeric substances, it was impossible to provide an explanation in other than extremely general terms. This did not happen until after the Carlsruhe Conference, when Cannizzaro's views were accepted.

Robert Siegfried (2002, p. 259) has more recently echoed the same theme.

> But the potentiality of molecular structure could not be successfully pursued until the atomic weights could be reliably known. This was not accomplished until fifty years later, by Stanislao Cannizzaro, through a rigorous application of Avogadro's hypothesis first published in 1811.

The common view that is illustrated by the above quotations can be summarized and expanded thus: Cannizzaro demonstrated, for the first time, how definitive atomic weights and formulae could be introduced into chemistry using Avogadro's hypothesis. This opened the way for structural formulae in organic chemistry by means of which isomers and polymers could be distinguished and soon led to the introduction of the notion of valency. Unambiguous atomic weights also laid the basis for the transformation of inorganic chemistry with the introduction of the periodic table.

This rendering of Cannizzaro's achievement is misleading, exaggerated and in some respects plain false. The structure of the argument of the *Sketch* is not adequately characterized in terms of a method for deriving atomic weights using Avogadro's hypothesis. This would have made those weights as questionable as Avogadro's hypothesis itself. Chemists of the first half of the nineteenth century had found plenty of reasons to doubt the truth of that hypothesis, as has been argued in detail for example in Brooke (1981) and Fisher (1982), whilst the kinetic theory of gases, which had Avogadro's hypothesis for physical molecules of gases as a consequence, was in its infancy. Cannizzaro explicitly argued for his atomic weights by noting the agreement of them with those derived from specific heat measurements, and he implicitly utilised their agreement with formulae that had emerged in organic chemistry.

Cannizzaro's method for atomic weight determinations involved an acceptance of the assumption that equal volumes of gases at the same temperature and pressure contain equal numbers of molecules. I will refer to this as the 'equal numbers hypothesis' rather than Avogadro's hypothesis. Avogadro's proposal involved both the equal numbers hypothesis and assumptions about the division of molecules into smaller parts. As Frické (1976, p. 298) clearly points out, Cannizzaro did not need to make the latter hypotheses because the degree of molecularity of molecules emerged as one of the consequences of his method.

Cannizzaro's utilization of the equal numbers hypothesis can be summarised as follows. Assuming that hypothesis, molecular weights can be compared simply by comparing the densities of substances in the gaseous state at the same temperature and pressure. So, with hydrogen taken as the reference substance, molecular weights relative to hydrogen can be estimated by comparing vapour densities with the density of hydrogen. The next step employs the relative weights of the elements in a compound as revealed by chemical analysis. This data can be used to split the relative weight of a molecule into the relative weights of the elements it contains. The atomic weight of an element is now determined by surveying the various instances of compounds containing that element and selecting the least value of the weight of that element in the molecules of those compounds. Knowledge of atomic weights enables the elemental constitution of compounds revealed by analysis to be converted into formulae.

This much leaves unsolved what was a major problem with atomic weights in inorganic chemistry. I refer to the atomic weights of metals. Their atomic weights could not be determined straightforwardly because they could not be vaporised in the laboratory. However, many of their chlorides could be vaporised, and the weight of metal in a molecule of chloride is either its atomic weight or a simple multiple of it. In the case of mercury, Cannizzaro refers to formulae for organic compounds containing it for further evidence concerning its atomic weight. The status and significance of appeals to organic formulae will be discussed below. As far as the majority of metals are concerned, as well as for further evidence concerning the atomic weight of mercury, Cannizzaro (1961, p. 22) turned to 'the law of the specific heats of elements and of compounds'.

Cannizzaro utilised the law of Dulong and Petit and its extension by Gaudin. He correlated the atomic weight measurements acquired through vapour density and specific heat measurements in inter-related ways. First, he was able to verify the laws of Dulong and Petit for those elements whose vapour densities had been measured, and of Gaudin for those compounds containing only elements whose vapour densities have been measured. As for the metals, we have noted that Cannizzaro was able to identify either the atomic weights of metals or simple multiples of them from the vapour density of their chlorides. It was then just a question of selecting from amongst the possibilities the value that satisfies the Law of Dulong and Petit for the measured specific heat of the metal. Cannizzaro could then double check this using Gaudin's law for compounds since he had established the atomic weights for each of the elements in the halides of the metals and the measurable specific heats of those halides. The way in which the verification of these mutual relationships give some support to the hypotheses employed (the equal numbers hypothesis and those of Dulong and Petit and Gaudin) as well as to the atomic weights proposed by Cannizzaro has been stressed by Clark Glymour (1980, 245–263) as an example of what he calls 'bootstrapping'.[6] I will have more to say about Glymour's analysis in the next section.

While it must be acknowledged that the overlap between atomic weight estimates based on vapour density and specific heat measurements gave some support to those atomic weights, and to the hypotheses on which they were based, the degree of that support and its significance for chemistry should not be overestimated. The identification of physical molecules, the particles of a gas, and chemical molecules assumed by Cannizzaro, was not entirely straightforward. The 'anomolous' vapour densities of elements such as sulphur, antinomy, phosphorous and arsenic and of compounds such as ammonium chloride were sufficient to show this. Thermal dissociation had already been suggested as a cause of these discrepancies, but the matter was far from settled in 1858. William Odling, in a Dictionary entry on atomic weights written in 1859 (Watts, 1879, p. 467), was clearly of the opinion that such issues posed problems for Cannizzaro. 'In the present state of knowledge', he wrote, 'it seems to us preferable to deduce the chemical atom or molecule of a body chiefly from chemical considerations, and to wait for further investigation to clear up the few anomalies which at present exist between the results of chemical and physical inquiry'.[7] The anomalous vapour densities were not mentioned by Cannizzaro in the *Sketch*. As we shall see in the next chapter, it was thermodynamics, not atomic theory, that led the way to an understanding of thermal dissociation and the removal of the anomalies.

Cannizzaro's arguments, then, were not entirely conclusive. However, the main points I wish to make go beyond that observation. They are associated with the achievements of organic chemistry that we have surveyed in Chapter 9. Cannizzaro's methods yield consistent atomic weights and formulas, but are they the right ones? Yes they are, because they agree with the ones already arrived at by organic chemists. That is my first point. The second one is that Cannizzaro owed a debt to the work of the organic chemists that was not adequately signalled. Thirdly, and perhaps most importantly, insofar as Cannizzaro's method yields formula, they are not the structured formulae that chemistry was in need of. Cannizzaro's method yields $C_2H_4O_2$ for acetic acid, not one that is able to distinguish between the differing ways of substituting hydrogen that we have discussed.

When Cannizzaro is discussing access to molecular and atomic weights via vaporizable chlorides he notes that, in the case of mercury, the vapour densities of its two chlorides suggest an atomic weight of 200, but he also notes that to verify this 'it would be necessary to compare the various quantities of mercury contained in all the molecules of its compounds whose weights and composition are known with certainty' (1961, pp. 21–22). He goes on to observe that 'there are some [compounds of mercury] in organic chemistry the formulae of which express well the molecular composition' and notes that the relative weight of mercury in a molecule of those organic compounds is 200. I presume that the mercury compounds Cannizzaro had in mind are mercury ethyl and mercury fulminate. Mercury fulminate, a white powder formed when mercuric nitrate is mixed with alcohol and nitric acid, had been known for some time. However, mercury ethyl was prepared only in 1850, and the path to its discovery by Frankland (1850) was via a debate about formulae, in particular the appropriateness of A. W. Kolbe's characterisation of alcohol as the ethyl radical plus OH. My point is that the ability of Cannizzaro to refer to this substance was dependant on moves to develop formulae in organic chemistry by a route quite different to that proposed by Cannizzaro.[8] Later in the *Sketch* Cannizzaro (1961, p. 51) uses the vapour density of zinc ethyl, another discovery arising from the quest by Frankland (1849, p. 265) for adequate chemical formulae, to help fix the formulae of organic compounds containing zinc.

The final third of the *Sketch* involves a discussion of formulae for organic compounds. Cannizzaro (1961, p. 40) makes the point that here one has the advantage that many of them are volatile, enabling molecular weights to be fixed via vapour density measurements in a way that is not possible for the majority of metallic compounds. 'This is the great benefit which the study of organic chemistry has rendered to chemistry in general'. In developing formulae for organic compounds Cannizzaro develops the notion of 'saturation capacity' or 'atomicity' of elements and radicals, the notion that corresponds to what has become known as valency, and which was already being employed by organic chemists to fix formulae, as we have seen. Cannizzaro also takes over from organic chemistry the device of writing what I have called structural formulae, rather than raw formulae, so as better to capture the chemical behaviour of organic compounds. To quote Cannizzaro himself:

> Sometimes I repeat in the same formula more than once the same symbol to indicate some difference between one part and another of the same element. Thus I write acetic acid $C_2H_3HO_2$, to indicate that one of the four atoms of hydrogen contained in the molecule is in a state different from the other three, it alone being replaceable by metals. Occasionally I write the same symbol several times to indicate several atoms of the same element, *only to place better in relief what occurs in some reactions.*

I have emphasized the last clause, which clearly shows that Cannizzaro is drawing on resources from organic chemistry that go beyond his considerations concerning vapour densities and specific heats. This is further in evidence when, for instance, Cannizzaro writes formulae such as Zn $(C_2H_5)_2$ for zinc ethyl. Cannizzaro(1961, p. 51) is not justified in saying that 'vapour densities demonstrate the accuracy of [such formulae]'. Such measurements justify raw formulae only. There is further evidence of Cannizzaro's debt to the chemists when he uses his means of constructing formulae to predict the asymmetric zinc ethylmethyl, which he represents with the structural formula Zn $C_2H_5CH_3$ (Cannizzaro, 1961, p. 52), an application of the device introduced by Williamson in his treatment of asymmetric ethers, as we have seen. It is worth recalling here that Williamson had produced powerful chemical arguments for writing C_2H_5OH and $(C_2H_5)_2O$ as the formulae for alcohol and ether respectively, formulae which were subsequently shown by Cannizzaro to be consistent with atomic weights derived from the vapour densities of alcohol and ether. There are two points to be stressed here. Firstly, Williamson's path to these formulae relied on chemical data and not measurements of vapour densities, although Cannizzaro was able to show that they were consistent with those measurements. Secondly, the formulae, insofar as they exhibit structure, do not follow from vapour density measurements alone.

I emphasise Cannizzaro's debt to developments in organic chemistry, not so much to downplay his achievement, but to better understand the structure and significance of his argument. Cannizzaro demonstrates that the atomic weights and formulae arrived at by way of vapour densities are consistent with, not only those derived from specific heat measurements, but also those arrived at independently by chemists in their endeavours to made sense of the chemical behaviour of organic compounds. Looked at in this way, the fact that the equal numbers hypothesis leads to the same values for atomic weights derived from specific heats and chemical formulae constitutes *evidence for the equal numbers hypothesis.*

There are two places in the *Sketch* where Cannizzaro himself describes the situation in these terms. The first place is in fact the opening of his paper. 'I believe that the progress of science made in these last years has confirmed the hypothesis of Avogadro, of Ampère, and of Dumas on the similar constitution of substances in the gaseous state; that is, that equal volumes of these substances, whether simple or compound, contain an equal number of molecules – '. Cannizzaro (1961, p. 5) makes the more specific claim that the work of the organic chemists serves to confirm the equal numbers hypothesis when he writes that 'the discoveries made by Gerhardt, Williamson, Hofmann, Wurtz, Berthelot, Frankland and others, on the constitution of organic compounds confirm the hypothesis of Avogadro and Ampère'. The fact that there was mutual support between the consequences of the equal numbers hypothesis, the laws of Dulong and Petit, and of Gaudin concerning specific heats and the novel views about structure and valency in chemistry meant that all of these assumptions were empirically supported to some degree. That is, the bootstrapping considerations suggested by Glymour (1980) can, and should, be extended to include the advances in organic chemistry.

It is interesting to note that two organic chemists, who were contributors to the developments in organic chemistry involved here, who in 1869 wrote histories of those developments, both saw formulae and atomic weights as being fixed by chemical considerations. A. Ladenburg explicitly claimed that it was organic chemistry that led to the fixing of formulae and atomic weights, and his detailed description of how that came about makes no mention of Cannizzaro. When Ladenburg (1900, pp. 298–304) came to write a second edition of his book in 1887 he added a chapter to cover developments that had taken place subsequent to his first edition. There Cannizzaro is mentioned mainly in the context of the problem of anomalous vapour densities and how they were eventually overcome, and not in a way that requires a qualification of Ladenburg's claims that atomic weights and formulae were settled by organic chemistry. Charles Adolphe Wurtz was the other chemist in question. In his account of the history, Wurtz (1869) referred to an implication of Cannizzaro's *Sketch*, namely, that many metals are bivalent, but noted that this conclusion, based on physical arguments in the *Sketch*, can also be established by chemical arguments. He went on to show how problems posed for the determination of molecular weights from vapour densities can be solved by appealing to investigations of thermal dissociation that took place after the writing of the *Sketch*. There is no suggestion that Cannizzaro's intervention was particularly decisive for chemists.

10.6 Was the Determination of Atomic Weights Important?

I suggest that singling out the problem of establishing definitive atomic weights and formulae as a central one faced by chemists in the first half of the nineteenth century is a misrepresentation of the historical situation that to some extent disguises the nature of the work that needed to be done to make experimental access to atoms possible. To make my point I take as my foil the, in many ways exemplary, discussion of the determination of atomic weights offered by Clark Glymour (1980, pp. 226–263).

Glymour presents a view on the confirmation of hypotheses in science that overlaps considerably with the position I have adopted in this book. He recognises that the testing of a hypothesis typically involves invoking other hypotheses, and demands that those other hypotheses be themselves testable. He uses the term 'bootstrapping' to refer to the advance of science as involving the stringent testing of a maze of inter-related hypotheses against the evidence, as opposed to accommodating them to the evidence. His case study involving atomic weight determination in the nineteenth century is intended as an illustration of his position. Glymour (1980, p. 229) characterises the situation in the following way.

> Every competent eighteenth-century physicist knew how to determine the masses of a great many things, but even if he believed there to be atoms, he did not know how to determine their masses, or how to count them. When the fundamental hypotheses of a theory cannot be tested but, for whatever reasons, the theory is appealing anyway, a reasonable requirement is that the properties the theory ascribes be determinable, and, if other hypotheses must be used in making such determinations, that these hypotheses be testable with respect to the theory in question. The scientific controversy from Dalton to the twentieth century over the atomic theory was about a great many things, but I believe a great deal of it can and should be understood to have been about whether or not the atomic theory could meet the requirement just stated. So long as that requirement remained unfulfilled, many responsible scientists found the hypothetico-deductive argument for the theory unconvincing; when, early in this century, the requirement had at last been met, even some of its most hostile critics reconciled themselves to the atomic theory.

Glymour proceeds to outline the way in which atomic weight determinations via vapour density and specific heat measurements were gradually made compatible, culminating in the case made by Cannizzaro in the *Sketch*. My quibble with this is the omission of significant reference to the formulae developed in organic chemistry Inclusion of this part of the story could have enhanced Glymour's general point concerning the mutual support of testable hypotheses.[9]

However, the mere omission of adequate reference to organic chemistry is not my main worry, which is more concerned with the way in which atomic weight determination is singled out as of key importance for understanding the epistemology of nineteenth-century chemistry. I suggest that there are grounds for arguing that it was not of central importance and that practising chemists recognised that. They were able to advance chemistry in spectacular ways by devising hypotheses that could more fruitfully be tested than those involving relative atomic weights. They did so by adopting a stance in line with what I have called chemical atomism or agnostic anti-atomism, rather than the physical atomism that is, in effect, Glymour's focus.

Glymour points out, rightly, that nineteenth century atomism presumed that atoms have weight and argues that it is reasonable to demand of those atomists that they provide means of establishing that weight. But it is also clear that atoms need other properties too. In chemistry they need properties responsible for chemical combination. What of the demand that these be determined too? How could claims about properties of atoms responsible for chemical behaviour be put in testable form? Could chemistry in the first half of the nineteenth-century best proceed by attempting atomic weight determinations along the lines described by Glymour, by persisting with the search for laws of chemical affinity, or in some other way? My discussion of the previous chapter suggests that the answer lies in the 'other way' that organic chemists were able to devise.

The issue, as I have described it, was a question of which kinds of hypotheses could be fruitfully pursued given the resources of the time. The answer implicit in Glymour's analysis is that it was the pursuit of atomic weight determinations that held the key to progress. My unease with this is not simply the point that such determinations were beyond the resources of chemists in the first half of the century, My point is more to do with why atomic weight determinations were important anyway. Atomic weight determinations combined with chemical equivalents fix formulae for compounds, at least empirical or raw formulae. One role of formulae is that they express, in a convenient way, facts about combining weights. But those facts could be expressed in other ways without necessitating formulae or atomic weights. More important is the way in which formulae were used in organic chemistry to reflect chemical properties, by introducing some structure into the formulae in the ways I have summarised in the previous chapter. An interesting feature of that endeavour is the extent to which progress was made using what were, from a modern point of view, the wrong formulae For example, in the papers referred to above Frankland writes C_4H_5Zn for zinc ethyl and HO for water, assuming 'atomic weights' of 6 for carbon and 8 for oxygen. Much progress in classifying organic compounds and predicting new ones could be and was accomplished in the decades following about 1830 using the 'wrong' formulae, since in many cases replacing them with the right ones makes no difference to the chemical argument. Where the choice of formulae did make a chemical difference, then chemists were able to exploit those differences to argue for the preferable formulae, as we have exemplified in the case of Williamson's work with mixed ethers.

The difference between Glymour and myself on the question of what kinds of hypotheses were testable and hence fruitful in the nineteenth-century chemistry is highlighted by a comparison Glymour makes between Gaudin and Dumas. According to his reading, Gaudin is to be praised for utilising theory and hypotheses and Dumas to be condemned for avoiding them. Glymour (1980, p. 254–255) cites what he describes as 'wonderful words' from an 1833 article of Gaudin's on the measurement if the specific heats of gases.

> For it is not facts which we are lacking; on the contrary, there are swarms of them, and they require no more in order to bring forth all of their fruits than a theory which will tie them together.

He contrasts this with Dumas' assessment of the prospects of arriving at atomic weights via vapour densities in 1837.

> Everyone will allow that one can acknowledge, if one wishes, equal numbers of molecular or atomic groups in equal volumes of gas; but up to the present such assumptions gave nothing useful to anyone. It will only be an hypothesis, after all, and on this subject there are already too many of those.

Compared with Gaudin's wonderful words, those of Dumas are portrayed by Glymour as those of 'a man who did not care about, nor much like, theorizing, and thought it a regrettable part of chemistry'. But it is quite wrong to characterise

Dumas' chemistry as untheoretical and free of hypotheses just because he was wary of commitments to some form of physical atomism. By the time he wrote the above, Dumas was pioneering the attempt to represent chemical properties by appropriate arrangements of terms in formulae, had introduced the notion of substitution and was freely using the device of tracing complex parallel reactions using equations. We have described the fruits of that endeavour. By contrast, Gaudin's attempt to deploy the equal numbers hypothesis involved his geometrical atomism, with point atoms at the vertices of regular solids, a programme that made little contact with experiments in chemistry and led nowhere. Dumas and other nineteenth-century organic chemists were able to formulate testable hypotheses by means of which they were able to bootstrap their way upwards. In doing so they opened a path to testable versions of atomism that had not existed prior to their innovations.

I conclude by pointing out that even if we focus on the eventual concurrence of the various ways of establishing relative atomic weights, it was possible for agnostic anti-atomists (such as Pierre Duhem and Wilhelm Ostwald) to avail themselves of those weights without committing to atomism. We have already seen in the previous chapter that 'atomic weights' can be understood in terms of combining portions rather than combining atoms. On that interpretation the specific heats law states an empirical relationship between those atomic weights and specific heats, whilst Avogadro's hypothesis can be interpreted as claiming that equal volumes of gases contain equal numbers of portions (whatever volume or weight of hydrogen is taken as the reference point for those portions.). This situation was to change, of course. I discuss how in the next two chapters.

Notes

1. See Avogadro (1923) for the introduction of the equal numbers hypothesis.
2. By the time Ampère reasserted his commitment to the hypothesis in 1835 he used the terms atom and molecule in something closer to the modern sense, but retained his geometrical account of the forms of molecules. See S. H. Mauskopf (1969, p. 65).
3. See Rocke (1984, pp. 107–109).
4. On this see Cole (1975, p. 346).
5. The first German edition of Ladenburg (1900), in which the quoted passage appears, was published in 1869. It was updated in the second German edition of 1887.
6. Glymour credits Joule with the extension of the law of Dulong and Petit to molecules in 1844. However, Gaudin made this extension over a decade before Joule. It is also the case that Gaudin is cited in Cannizzaro's *Sketch* as one of his sources, whereas there is no mention of Joule. Cole (1975, pp. 345–348) gives details of Gaudin's extension of Dulong and Petit's law to compounds.
7. Footnotes to the Dictionary entry on atomic weights indicate that it was written in 1859 and that its author (William Odling) subsequently altered his opinion on a number of basic issues.
8. Incidentally, having noted the scarcity of mercury compounds of known composition, Cannizzaro turns to specific heats to supply further evidence for an atomic weight for mercury of 200, overlooking the fact that he could have employed the vapour density of mercury ethyl to bolster his case.
9. Glymour (1980, p. 263) follows his discussion of atomic weight determinations via vapour densities and specific heats with the remark that the remainder of Cannizzaro's paper 'is de-

voted to solving outstanding questions of the day regarding the molecular formulas of various compounds, especially organic compounds'. I hope it is clear why I regard that remark as totally inadequate. I should stress that the work by Rocke, Klein and others that have emphasised the importance of formulae determination in organic chemistry, which I have heavily drawn on in my own analysis, post-dates Glymour's book.

Chapter 11
Thermodynamics and the Kinetic Theory

Abstract From 1859 the kinetic theory gained considerable support from experiment, yielding a range of known phenomena such as the gas laws and predicting new phenomena such as the independence of the viscosity of a gas from its density. Alongside these developments was the rise of thermodynamics, which explained a range of phenomena without any assumptions about the underlying structure of matter and which also received considerable experimental support. Thermodynamics yielded two results, an account of thermal dissociation and a measure of chemical affinities, in areas that had troubled atomists. Two basic problems faced the kinetic theory, its clash with measurements of the specific heats of gases and the problem posed by irreversible processes implied by the second law of thermodynamics. The latter problem was solved by appeal to statistical fluctuations, so that the inverse of apparently irreversible processes became unlikely rather than impossible. There was no independent evidence in support of this move in the nineteenth century.

11.1 Introduction

I have argued that nineteenth-century chemistry made less experimental contact with atoms than is typically supposed. It is time to turn our attention away from chemistry to nineteenth-century physics. Two approaches to the study of heat made considerable progress in the second half of the nineteenth century, but they were very different in kind and the relationship between them was problematic. One approach became consolidated into what has become known as phenomenological thermodynamics. The term 'phenomenological' is designed to capture the extent to which the theory dealt with measurable entities such as temperature and quantity of heat in a way that was independent of any theory about the underlying structure of matter. The approach involved in the kinetic theory of gases, by contrast, was just about the opposite, insofar as that theory was based on quite specific assumptions about molecular motions and collisions.

As we shall see, both thermodynamics and the kinetic theory were borne out by a range of experiments. The former theory was able to make progress in two areas that had posed problems for the atomic programme, namely, the measurement of chemical affinities and understanding the 'anomalous' vapour densities of some

A. Chalmers, *The Scientist's Atom and the Philosopher's Stone*, Boston Studies in the Philosophy of Science 279, DOI 10.1007/978-90-481-2362-9_11,

gases. The kinetic theory also had its early successes, but there were also some very basic problems, recognised as such by the proponents of the theory, that it took twentieth-century developments to solve. Both theories had ramifications for chemistry, chemical thermodynamics making possible the study of chemical affinity, speeds of reaction and thermal dissociation and the kinetic theory gave evidence for the molecules that atomic chemistry needed and yielded Avogadro's hypothesis and the di-atomicity of common gases such as oxygen, hydrogen and nitrogen.

Some nineteenth-century scientists, such as Rudolph Clausius, Clerk Maxwell and William Gibbs, worked with and contributed to both theories, recognising and struggling with the tensions between them. Others, such as Pierre Duhem, Wilhelm Ostwald and a young Max Planck, took phenomenological thermodynamics to be a paradigm of good science and ruled out, or were suspicious of, theories such as the kinetic theory that invoked atoms or molecules lying behind the measurable phenomena. Those that took a stand against the atomic theories of the time took strength from their perception that in the last decades of the nineteenth century the kinetic theory was not making headway with the problems that beset it and was proving unable to emulate its early successes. The basic problems faced by the kinetic theory were overcome early in the twentieth century as we shall see in the next chapter. By that time it was possible to make experimental contact with atoms, and the achievement marks the end of the story as far as my book is concerned.

11.2 The Rise of Thermodynamics

Those who formulated and contributed to the spectacular success of thermodynamics in the second half of the nineteenth century, from Clausius and Maxwell through to Gibbs and then Planck, quite explicitly saw themselves as developing a theory that abstracted from and was independent of any views about the underlying structure of matter or heat. Maxwell (1965, Vol. 2, pp. 664–665) for instance, defined thermodynamics as 'the investigation of the dynamical and thermal properties of bodies, deduced entirely from what are called the First and Second laws of Thermodynamics, without any hypotheses as to the molecular constitution of bodies'. These formulations drew on a century of work on heat which had involved various hypotheses, involving such things as caloric or atomic motions, but consciously abstracted from those earlier theories. The aim was to construct a general theory that could learn from the success of earlier theories but jettison from those earlier theories hypotheses that were unsubstantiated and unnecessary. The resulting theory, based on the conservation of energy (the first law) and the increase of entropy (the second law), was remarkably successful and led to a range of novel discoveries.

Joseph Black, in the second half of the eighteenth century, formulated the notions of specific heat and latent heat in the context of a caloric theory. Specific heat is a property of a substance measuring the amount of heat necessary to raise the temperature of unit mass of a substance one degree. Latent heat is a measure of the amount of heat required to change the state of unit mass of a substance without change of temperature. The heat that enters a liquid to vaporise it is latent (hidden) in the sense that it is absorbed without a corresponding rise in temperature. This was commonly understood in terms of a chemical combination of the heated substance with heat (caloric). Lavoisier, a proponent of the caloric theory, listed caloric as one of his chemical elements. Black's innovations, and the clear distinction between amount of heat and degree of temperature that they involve, became the basis of calorimetry. Changes that involve a mere redistribution of heat in an isolated system are such that the amount of heat in conserved. If a known mass of a hot liquid with a known specific heat and specified temperature is added to a given mass of water with a known specific heat and at a specified temperature, then the temperature of the resulting mixture can be calculated from the fact that the heat lost by the liquid must equal the heat gained by the water.

Early defenders of the caloric theory, such as Laplace and, as we have already had occasion to mention, Dalton, specified much detail. They assumed that particles of a gas are surrounded by atmospheres of caloric and attributed the pressure of a gas to the repulsive effects of the caloric. It was this repulsive effect that accounted for the expansion of a gas when it is heated, that is, when more caloric is added. It is not difficult to see how calorimetry and other common heat phenomena can be captured with the idea that heat is a something that is conserved in a wide range of changes and which flows from a hot to a cold body. Joseph Fourier developed an account of heat conduction that abstracted from the caloric theory in this way. Sadi Carnot also went some way towards abstracting from details of the caloric theory in his theoretical treatment of heat engines that was a major step towards the formulation of thermodynamics.

Carnot understood the performance of work by a heat engine as resulting from the falling of heat from a high to a low temperature. He imagined the quantity of heat (caloric) to be conserved in the process. He introduced the device of considering infinitesimal cyclic processes which involve the returning of a system to its original state, in which case any internal changes in the working substance can be ignored. He considered an ideal, reversible sequence of infinitesimal changes involving two isothermal and two adiabatic transformations (transformations taking place at constant temperature and without heat loss or gain respectively) that brought the working substance back to its original state. Carnot showed that the work performed in the execution of the cycle represented a maximum that could not be exceeded by any real, and hence irreversible, change. Otherwise, an indefinite amount of work could be extracted from a cyclic process by having the ideal engine drive a real one around the cycle. Carnot's considerations also led to a straightforwardly testable prediction, that the difference between the specific heat at constant pressure and the specific heat at constant volume of a gas be a constant.

Carnot's results were published in 1824. They were viewed in a somewhat different light a quarter of a century later, because, by that time, the law of conservation of energy had been formulated and generally accepted. There was a tension between Carnot's analysis and the conservation of energy. An understanding of work as the result of a mere transfer of heat from a high to a low temperature suggested to Rudolph Clausius that there would be a steady loss of working power in nature, which he regarded as implausible. In the 1850s he transformed Carnot's theory by dropping the idea that heat is conserved and recognising the conversion of heat into work (mechanical energy) for which there was mounting experimental evidence, exemplified, for example, in the work of James Joule. Clausius's reworking of Carnot's analysis of his cyclic changes led to a version of the second law of thermodynamics, 'heat can never pass from a colder to a warmer body without some other change connected therewith occurring at the same time'[1].

The 'mechanical theory of heat' as formulated by Clausius was transformed into what became known as 'phenomenological thermodynamics' by casting principles of conservation of energy and increase of entropy in a general form free of any assumptions about hidden structures of the systems possessing energy and entropy. Gibbs and Planck were the major contributors to that move. The former's work, in particular, made possible the extension of thermodynamic considerations to chemistry.

Thermodynamics was confirmed in a way that conforms to the view of confirmation adopted in this book. A variety of experimentally testable predictions were made in an uncontrived way and vindicated. However, there is a qualification to be made about the extent to which thermodynamics led to natural consequences. Because of the very general form of the laws of thermodynamics those laws alone had scarcely any testable content. Predictions were extracted from them only by adding known empirical laws. This can be seen in the case of two early successes of thermodynamics, the prediction that the difference between the two specific heats of a gas is a constant, and the prediction that the freezing point of water decreases with pressure. Derivation of the first requires the addition of the gas laws to the laws of thermodynamics, whilst thermodynamic reasoning yields the second given the empirical observation that ice is less dense than water at the same temperature. Clark (1976) has pointed out that what he calls the heuristic of thermodynamics was weak. It yielded predictions only by way of the empirical regularities fed into it. Nevertheless, the regularities predicted in this way were natural consequences of thermodynamics plus the independently vindicated empirical knowledge fed in, so there is a strong sense in which the phenomenological theory was confirmed.

11.3 Thermal Dissociation and Affinities

Experiments on a number if gases reveal that they do not obey the gas laws in a straightforward way. Nitrogen peroxide is an example. As was suspected from the early 1860's, the 'anomalous' behaviour is due to the fact that the gases exist in more than one form, the relative proportions of each form depending on the temperature. From 1875 Gibbs treated this problem from the point of view of thermodynamics. He assumed that each component separately obeys the gas laws, used the equations of the component substances to deduce their relative densities, (so that NO_2 is half as dense as N_2O_4) and found the temperature-dependence of the relative proportions of each gas in the mixture from the fact that the equilibrium condition will correspond to maximum entropy. The argument was entirely phenomenological. (We have already seen that the use of formulae need not involve a commitment to atomism.)

Gibbs' predictions were confirmed by experiments by Henri Deville and his student L. J. Troost. In this way, a problem that stood in the way of the determination of relative atomic weights via vapour densities was removed. Clark (1976, pp. 71–72) notes that Boltzmann attempted to derive the equilibrium conditions within the kinetic theory, and succeeded in 1896, but only in an ad hoc way by adapting his assumptions to the known results.

Chemical affinity is a measure of the facility with which chemical substances combine. A theory of affinities should be able to determine the rate and direction of a chemical reaction. Many spontaneous chemical reactions involve the release of heat. A natural assumption informing early thermodynamic treatments of affinity was that chemical reactions take place spontaneously in a way that involves a maximum release of heat. This is not borne out empirically. There are spontaneous chemical reactions that result in cooling and there are reactions involving the release of heat that do not take place spontaneously. Helmholtz, in 1882, followed by van't Hoff and Duhem, developed a theory of affinity and rates of reaction that took account of entropy as well as energy changes. As with the case of thermal dissociation, the arguments were purely thermodynamical and they were borne out by experiment.

In the second half of the nineteenth century, then, thermodynamics was making impressive progress, some of it in territory that had proved problematic for atomists. It is time to switch our attention to developments in the kinetic theory that were taking place concurrently. Did atomism in the form of the kinetic theory lead to experimentally-confirmed knowledge in an especially significant way?

11.4 Early Versions of the Kinetic Theory

An early version of the kinetic theory of gases was proposed by Daniel Bernoulli in 1738. His description of the theory occupied just 5 pages in his book on hydrodynamics, published in Latin.[2] The basic idea that gas pressure is due to the elastic impact of gas particles moving freely and bodily through the volume of the gas was present and Bernoulli in effect showed that his basic assumptions could explain the gas laws. Indeed, insofar as the theory predicted the dependence of volume and pressure on temperature, Bernoulli's theory actually *predicted* the gas laws.

Bernoulli noted that the number of particles striking a piston bounding the surface of a gas, and hence the gas pressure, will be proportional to the gas density. That is, the theory explains Boyle's law. Bernoulli further reasoned that the contribution to pressure of the impact of a particle would be proportional to the mass of the particle and to the square of its velocity. Velocity affects the pressure in two ways. The greater the velocity the greater the impact and the greater the velocity the faster the particle will traverse the gas volume and so the more frequently will it strike the piston. Hence we have the dependence of pressure on the square of the velocity. The further assumption, that temperature is related to the square of the velocity, yields the proportionality of both pressure and volume to temperature.

There are a number of reasons why Bernoulli's version of the kinetic theory cannot be considered to have been well-confirmed at the time. Thermometry was in its infancy in the early decades of the eighteenth century, and it was to be some time before the laws relating the volume and pressure of a fixed quantity of gas to its temperature were to be regarded as well confirmed experimental laws. If Bernoulli's theory is accepted then the persistence of the state of a well insulated gas implies that the mean velocity of the particles persists undiminished. This in turn implies that their collisions are perfectly elastic and that they experience no resistance to their motion through the body of the gas. These are strong assumptions far from obviously true, and an atomist who favoured a Newtonian model involving a static array of atoms repelling each other with short range forces inversely proportional to their separation and perhaps vibrating about equilibrium positions, must have seemed more plausible at the time. While the basic idea of the cause of pressure as presented by Bernoulli seems straightforward enough, there are problems in the detail. Not all particles can be moving at the same velocity nor can they be moving always normally to the surface on which they impact because of the results of collisions of particles with each other and with the walls of the containing vessel. The fact that particle motions will yield stable averages and produce a steady pressure obeying the gas law was something that needed to be argued for. Further, there is something ad hoc about the assumption that temperature is related to the square of the velocity. That relation is chosen in order to fit the known phenomena to some extent. Further, precisely which function involving the square of the temperature to choose is under-determined in Bernoulli's theory. From Bernoulli's discussion it is natural to identify temperature T with the average value of v^2. Later it was found necessary to identify it with the mean value of the kinetic energy, $1/2mv^2$.

Versions of the kinetic theory similar to that of Bernoulli appeared in the first half of the nineteenth century, in papers by John Herapath and John Waterston. Like Bernoulli, they were able to show how the theory yielded the gas laws, which by then were well established experimentally. However, the theory presented by Herepath and Waterston were not free of the difficulties I have indicated in connection with Bernoulli's version proposed a century earlier. I suggest there is an analogy between the situation with respect to these early versions of the kinetic theory and Dalton's atomic theory. Just as Dalton's theory could explain the laws of proportion so the kinetic theory could explain the gas laws. But (i) were they explanations that could be vindicated by experiment in the way that had come to be demanded of scientific theories and (ii) could the respective theories be defended in the face of obvious objections (for example, how can two like atoms combine to form a molecule, in the Dalton case, how can atomic motions persist and stable averages emerge in the case of the kinetic theory)?

A paper by Karl Kronig summarising the simple version of the kinetic theory was published in the *Annalen der Physik* in 1856 and attracted much more attention than those of Herepath or Waterston. His version did not progress beyond the earlier ones, but by 1856 the scene had changed in a significant way. By then the conservation of energy, and the appreciation that a quantity of heat is equivalent to a definite amount of mechanical energy, was well appreciated, and this made the identification of heat with the mechanical energy of moving molecules more plausible and attractive than it had been hitherto.

Rudolph Clausius and James Clerk Maxwell were soon to produce versions of the kinetic theory that were significant improvements on the simple version. Clausius recognised that most gas molecules, being complexes of atoms, could not be smooth spheres and so would be set in rotation by collisions. Taking into account the rotational and translational energy, he arrived at a novel empirical prediction of the kinetic theory, namely a value for the ratio of the principal specific heats of a gas, the specific heat at constant pressure and the specific heat at constant volume. (The fact that the prediction did not altogether match the empirical data was to prove one of the chief difficulties for the theory as is discussed below.) He also established a method of estimating the mean velocity of molecular motions, the result being velocities on a par with the velocity of sound. These high speeds posed a difficulty for the theory. If molecules in a gas move so quickly why does it take so long for a gas to diffuse through another? Why does it take a few minutes before the students at the back of a lecture room can smell the ammonia emerging from a freshly opened bottle at the front? Clausius responded by introducing the notion of the mean free path. A molecule progresses slowly in spite of its high velocity in between collisions because of the many changes in direction occasioned by those collisions. From Bernoulli onwards, supporters of the kinetic theory had assumed molecules to be of a volume negligible compared with the volume of gas which were aggregates of them. The mean free path is a function of their size, and it was necessary for Clausius to assume a size consistent with observed diffusion rates. There is a strong sense in which, at this stage, the kinetic theory was accommodated to rather than supported by observable rates of diffusion.

11.5 The Statistical Kinetic Theory

From 1859 onwards Clerk Maxwell set about investigating the scope and merits of the kinetic theory and was soon to considerably improve it. A reflection of the fact that, in 1859, the theory, though promising, was far from confirmed, is evident in Maxwell's early attitude to it. Writing to Stokes about his early work on the theory, inspired by Clausius's efforts, Maxwell made it clear that he regarded the theory as speculative and that it might well be refuted by experiment. He clearly recognised that the early form of the theory 'is wrong' but set about investigating it anyway 'as an exercise in mechanics'. Here are some of his own words:

> I do not know how far such speculations may be found to agree with facts, even if they do not it is well to know that Clausius' (or rather Herepath's) theory is wrong and at any rate as I found myself able and willing to deduce the laws of motion of systems of particles acting on each other only by impact, I have done so as an exercise in mechanics. Now do you think there is any so complete a refutation of this theory of gases as would make it absurd to investigate it further so as to found arguments upon measurements of strictly 'molecular' quantities before we know whether there be any molecules?[3]

Maxwell's doubts about the kinetic theory, and even about the existence of molecules, notwithstanding, he was soon able to greatly improve the status of the theory and invoke more empirical support for it. A key move was the introduction of statistics to deal with the net effect of the motions of systems of moving and colliding molecules. In Maxwell's 1859 version of the theory the velocities of molecules, constantly changing through collisions, were randomly distributed about a mean. Using his statistics and the notion of mean free path Maxwell was able to analyse how local groupings of molecules and their velocities would spread through the body of a gas in equilibrium, and in this way was able to offer explanations of diffusion, viscosity and heat conduction. His first paper on the kinetic theory made possible a striking confirmation of it. The statistical theory yielded the result that the viscosity of a gas is independent of its density, so a pendulum swinging in a region evacuated by an air pump would experience no less resistance than one swinging in atmospheric air at the same temperature! Counter-intuitive or not, the prediction was confirmed by experiment. It is possible to comprehend, in a qualitative way, why resistance to motion in a gas should be independent of its density according to the kinetic theory. An object moving through a gas experiences resistance because its motion needs to set in motion the adjoining gas. The *more* dense the gas, the more molecules there are adjacent to a moving object, and so the greater the resistance to the motion spreading. However, on the kinetic theory the moving molecules are in random motion, and molecules set in motion by a moving object will have a tendency to migrate in a direction transverse to the moving object by virtue of the randomising effect of molecular collisions. This tendency for velocity to be transmitted through a gas will be greater the greater the mean free path, and that in turn is greater the *less* the density. The two factors affecting the spread of velocity, the one increasing and the other decreasing with density, cancel each other out. Incidentally, on the rival theory that construes a gas as a static array of molecules held together by forces, it is difficult to see how viscosity could fail to increase with density. So the experimental test was a crucial one in favour of the kinetic theory.

Early versions of the kinetic theory were idealisations that assumed that the volume occupied by the molecules themselves is vanishingly small compared to that of the gases they form, that the time spent in collision is negligible and that molecules interact only when colliding. These assumptions needed to be relaxed in order to predict the detailed behaviour of real gases. The kinetic theory met with some success in this direction. A notable one was due to the work of J. D. Van der Waals. He took into account intermolecular forces and the finite volume occupied by the molecules themselves to arrive at a modification of the ideal gas law. The modified law took the form

$$(\mathrm{P} + \mathrm{a}/\mathrm{V}^2)(\mathrm{V} - \mathrm{b}) = \mathrm{RT}$$

This equation proved to be in conformity with detailed experiments on carbon dioxide at temperatures near and far from its point of liquefaction, although some other gases did not fit so well, especially under high pressure.

Phenomena predicted by the kinetic theory did not permit the determination of the absolute dimensions, weights and numbers of the molecules that it presupposed. The mean free path of molecules could be calculated, for example from the rate of diffusion of gases or from their viscosity. Various ways were devised to add one other consideration to enable the absolute magnitudes to be determined. Loschmidt added such a relationship by supposing, for example, that the volume occupied by the molecules in a gas equals the volume occupied by that gas when it is liquefied. This yielded a value for NV, the number of molecules in a volume times the volume of each molecule. This relationship in conjunction with the expression for mean free path derived from the kinetic theory enabled the absolute magnitudes to be estimated. It yielded, for example, a value for Avogadro's number, the number of molecules in a gram molecule of gas. The constant b in Van der Waals equation was also related to molecular volume and measuring it gave another way of estimating Avogadro's number. Yet a third way was opened up from a study of thin films, the measurable thickness of which put an upper limit on the size of molecules. The fact that these various methods yielded values for Avogadro's number that were of the same order of magnitude was evidence in favour of the kinetic theory. The values for Avogadro's number were not contrived in the sense that they were consequences of the results of measurements, of the volume of a sample of gas when liquefied, of the thickness of an oil film and of the constant b in Van der Waals equation derived from measurements of the pressure and volume of a gas as a function of temperature. The approximate agreement of the numbers calculated from basic assumptions of the kinetic theory was genuine evidence, if not conclusive evidence, for those assumptions.

Maxwell stressed the central role in the kinetic theory played by the equipartition of energy. The interchange of energy between colliding molecules leads, on average, to an equalisation of kinetic energy. So, if velocities in one direction should happen to exceed those in another direction, collisions would soon ensure that the slower ones gain more energy than the faster ones lose in collisions, thus equalising the average velocity in each direction. Similarly, in a mixture of two gases, collisions will lead to the average kinetic energy of the molecules of each gas becoming equalised. Avogadro's hypothesis follows straightforwardly from this. Our analysis of developments in organic chemistry has shown how support for Avogadro's hypothesis followed from the formulae and molecular weight determinations that those developments made possible. The extent to which the kinetic theory was able to yield that result in an uncontrived way therefore constituted support for the theory. However, the equipartition of energy ran into trouble by clashing with specific heat measurements, as we shall see in the next section.

11.6 Problems with the Kinetic Theory

The basic assumptions of the kinetic theory were borne out through their ability to yield both known laws and novel ones in a natural way. They yielded the gas laws, Avogadro's hypothesis, the independence of the viscosity of a gas and its density and various interconnections between macroscopic properties such as diffusion, viscosity and heat conductivity, where estimates of mean free path by reference to one of these phenomena made possible quantitative predictions in one of the others. There is no doubt that the basic assumptions of the theory received important empirical support. However, as we have noted in our discussion of the contribution of Van der Waals, the basic kinetic theory involved assumptions, such as the small size of molecules and the extremely short range and elastic character of their interactions, that were known to be violated by real gases. Progress of the theory beyond its initial qualitative successes required its development in the direction of being able to cope with the more complicated situations present in real gases.

Some of the moves to deal with the complications met with some success. The contribution of Van der Waals is a case in point. Others were not so clear-cut. Consider, for example, the variation of the viscosity of a gas with temperature. Experiment shows viscosity to be proportional to absolute temperature. The basic theory predicted that the viscosity be proportional to average velocity and hence to the square root of the absolute temperature. Maxwell attempted to remove the difficulty by introducing an effect of temperature in addition to its effect on the mean free path through an increase in velocity. He argued that the effective diameter of a molecule would also be a function of temperature, since the faster a molecule was moving the more effectively it could counter the repulsive forces of a neighbouring molecule and so the closer it could approach it. Maxwell found that by assuming a repulsive force between two molecules to be inversely proportional to the fifth power of their separation the kinetic theory yielded a value for the viscosity that was proportional to temperature, in accordance with experimental results. This should be classified as an accommodation to rather than confirmation by experiment.

A serious problem with the kinetic theory was posed by measurements of the two specific heats of gases. As Maxwell, most notably, appreciated, these measurements threatened to undermine the equipartition of energy, an assumption at the basis of the theory. It was known from chemistry, and indeed, from the implications of Avogadro's hypothesis now derivable from the kinetic theory, that very few molecules consisted of one atom and so had a structure that could not be spherically symmetric. Molecules made up of atoms must be able to rotate and to be subject to internal vibrations. As far as the latter are concerned, atomists were forced to conclude from the line spectra of gases that their molecules most be capable of a range of modes of vibration. Once molecules are capable of rotation and vibration, collisions of two of them cannot in general be perfectly elastic in the sense that the sum of their kinetic energy is equal before and after collision. If, for instance, a colliding molecule encounters a component atom of a molecule with which it collides as that atom moves towards it, because of rotation or vibration internal to the molecule, then it will receive a boost to its kinetic energy at the expense of the rotational or vibrational energy of the molecule it has struck. This was dealt with within the kinetic theory by assuming the setting up of an equilibrium between the three modes of energy, translational, rotational and vibrational, with equal amounts of energy, on average, being possessed by each degree of freedom of a molecule. This in general will consist of three translational modes along three mutually perpendicular directions, three rotational modes about three mutually perpendicular axes, and a range of vibrational modes. On the assumption of equal partition of energy amongst the degrees of freedom the kinetic theory yields a value of $(n+2)/n$

for the ratio of the specific heat at constant pressure to the specific heat at constant volume for a gas whose molecules have n degrees of freedom.

This formula was not in general borne out by measurements of the specific heats of gases. The formula predicts 1.33 for the ratio for diatomic gases if we take into account the three rotational modes in addition to the three translational modes *and ignore the vibrational modes*, conflicting with the value of 1.4 measured by experiment. As John Nyhof (1988) has stressed, things were not quite as bad as this result suggests. Both R. Bosanquet (1877) and Ludwig Botlzmann (1877) independently proposed that if diatomic molecules involved the combination of two perfectly smooth spheres then any rotation that they possessed about an axis joining the two spheres could not be changed by collisions so that this degree of freedom could be ignored in calculating specific heats. A diatomic molecule on this picture has five degrees of freedom, yielding a predicted value of 1.4 for the ratio of specific heats, conforming precisely to the measured value. Experiments on mercury vapour, a monatomic gas, lent support to this point of view. If atoms are perfectly smooth spheres then molecules of a monatomic gas should have no rotational degrees of freedom, leaving just three translational degrees of freedom if vibrations are ignored. This yields 1.67 for the predicted ratio of specific heats, and this was precisely the value for the specific heat of monatomic mercury vapour, measured by Kundt and Warburg (1876).

However promising this resolution of the specific heats problem may have appeared it remained deeply problematic, as Maxwell for one was well aware. The assumption of perfectly smooth atoms had no independent support so there was a degree of ad hocness about it. More serious is the ignoring of degrees of freedom associated with vibration. As we have mentioned, line spectra indicate that atoms and molecules vibrate in specific modes (and mercury vapour is no exception), and so the degrees of freedom involved in these modes should be taken into account in calculations of specific heat. Once this is done, and once it is recognised that molecules must possess a number of vibrational modes, the predicted value for the ratio of the specific heats of all gases approaches unity. Maxwell frequently expressed his awareness of the seriousness of the problem from his 1860 paper onwards. Maxwell (1877, p. 245) put the point in no uncertain words when, discussing the results of specific heat measurements, he wrote:

> Some of these, no doubt, are very satisfactory to us in our present state of opinion about the constitution of bodies, but there are others which are likely to startle us out of our complacency, and perhaps ultimately drive us out of all the hypotheses in which we have hitherto found refuge into that state of thoroughly conscious ignorance which is the prelude to every real advance in knowledge.

We now know that the classical kinetic theory cannot remove the specific heats problem. The reason why vibrational modes cannot be activated at low temperatures is fundamentally quantum mechanical. The quotation from Maxwell can be seen to be quite prophetic in retrospect.

A second central problem for the kinetic theory stems from the irreversibility of most processes in nature. Newtonian systems, provided they do not involve forces such as frictional forces that are themselves asymmetrical under time inversion, are invariant under time reversal. The time inverse of any such Newtonian system is also a Newtonian system. For instance, a system of colliding billiard balls free of frictional forces is time-reversible. A film of such a system played backwards shows motions that would indeed occur if all the moving billiard balls could be turned back on their tracks. Macroscopic phenomena are not in general like that. The inverse of the wave radiating out from the point of entry of a stone in a pond, the diffusion of sugar in one's coffee, the cooling of a hot body are all processes such that their time-inverses run counter to what occurs in nature. In general, it is quite easy to discern that a film is being run backwards. However, if the world consists entirely of molecules executing friction-free motions governed by Newton's laws, then the time-inverse of any happening should also be a possible happening. The time asymmetry in evidence in the world clashes with the time-symmetric kinetic theory.

Proponents of the kinetic theory such as Maxwell and Boltzmann were able to rise to this challenge to a considerable degree. They could appeal to statistics to explain the asymmetry. There are innumerable ways of distributing the molecules in a cubic centimetre of ammonia at atmospheric pressure around a laboratory, and considerably less ways of arranging the molecules so that they are confined to a small region near the demonstration desk at the front. Given the basic assumption that molecular motions are random, the probability of the ammonia staying near the desk once released is negligible. Its diffusion through the room is virtually inevitable. This explanation of irreversibility is not ad hoc. It hinges on the randomness of the molecular motions that formed part of the basic assumptions of the theory from Maxwell's 1860 paper onwards. However, this did not dispose of the problem to the satisfaction of those that were most worried by it. An implication of the statistical explanation of time irreversibility is that that irreversibility is merely an improbability rather than an impossibility. This circumstance clashes with the second law of thermodynamics taken literally. The spontaneous congregation of ammonia dispersed throughout a laboratory into a small region near the demonstration desk, with a resulting increase in entropy, is an improbability rather than an impossibility according to the kinetic theory. The power and success of thermodynamics based on the first and second laws that we have already described provided a strong case for taking those laws as fundamental truths. Explanation of time-irreversability by the kinetic theory clashed head-on with that position. It is important to appreciate, in this connection, that there was no evidence available in the nineteenth century for the statistical nature of apparently time-irreversible processes such as diffusion. This became available only with Jean Perrin's experiments on Brownian motion in 1908.

I have highlighted specific heats and irreversibility as the source of two key difficulties for the kinetic theory. A third problem stems from the mathematical complexity of the statistical theory. Progression from the basic theory, that treated gases in equilibrium whose molecules interacted elastically in negligible time whilst occupying a negligible amount of the volume of the gas, needed to be relaxed in order to deduce predictions that could stand comparison with experiments on real gases. In a wide variety of cases the problems proved intractable. Various approximations were made, yielding conflicting results for such things as diffusion and heat conduction, and with no way of testing the superiority of one set of approximations over another. Clark (1976, pp. 86–88) has characterised the problem in this way and described Boltzmann's ineffective struggles with it, and cites contemporaries of Boltzmann such as Planck who read the problem in just this way. Boltzmann's frustration with these problems was probably a contributing cause to his suicide in 1906.

11.7 The Status of the Kinetic Theory in 1900

In the second half of the nineteenth century the kinetic theory received strong support. It yielded at least approximate predictions of known laws such as the gas laws, successfully explained and exploited departures from them, predicted novel phenomena such as the independence of the viscosity of a gas and its density and gave a natural explanation of many of the time-asymmetries in nature. There was genuine experimental support for the theory that was not contrived. There surely had to be something right about the theory. What is more, the success of the kinetic theory could not be reproduced by bracketing off assumptions about the existence of molecules in the way in which the success of nineteenth-century chemistry could be retained by bracketing off atoms. There is no doubt, then, that the development and testing of the kinetic theory improved the case for atoms and molecules.

However, it must be recognised that the theory had basic problems, problems that had become recalcitrant by the last decade of the nineteenth century. The specific heats problem could be offset only by restricting the application of the equipartition of energy, a principle which was otherwise a central and crucial part of the theory. The claim that the second law of thermodynamics was only statistically true had not been supported by independent evidence and showed no prospects of being so. Partly because of formidable mathematical difficulties, little new evidence for the kinetic theory came to light in the last two decades of the nineteenth-century.

It is not inconceivable that, given the state of play in the late nineteenth-century, molecules might have gone the same way as the aether. From Fresnel to Maxwell and beyond the aether had figured prominently in well-confirmed theories. Most scientists saw a range of phenomena involving diffraction, interference and polarisation of light as establishing that light is a transverse wave in an elastic aether. Maxwell was encouraged to reduce electromagnetism to the mechanics of that aether and succeeded in giving an electromagnetic theory of light that successfully predicted that the ratio of the electromagnetic unit of charge to the electrostatic unit be equal to the velocity of light. The eventual confirmation of Maxwell's theory by Hertz in 1888, with the production of the radio waves that it predicted, was seen by many scientists as establishing the existence of the aether. FitzGerald made the point emphatically when reporting on Hertz's experiment to a meeting of the British Association in Bath in 1888.

> Henceforth I hope no learner will fail to be impressed with the theory – hypothesis no longer – that the electromagnetic actions are due to a medium pervading all space, and that it is the same medium as the one by which light is propagated.[4]

In spite of the evidence in its favour the aether was, in a sense, banished from science in the first decade of the twentieth century. But only in a sense. The electromagnetic field pervading all space has persisted. The aether has been rejected, but only by being replaced by something that can duplicate its role in the successful predictions of the nineteenth century that had supported it. The molecule of the nineteenth-century kinetic theory may not have survived. After all, as we have seen, the theory faced fundamental difficulties of various kinds. But any replacement would have been required to account for the empirical success of that theory, success that was too diverse to have come about by accident.

Much of the literature on attitudes to assumptions about atoms and molecules in the late nineteenth century assumes that avoidance of or refusal to commit to such notions amounted to positivism. So, for instance, Einstein attributed the opposition to atomism of scientists such as Ostwald and Mach to 'their positivistic philosophical views' a position echoed by Stephen Brush.

> Those scientists who did suggest that the kinetic theory be abandoned in the later 19th Century did so not because of empirical difficulties but because of a more deep-seated purely philosophical objection. For those who believed in a positivist methodology any theory based on invisible and undetectable atoms was unacceptable.[5]

There is no doubt that some scientists who were suspicious of or inclined to reject the claims of atomism invoked philosophical positions that are positivist in some reasonably strong sense. Take, for instance, the following remark made by Mach in 1872 in the context of a denial that it has been established that heat can be attributed to the motion of molecules.

> One thing we maintain, and that is, that in the investigation of nature, we have to deal only with knowledge of the connections of appearances with one another. What we represent to ourselves behind the appearances exists *only* in our understanding, and has for us only the value of a *memoria technical* or a formula, whose form, because it is arbitrary and irrelevant, varies very easily with the standpoint of our culture.[6]

Insofar as objections to the kinetic theory appeal to a positivist philosophy they can be rejected along with that philosophy. Appeal to the conservation of energy or the undulatory character of light, endorsed by Mach himself, go beyond the appearances, as do common sense assumptions about tables and chairs. Science in general, not just atomism, clashes with positivism. The clash gives no specific reason to reject atomism and perhaps every reason to reject positivism.

Other philosophical objections to atomism appealed to considerations that are best described as instrumentalist rather than positivist.[7] Pierre Duhem raised objections of this kind. An instrumentalist can endorse theory in science and can recognise that theories involve conceptions, such as energy and mass, that abstract from and go beyond the deliverances of the senses. However, an advocate of this position, like Duhem, insists that scientific theories are adequate to the extent that they adequately order phenomena. There may be a reality behind the phenomena accessible to observation and measurement, but science is not capable of revealing it. Duhem denied that science can 'strip reality of appearances covering it like a veil, in order to see the bare reality itself'. A physical theory, according to Duhem (1962, p. 19), is not an explanation of the phenomena. Rather, it is 'a system of mathematical propositions which aim to represent as simply, as completely, and as exactly as possible a set of experimental laws'.

As with positivism, it can be argued that this philosophical position clashes with science. Indeed, the general claim that science is not explanatory clashes with Duhem's own science. Duhem was a pioneer of the extension of thermodynamics into chemistry and, as we have seen, one of the merits of thermodynamics was that it could *explain* such phenomena as anomalous vapour densities where atomism had failed.

Objections to the kinetic theory based on positivist or instrumentalist philosophies stand or fall with those philosophies. A scientist impressed with the progress of the kinetic theory might well take that success as one reason among others for rejecting positivism or instrumentalism. If a philosophy cannot accommodate the kinetic theory so much the worse for that philosophy.

As we have seen, then, objections to atomism by the likes of Mach, Ostwald, Duhem and many lesser figures were sometimes couched in general philosophical terms and can be rejected along with the philosophies they presuppose. However, I think it would be a mistake to leave matters there. There were a range of other kinds of objections to atomism in the late nineteenth century that were more closely tied to science itself and not to some positivist or instrumentalist philosophy. Many of them were explicitly invoked by critics of atomism like Mach and Duhem. In the following I extract some of the lines of argument that had something going for them.

All three of the critics of atomism mentioned in the previous paragraph frequently linked their opposition to atomism to an opposition to mechanism. There are aspects of this criticism that cannot be readily dismissed along with positivism. From the mechanical philosophers of the seventeenth century through to many practising scientists of the nineteenth century there was a widespread assumption that the material world is in some strong and narrow sense mechanical. One version of mechanism identifies the mechanisms assumed to underlie phenomena as atomic mechanisms, either systems of atoms with shape, size and motion like Boyle's or as centres of Newtonian force like Newton's or those of Boscovich. After the introduction of the conservation of energy, a weaker version of mechanism assumed all energy to be mechanical energy, without necessarily assuming that energy be borne by discrete moving atoms. (Maxwell, for example, who was a mechanist in the weaker sense, assumed electromagnetic energy to be located in the motions and strains of a continuous aether.) Mach, Duhem and Ostwald all objected to mechanism in this sense, arguing that it went beyond what was implied by or necessary for science.

From a scientific point of view, the possibility of reducing a branch of science to mechanics is to be supported by achieving such a reduction rather than by philosophical decree. Not only had this not been accomplished in several areas of science but it could be argued that things were moving in a contrary direction. For instance, mechanists assumed light to involve transverse vibrations in an aether. After Maxwell, light became understood as fluctuating electro-magnetic fields. So the first major substantiated reduction in physics was the reduction of optics to electromagnetism. It wasn't a mechanical reduction at all! It is true that Maxwell and his followers believed electric and magnetic fields to be the manifestations of motions and strains in a mechanical aether. But as opponents of mechanism such as Duhem and Ostwald were able to point out, the mechanical models involved were contrived and became increasingly implausible and unsubstantiated as the nineteenth century progressed. It can be argued that not only did progress in electromagnetism move away from mechanical reduction but also that that progress owed nothing to mechanism.[8]

Chemistry was another area providing fertile ground for opposition to mechanism. We have seen in detail that the success of nineteenth-century chemistry could be made sense of via an interpretation of chemical formulae that dispenses with atoms, a point made in detail by Duhem around the turn of the century. Even if the hypothesis of chemical atoms is entertained, as most chemists of the late nineteenth century admittedly did, those atoms could not be mechanical atoms as traditionally conceived. Those atoms needed to possess valency and to combine in various ways that no mechanical atomic theory of the time was able to come close to explaining. Chemical atomism could not be regarded as a vindication of mechanical atomism.

As we have seen, there were specific problems with the kinetic theory stemming from specific heats and irreversibility. They pointed to genuine problems that the theory needed to overcome. However, to take the difficulties as sufficient reason to reject the theory was inappropriate and ignored the significant successes of the theory. So, for instance, the rejection by Ostwald (1896, pp. 345–346) of the kinetic theory on the grounds that it clashes with irreversibility paid inadequate attention to the resources that the theory had for meeting the objection. Those opponents of atomism who favoured outright rejection of it were ill-advised and had at least failed to appreciate the successes of the kinetic theory. On the other hand, in denying that the successes of atomism were a victory for mechanism, questioning the extent to which chemical atomism had been vindicated, stressing the non-mechanical aspects of chemical atoms and arguing that developments in electromagnetism were moving in a direction away from mechanism and atomism towards a theory involving continuous fields and electric charge as primitives, the anti-atomists were on defendable ground.

In the next chapter we will see how new experiments were to leave little scope for a denial of atoms, and those who claimed that atoms could have no place in science in principle were simply shown to be wrong. However, we shall also see that some important aspects of the case against mechanical atomism were vindicated rather than undermined by these developments.

Notes

1. As cited by Clark (1976, p. 65).
2. An English translation of the key part of Bernoulli's paper is in Brush (1965, pp. 52–65).
3. See Brush (1965, pp. 26–27).

4. As cited by Hunt (1991, p. 160).
5. Both quotations are cited in Clark (1976, p. 42).
6. As cited in Nyhof (1988, p. 88). The emphases are Mach's.
7. Nyhof (1988) makes this distinction.
8. I have argued the case that Maxwell's progress in electromagnetism owed little to mechanism in Chalmers (2001). Ostwald (1896) explicitly invoked the history of electromagnetism and optics as part of his case against mechanism.

Chapter 12
Experimental Contact with Molecules

Abstract Perrin's experiments on Brownian motion left little room for reasonable doubt about the existence of molecules. They provided independent support for the statistical fluctuations that had been postulated by proponents of the kinetic theory and they gave direct support to two basic assumptions of that theory, the randomness of the molecular motions via the motions of the Brownian particles that they jostled, and the equipartition of energy for motions of translation and rotation. Three methods for calculating Avogadro's number were in agreement with each other and with estimates acquired by other means. The kinetic theory was not confirmed in an unqualified way. It became increasingly clear that violations of the equipartition of energy constituted a problem that could not be solved without a drastic modification in the fundamentals of the theory. The challenge around the time of Perrin's experiments was to clarify what the kinetic theory got right and what it got wrong. Perrin was able to show that it got things right to an extent that it became unreasonable to doubt the existence of molecules.

12.1 Introduction

We have seen that those who were inclined to be skeptical about the existence of atoms and molecules had some grounds for their position and profitable ways of pursuing their science without them. Chemical formulae and the atomic and molecular weights to which they led could be exploited without commitment to atoms and thermodynamics was able to guide research in a progressive way without commitment to any matter theory. The work of Ostwald and Duhem in the late nineteenth century bares witness to this. The strongest case for atomism stemmed from the successes of the kinetic theory, but that theory had its problems, as we have seen, and Maxwell, for one, appreciated that the theory could not survive without some major transformation in its fundamentals.

Whatever room there was for doubting the existence of a granular structure of matter it was effectively removed in the light of experiments performed late in the nineteenth and early in the twentieth century. Jean Perrin's experiments on Brownian motion in 1908 made possible a case for some basics of the kinetic theory that was difficult to deny. They supplied the first unambiguous evidence for statistical

A. Chalmers, *The Scientist's Atom and the Philosopher's Stone*, Boston Studies in the Philosophy of Science 279, DOI 10.1007/978-90-481-2362-9_12,

fluctuations that violate an unqualified form of the second law of thermodynamics. Preceding Perrin's experiments by a decade were experiments of a different kind, and coming from a quite different direction, that involved the detection of what we now refer to as the electron. The granular structure of matter was established by identifying sub-atomic particles, an irony given that atoms were first introduced as the permanent entities underlying change.

An analysis of Perrin's experiments is a natural sequel to the discussion of the kinetic theory in the previous chapter so I will depart from a chronological order and discuss them and their significance in this chapter. Then, in the following chapter, I will focus on experiments that revealed the ubiquitous presence of electrons.

12.2 Brownian Motion

The agitated motion of minute particles suspended in a fluid was observed through a microscope by the British naturalist Robert Brown in 1827. The phenomenon was exhibited by inorganic as well as organic matter, undermining any notion that the moving particles were alive and self-moving for that reason. Brownian motion is striking because of its spontaneous and permanent character. From the early 1860's some scientists suspected a connection between the motion of the Brownian particles and the thermal agitations of the molecules of a fluid postulated in the kinetic theory. Those wishing to support such a connection could point to the independence of the motion from extraneous factors. The fact that neighboring Brownian particles moving in opposite directions could freely pass each other told against any suggestion that the motions were due to convection currents, whilst the motions persisted in the absence of vibrations. The apparently chaotic nature of the motions made it readily distinguishable from the co-ordinated motions known to arise from vibrations and convection currents. These features, together with the persistence of the motion, exhibited, for example, by the fact that the motion could be observed in liquid bubbles that have been trapped in solid quartz for thousands of years, lent support to those, like C. Wiener (1863), who wished to attribute the source of the Brownian motion to impacts with moving molecules. L. G. Gouy (1888) even saw the spontaneous upward motion of many Brownian particles as conflicting with the second law of thermodynamics.

R. Maiocchi (1990, pp. 258–263) is right to point out that there was considerable disagreement over just what the observable features of Brownian movement were prior to Perrin's experiments, so that to pick out those observations that seemed to favor their interpretation in terms of the kinetic theory, as Perrin (1990, p. 83–86) did, is to be unduly selective. A particular source of confusion and disagreement involved estimates of the velocity of Brownian particles calculated by dividing apparent distance moved by time taken. On a sub-microscopic scale an apparent linear displacement of a Brownian particle is in fact a highly irregular movement with the particle changing direction many times. Estimating velocities by dividing observed displacements by the time is in fact as inappropriate as measuring the velocity of flight of a bee by noting the separation of its position in the swarm at two separate times and dividing by that time. As we shall see, Perrin's experiments were to remove such confusions and confirm the kinetic theory in striking ways.

12.3 The Density Distribution of Brownian Particles

From the point of view of thermodynamics a solution is categorically distinct from a suspension of Brownian particles, the former being regarded as a single liquid phase and the latter as a solid phase (the particles) together with a liquid phase (the suspending liquid). However, from the point of view of the kinetic theory this is not the case. From that point of view the only significant difference between the solute molecules in a solution and suspended Brownian particles is one of size. Consequently, properties exhibited by systems of molecules and by systems of Brownian particles should differ quantitatively but not qualitatively. Both Einstein and Perrin exploited this idea in their application of the kinetic theory to Brownian motion.

Osmotic pressure is a phenomenon that makes its appearance in the functioning of living cells. The membrane constituting the walls of the cells allows water to pass freely through it, but is impervious to various substances dissolved in the water. If the walls of a cell containing a solution of solute in water cannot be penetrated by that solute, then, if the cell is situated in water, water will pass freely through the walls until the concentration of water is the same either side of the cell walls. The difference in pressure on either side of the membrane that results is the osmotic pressure of the solute. Work by Raoult and Van't Hoff in the 1890s established that the osmotic pressure of weak solutions of non-electrolytes obey the gas laws. The behavior of the solute, as far as the interdependence of pressure, volume and temperature is concerned, mimics that of a gas. The idea that Brownian particles in a liquid differ from solute molecules in a solvent only in size suggests that Brownian particles should exhibit an osmotic pressure that obeys the gas laws. This, in effect, is what Perrin explored with his investigation of Brownian motion.

In his first paper on Brownian motion, published in 1908, Perrin applied the idea that Brownian particles behave like molecules of a gas to explore the variation of the density of suspended Brownian particles with height. Perrin's own presentation of the density distribution makes intuitive sense and appeal in a way that does not require mastery of the algebraic details. The variation in density of Brownian particles with height is explained in just the same way as the variation with height of the density of a gas or of the distribution of solute in a solvent. If the molecules of gas or solute or the Brownian particles are in random motion, and if their density decreases with height, then the number striking the bottom surface of a thin horizontal area will exceed the number striking the top surface in the same time. The net result of the difference in number of impacts is an upward pressure that counters the downwards force of gravity due to the weight of the molecules or Brownian particles once equilibrium has been reached.

I give the details of the calculation of the formula representing the density distribution as a function of height in a footnote.[1] It is arrived at by deriving the difference in pressure across a thin horizontal layer containing Brownian particles utilizing the expression for the pressure on a surface assuming it to be caused by the impact of molecules or particles on that surface. The expression for the pressure, P, basic to the kinetic theory, is $P = 1/3\,nmc^2$, where c^2 represents the average of the squares of the velocities of the molecules. It can be readily understood why the pressure on the walls of a container of a gas should be proportional to n, the number of molecules per unit volume, and the mass, m, of each molecule colliding with the walls. The velocity appears squared because pressure depends on that velocity in two ways. The greater the velocity of a molecule the greater the impact and the greater the velocity the more often it will collide with the walls of the container. This expression is assumed to apply alike to gases and to systems of Brownian particles. Using this expression to calculate the upward force due to pressure difference and equating it to the downward force due to the weight of the Brownian particles in the suspending liquid yields the expression

$$W.\log(n_0/n) = 2\pi r^3 \Delta.g.h$$

This expression shows the density of Brownian particles falling off exponentially with height, h. Here, n_0/n represents the concentration of particles at some reference level divided by the concentration at height h above that reference level, r is the radius of the Brownian particles, Δ represents the difference between the densities of the material of the particles and the suspending liquid, g is the acceleration due to gravity and W is the mean kinetic energy of the particles. As we shall see, Perrin was able to determine all of the quantities in this expression other than W, which could therefore be calculated.

If the basic equation for pressure is combined with the gas law, $PV = RT$ for a gram-molecule of gas, W can be shown to be equal to (3/2)RT/N, where N is Avogadro's number. Consequently, knowledge of W enabled Perrin to calculate a value for Avogadro's number. This could be done because the ratio n_0/n, r and Δ could be measured for Brownian particles because they are visible in a microscope in a way that the molecules of a gas are not.

12.4 Experimental Details

A key piece of technology that helped to make Perrin's experiments possible was the ultra-microscope, invented by Siedentopf and Zsigmondy in 1903.[2] Perrin illuminated a narrow layer of Brownian particles he wished to observe by a horizontal pencil of light and viewed the light scattered vertically through a travelling microscope whose height could be varied and accurately determined. The advantage of the transverse illumination involved in the ultra-microscope over direct illumination was that only light from the particles and not from the suspending liquid reached the microscope, the light scattered from the molecules of the liquid being negligible compared with that scattered by the Brownian particles.

Treating a system of Brownian particles as a gas required that the particles be of the same size. Perrin was also able to use a state-of-the-art centrifuge to separate particles of differing size from suspensions of gamboge or of mastic, the two resins he employed as the material of his suspended particles. The elaborate procedures involved in the latter process typically took him a few months! They resulted in Perrin being able to prepare suspensions of resin particles of similar size and observe them using the ultra-microscope.

Perrin needed to know the density of the resins forming the Brownian particles. He devised several methods for measuring it, checking the respective results for consistency. The first involved solidifying a portion of resin and then varying the density of water by adding potassium bromide until the resin would just float in it. The easily measurable density of the fluid then gave the density of the resin. In a second method, a known volume of a suspension of resin in water was weighed. The water was boiled off and the remaining resin weighed. The difference between the weights gave the weight of the water boiled off and hence its volume. The difference between this and the original volume gave the volume of resin. Knowledge of both the weight and volume of the resin thus yielded its density. Perrin subsequently added a third method. He varied the density of liquid in which the particles were suspended until centrifuging failed to separate them. Perrin demanded that the measurements achieved by the various methods be mutually supportive.

Perrin also needed to measure the radius of the resin particles. Again he used three different methods which were required to be mutually supportive. One of them involved appeal to Stokes' law, which gives the force on a sphere moving through a fluid as a function of its speed and radius and the viscosity of the fluid. The steady velocity of fall of a particle will occur when this force equals the weight of a particle, itself a function of its radius and density. By measuring the velocity of fall of a cloud of particles sufficiently dense to swamp the effects of Brownian motion, Perrin calculated their radius since the density of the material of the particles and the viscosity of the liquid (usually water) were known. A second and third method utilized the fact that resin particles tended to congregate together near the edge of the container holding the suspending liquid. When a few particles fortuitously arranged themselves in a line, the length of the line could be measured using a travelling microscope. Dividing by the number of particles yielded the average diameter of each. Repeating this measurement for a variety of linear arrangements gave Perrin a way of checking that the particles were in fact uniform in size. Collecting and weighing a volume of resin containing a countable number of particles yielded a third measure of the radius assuming the particles to be spheres. The extent to which repetitions of these measures on different samples gave the same answer indicated the extent to which the particles were indeed all of the same size. Once again, Perrin demanded that measurement of radii by the three methods yield consistent results.[3]

There is a rationale behind this duplication of methods of measurement. Any one method can be prone to errors from known or unknown causes. Estimating particle size by measuring the length of a row of them will be erroneous to the extent that the arrangement is not strictly linear and may be subject to systematic error stemming from some miscalibration of the microscope. Measurements of radius based on Stokes' law presumed the truth of that law which had only been verified empirically for much larger particles and the theoretical derivation of which treated the forces resisting the motion of a sphere to be distributed continuously over it, an assumption that must break down at sizes comparable to molecular sizes. The remaining method was subject to errors in counting of particles and in the weighing techniques involved. The fact that the three different methods yielded mutually supportive results in spite of being subject to different kinds of errors was a strong indication that the errors were insignificant.

Perrin needed to measure the variation of the density of Brownian particles as a function of height. With some ingenuity, Perrin accomplished these measurements with a good degree of accuracy. With the ultra-microscope focused on particles at a particular height, Perrin restricted the field of view by means of a diaphragm so that only a very small number of particles were visible through the window. Interrupting the light beam intermittently with a shutter, Perrin noted the number of particles visible in the window on each exposure, repeating the process two hundred times. The sum of the two hundred numbers resulting, which were subject to statistical fluctuations and ranged from 0 to 6, was proportional to the density of particles at that height. Repeating the measurements at various heights gave Perrin the variation of relative density of particles with height.

12.5 Support for the Kinetic Theory

In my opening chapter I discussed in general terms what conditions need to be satisfied before it can be confidently asserted that a theory has been tested against and supported by empirical evidence as opposed to being merely accommodated to that evidence. Prominent in my discussion was the idea that a theory is supported only if the evidence conforms to predictions following in a natural rather than a contrived way from the theory in conjunction with subsidiary assumptions that themselves have independent support. The support Perrin offered for the kinetic theory by measuring the density distribution of Brownian particles satisfied this kind of demand with flying colors. The equation for the density distribution follows from the basics of the kinetic theory with no need for additions or adjustments. The measurements of particle density were derived by summing two hundred measurements, as we have seen. Perrin did not assume that the density indicated by these sums was constant. He repeated the measurements, finding that the density did remain constant after a time of up to three hours had elapsed, which proved to be sufficient for equilibrium conditions to be attained. Perrin even disturbed the equilibrium conditions deliberately, by cooling the lower part of the suspension. When the temperature was allowed to return to its original uniform value so did the measured densities return to their original values. Perrin could not choose the densities that he measured as a function of height. They were determined for him by the situation he was investigating. The measured densities varied exponentially with height in exactly the way predicted by the kinetic theory. Subsequently, Perrin (1990, p. 106), with the help of Bruhart, showed that the temperature-dependence of the density distribution conformed to the theoretical prediction.

As we have seen, the density distribution measured by Perrin, together with values for the radius and density of individual particles, each of which were measured by three independent methods, could be fed into Perrin's theoretically-derived formula to yield a value for Avogadro's number. Perrin himself dramatized the result of his doing this. He posed the question of the range of possibilities that might have been expected for the density distribution by a scientist not wedded to the kinetic theory. Given that the Brownian particles are composed of a material denser than the liquid in which they are suspended, a reasonable expectation would be that they would settle to the bottom of the container. Substituting that density distribution into Perrin's equation yields infinity for Avogadro's number. Given that this does not happen and particles do in fact remain indefinitely in suspension, one can pose the question of how one might expect the density of particles to vary over a height of less than a tenth of a millimeter.[4] A reasonable answer would be that there is no detectable variation. Had this proved to be the case a value for N of zero would have eventuated. The range of plausible possibilities for calculated values of N could hardly be larger. The value for N that in fact resulted from Perrin's measurements was about 7×10^{23}, close to the values previously estimated by Loschmidt and others using alternative methods unconnected with Brownian motion. One can understand why Perrin (1990, p. 104) greeted this result 'with the liveliest emotion'. There is no doubting that Perrin's experiments on the density distribution of Brownian particles, the first version of which were conducted in 1908 and duly published in Perrin (1908a), provided significant support for the kinetic theory.[5] But Perrin had not finished yet by any means.

12.6 The Mean Displacement and Mean Rotation of Brownian Particles

Perrin's first study of Brownian motion via the density distribution seems to have been carried out in ignorance of, or at least uninfluenced by, theoretical treatments of Brownian motion that had appeared in the literature from 1905. Einstein had derived an equation for the diffusion of particles through a fluid in 1905 and speculated that Brownian motion might be a manifestation of the phenomenon. As far as the equation itself is concerned, Einstein had in fact been anticipated by my fellow-countryman William Sutherland (1904, 1905). Sutherland derived his version of what Abraham Pais (1982, p. 92) has called the 'Sutherland-Einstein relation' for the diffusion of large solute molecules such as albumen in a solvent. He did not connect the phenomenon with Brownian motion. The 1905 papers were soon followed by further papers by Einstein (1956) in the next couple of years and papers by M. Smoluchowski (1906) and P. Langevin (1908). There is no evidence that Perrin took heed of Sutherland's work, but his exploration of Brownian motion subsequent to his first publication certainly made use of the other work I have cited, especially that of Einstein.

Einstein investigated the diffusion of spherical particles through a liquid, as they were jostled by the random motions of the liquid molecules and moved against the viscosity of the liquid. Assuming the particle motions to be a random walk and the resistance offered by the viscosity to be governed by Stokes' law (see note 3) Einstein derived an expression for the mean displacement of a particle in time t. This expression could be written in a form relating the mean displacement to Avogadro's number once it was assumed that the mean kinetic energy of the Brownian particles was equal to the mean kinetic energy of the liquid molecules jostling the former, the equipartition of energy basic to the kinetic theory. The resulting expression for the mean displacement, d, in a horizontal direction in time, t, is

$$d^2 = (RT/N)(1/3\pi r\zeta)t$$

Here, r is the radius of a particle and ζ is the viscosity of the suspending liquid. The assumptions lying behind this equation are that the moving particles are spheres obeying Stokes' law, that their horizontal motion, uninfluenced by gravity, is random and that their mean kinetic energy is equal to that of the liquid molecules.

We have seen that Perrin was able to prepare emulsions of Brownian particles consisting of spheres of constant size. He had only to measure the mean displacement, d, to make possible an experimental test of Einstein's formula and another means of calculating Avogadro's number. Perrin (or his research students) did indeed carry out the necessary determinations. The motion of a particle could be tracked and its horizontal position in successive equal time intervals (of the order of $^1/_2$ to two minutes) noted using a *camera lucida*. Series of measurements of this kind involving particles of various sizes and differing materials suspended in liquids of a range of viscosities all showed Einstein's formula to be confirmed and yielded values for Avogadro's number within a few percent of 7×10^{23}.

Perrin was able to independently test most of the assumptions on which his calculations were based. We have already seen (note 3) that the fact that determinations of particle radius assuming Stoke's law agreed with determinations of that radius by other methods amounted to experimental evidence for the applicability of that law to the motion of Brownian particles in the suspending liquid. Perrin, following the suggestion of his research student, Chaudesaigues, also put to direct test the random character of the displacement of the Brownian particles. In fact, he carried out a number of independent tests of that assumption in a way that Deborah Mayo (1996, pp. 228–229) has described as 'statistical overkill'.[6] A further assumption involved in the derivation of the equation for the mean displacement was, in effect, that no forces other than those due to gravity and viscosity act on the particles. (Gravity acts vertically and so had no effect on the horizontal displacements measured by Perrin.) At the time there was a concern that electrostatic forces due to charges at the liquid/particle boundaries might have a significant effect on the motions. Perrin was able to rule out significant influence of such forces. He showed that, provided particles were remote from the walls of the container, the displacements of particles were unaffected by additions of a small amount of acid, sufficient to reverse the polarity of the charges on the particles but causing only a negligible change in the viscosity of the suspending liquid.[7] The fact that the values for Avogadro's number agreed with those arrived at by other means constituted fairly direct evidence for the equipartition of kinetic energy, the key theoretical assumption in the derivation of the displacement equation.

As Einstein showed, extension of the equipartition of energy to rotations as well as linear translations of particles leads to an expression for the mean rotation of a particle in unit time similar to that for mean displacement. The formula for the mean rotation, α, about an axis is

$$\alpha^2 = (RT/N)(1/4\pi r^3 \zeta)$$

Once again, Perrin was able to rise to the occasion, put this equation to experimental test and determine Avogadro's number in yet another way. Support for the equipartition of energy was thus extended to rotational modes.

To make observations of the orientation experimentally feasible Perrin needed to generate relatively large particles, not only because of the improved visibility of the larger particles but also because only the larger ones rotate sufficiently slowly to make measurement of their average rotation possible. To eliminate motion under gravity Perrin introduced additives to the suspending water with the aim of rendering the density of the mixed fluid equal to that of the material of the Brownian particles. A difficulty was that such additions caused the particles to coagulate. Fortunately Perrin found that such coagulation did not occur when the additive was urea. A second fortuitous practical outcome was that some of the large spheres that Perrin found he could prepare by the slow introduction of water into an alcoholic solution of his resins had inclusions which made it possible to track the orientation of the otherwise transparent particles.

Perrin's measurements confirmed Einstein's equation for the mean rotation, including the implication that it be independent of the density of the material of the particles. The new estimates for Avogadro's number were in line with the earlier estimates within a few percent. This, in effect, confirmed the legitimacy of extending the equipartition of energy to include rotational as well as translational motions.

12.7 The Kinetic Theory Confirmed? – A Nuanced Discussion

By 1911 Perrin had amassed substantial evidence for the claim that Brownian motion exhibited by dilute emulsions is caused by the random motions of the molecules of the suspending liquid and that, consequently, the particles mimic the behavior of an ideal gas. He had shown that detailed predictions of the kinetic theory are borne out by the density distribution and mean displacement and rotation of the particles and that the three sets of phenomena yield values for Avogadro's number that agree within a few percent with each other and with estimates acquired by other means. The verifications were carried out in a wide variety of circumstances. They involved particles of two different materials, gamboge and mastic, and a variety of suspending liquids, including pure water, pure glycerine, and water with various proportions of glycerine, sugar or urea added. Temperatures varied between −9 and +58 degrees Centigrade, the viscosity of the liquid varied by a factor of 330 and the mass of the particles varied by a factor of 70,000. Surely it was with considerable justification that Perrin (1909, p. 599) wrote:

> I think it impossible that a mind, free from all preconception, can reflect upon the extreme diversity of the phenomena which thus converge to the same result, without experiencing a very strong impression, and I think that it will henceforth be difficult to defend by rational arguments a hostile attitude to molecular hypotheses, which, one after another, carry conviction, and to which at least as much confidence will be accorded as to the principles of energetics.

One reason for being skeptical about Perrin's enthusiasm has been expressed by Stephen Brush (1968/9, p. 34), who suggests that Perrin's experiments were of marginal significance and made only a quantitative rather than qualitative change in adding to evidence for the kinetic theory that was already strong.

> The evidence provided by the Brownian movement experiments of Perrin and others seems rather flimsy, compared to what was already available from other sources. The fact that one could determine Avogadro's number and the charge on the electron by one more method seems hardly sufficient to justify such profound metaphysical conclusions. Several independent methods of demonstrating these parameters had been known since 1870 or before, to say nothing of the many successes of kinetic theory in predicting the properties of gases.

Another reason that might be invoked to qualify Perrin's enthusiasm is that the kinetic theory in the form that Perrin utilized it is false. Because of quantum effects, energy is not equally distributed among degrees of freedom as presupposed in the classical kinetic theory. We noted in the previous chapter that Maxwell, for one, was well aware that something was seriously wrong. Hints that a solution of the problem lay in the direction of quantization of energy were already on the table by the time Perrin conducted his experiments and he was able to discuss them in his 1913 book.

I suggest that appraising the case for the kinetic theory from the point of view of truth or falsity, confirmation or disconfirmation, is inadequately nuanced and misconstrues what is typically involved in the testing of a reasonably general scientific theory. Theories are speculative and general. They go beyond phenomenological laws to explain them. Just because of the speculation and generality involved they typically get at least something wrong. Newton's mechanics, in spite of two centuries of significant success, proved to have its limits. Contemporary physicists are aware that General Relativity and quantum field theory do not fit well together and eventually something will need to give. Given that theories, however successful, are likely to need to be productively replaced or modified, there are two kinds of information one needs to facilitate such a step. One needs to know what a theory gets right and how and what it gets wrong and how. It is knowledge of this kind that sets the scene for future progress.

I believe that the unsophisticated thoughts expressed in the previous paragraph provide the appropriate framework for understanding work by physicists on the kinetic theory early in the twentieth century. When the leading physicists gathered at the Solvay Conference in 1911, with Perrin among them, they were well aware that classical physics was in disarray and in need of replacement. The specific heats problem was just one of a range of issues that was pushing them towards the quantum theory. What they needed, and got, from Perrin, was a nuanced account of what the kinetic theory gets right and how.

One further set of useful platitudes will set the scene for my appraisal of Perrin's work on Brownian motion. When the predictions of a theory are borne out and when they fail it is important to know which portions of the maze of claims implicated in the test are responsible for the success or failure. This involves discriminating between the assumptions involved in an argument, both high level and low level, and, wherever possible, subjecting them to independent tests. So, for instance, once it is appreciated that Einstein's General Theory of Relativity involves curved space-time plus a specific account of the cause and degree of the curvature, then a test, such as detection of the red shift, that depends only on the first assumption should not be taken as confirming the theory as a whole. On the other hand, when it appeared that one set of Eddington's observations of relative star positions at the time of an eclipse favored Newton's theory of gravity rather than Einstein's, it proved possible to trace the fault to a telescope distorted by the heat of the sun rather than to Einstein's theory.[8]

In the light of these preparatory remarks, we are in a position to appreciate that a key feature of Perrin's case for the molecules of the kinetic theory was the extent to which he was able to partition the various assumptions in his case and subject them to independent test. The fundamental assumptions of the kinetic theory were the random distribution of the motions of the molecules as expressed in the Maxwell distribution and the equipartition of energy amongst the degrees of freedom of the colliding particles. As we have seen, Perrin directly tested and confirmed by a number of methods the assumption that displacements of the observable Brownian particles do indeed conform to the Maxwell distribution. As far as the equipartition of energy is concerned, the fact that kinetic energy is equally distributed amongst Brownian particles of varying size was confirmed directly by Perrin since he was able to calculate that energy from measurable magnitudes. The fact that the measurable displacements and rotations of particles led to the same value for Avogadro's number fairly directly indicated that energy was equally distributed amongst translations and rotations of the particles. The fact that the mean kinetic energy of Brownian particles varies with absolute temperature according to the dictates of the kinetic theory was also subject to direct experimental test.

Other aspects of Perrin's case involved showing that derivations from the kinetic theory were borne out by experiment. A strength of these arguments was the extent to which there were a variety of them. The theory correctly predicted the density distribution, mean displacement and mean rotation of particles and enabled three methods for calculating Avogadro's number that were mutually supportive. It is also worth stressing that the experimental results against which Perrin tested the predictions of the kinetic theory were established independent of the theory itself. We have seen how they involved such things as counting particles, noting their positions after successive equal time intervals, measuring their radius and density by a variety of independent and mutually supportive methods and so on. The concordance of a variety of indisputable evidence with the predictions of the kinetic theory amounted to a powerful argument from coincidence. How could the theory get things so right if it were not at least roughly true?

Aware that showing predictions of a theory to be correct does not amount to a logically compelling case for it, Perrin explored an attempt to derive some basic features of the theory 'from the phenomena'. A premise of his argument was that the cause of the motion of the Brownian particles lies in the liquid. The evidence for this was considerable. Convection and vibration could be ruled out as a cause because the motions of neighboring particles are not co-ordinated in the way they would be if their motion was due to such macroscopic causes. Perrin's demonstration that the Brownian motion is random in a technical sense ruled out all macroscopic causes. Incident light could also be ruled out because of the independence of the phenomena of the wavelength and intensity of the illuminating light. As Perrin pointed out, motions due to convection currents caused by the heating accompanying such illumination, was distinct, and easily distinguishable from, the chaotic Brownian motion. I should stress that it is not all logically possible causes of the motion that is at issue here, but causes known through the science of the time. The importance of systems shielded from all known energy inputs (adiabatic systems) were freely utilized in thermodynamics by opponents of the kinetic theory. In 1908 the claim that the source of Brownian motion lay in the liquid was a powerful if not irresistible one.

Here is my rendering of the argument 'from the phenomena' put forward by Perrin (1909, pp. 513–515). A drop of water introduced into a larger mass will soon disperse throughout that larger mass. (This can be witnessed up to a point if the added mass is dyed.) When first added to the mass of water the motions of the parts of the added drop are coordinated. The drop moves as a whole. But the motion of the parts of the drop soon become decoordinated as the drop disperses. Suppose we focus on a sub-drop of the partially dispersed drop. The motion of this sub drop will be coordinated insofar as it moves as a whole. But it too will gradually disperse and the motions of its parts will be decoordinated. We now focus on a sub sub drop and so on. We raise the question of whether the process of decoordination proceeds indefinitely. Perrin argues that the existence of Brownian motion indicates that this cannot be so. If the motion of the liquid were completely decoordinated then it would be incapable of causing a Brownian particle to move. Such a particle cannot go with the flow if there is no flow. Spread of the added drop is an indication of decoordination but that decoordination must have a limit for there to be sufficient coordination to cause Brownian motion. As Perrin puts it, Brownian motion signals a balance between coordination and decoordination of the motion of the water. There must exist elements of water that move as a whole and which are such that their motion cannot be redistributed amongst their parts. Water must have least parts in this sense. (Nothing hinges on the suspending liquid being water, of course. The argument applies to any liquid in which Brownian motion occurs.)

This argument proceeds from observable phenomena to the existence of least parts that move as wholes and, to the extent that it is successful, it confounds the sceptic who doubts that science can establish the existence of such parts. This argument alone falls short of giving the kinetic theory the molecules that it needs, of course.

There is no sensible doubting of the claim that Perrin had succeeded in making experimental contact with molecules. But let us not forget that the kinetic theory was nevertheless strictly false. Its fundamental assumption, that energy is equally distributed about the degrees of freedom of a molecule does not in general hold. The specific heats of gases are an indication of how equipartition breaks down for vibrational modes at normal temperatures, as Maxwell had sensed. By the time Perrin (1990, pp. 73–74) wrote his book *Atoms* he was able to discuss how equipartition breaks down, at low temperatures, for rotational modes also. But given the extent to which Perrin had subjected the various claims of the kinetic theory to rigorous test it was possible to isolate the problem area without jettisoning the theory as a whole.

Over a decade prior to Perrin's analysis, Boltzmann (1895, pp. 413–414) had made the point that the kinetic theory was sufficiently confirmed to have already shown its worth in spite of difficulties.

> Every hypothesis must derive indubitable results from mechanically well-defined assumptions by mathematically correct methods. If the results agree with a large series of facts, we must be content, even if the true nature of the facts is not revealed in every respect. No one hypothesis has hitherto attained this last end, the Theory of Gases not excepted. But this theory agrees in so many respects with the facts that we can hardly doubt that in gases certain entities, the number and size of which can be roughly determined, fly about pell-mell.

Perrin's experiments were a significant addition to the 'indubitable results' supporting the kinetic theory to a degree that made it difficult to avoid the conclusion that most of its claims were at least roughly correct. They also enabled the problem of specific heats to be pinpointed as a basic problem for the theory, and added to the range of phenomena that, by the time of the 1911 Solvay Conference, were indicative of a fundamental change in the 'mechanically well-defined assumptions' that Boltzmann invoked in the above quotation.

Notes

1. The equation relating pressure to the momentum changes due to impacts of molecules or particles is $P = (1/3)\,nmc^2$. The pressure change, dP, over a height change dh involving a density change of dn is thus $dP = (1/3)\,dn.mc^2$ which becomes (2/3) dn.W when we write W for the average kinetic energy of a molecule or particle. This force per unit area upwards must balance the downwards weight of the particles. The number of particles in a cylinder with unit area and height dh is n.dh. Each of these particles will weigh $(4/3)\pi r^3\Delta.g$, where Δ is the excess of the density of the material of the Brownian particles over that of the suspending liquid. We can thus represent the balance of the forces per unit area by writing $(2/3)\,dn.W = (4/3)\,\pi r^3.\Delta.g.n.dh$ or $(2/3)\,W\,(dn/n) = (4/3)\,\pi r^3.\Delta.g.dh$. Integration yields $W.\mathrm{Log}\,n_0/n = 2\pi r^3.\Delta.g.h$. W is thus expressed as a function of factors that Perrin knew or was able to measure. This in turn gives a path to Avogadro's Number via the equation $W = 3RT/2N$, which results from a combination of the standard expression for pressure in the kinetic theory and the gas law, $PV = RT$ for a gram molecule of gas.
2. See Siedentopf (1903) and Siedentopf and Zsigmondy (1903).

3. Use of Stokes law to supply the resistance to motion of Brownian particles was controversial. Einstein assumed it in his 1905 paper and was taken to task on that score. Sutherland (1905) did the same. The problems were, firstly, there was no empirical evidence for the truth of Stokes' law for particles as small as Brownian particles, and secondly, Stokes' theoretical derivation of it assumed the motion and the resisting force to be continuous, an assumption that clearly breaks down to the extent that the kinetic theory is correct. Perrin's first experiments relied on only one method for measuring the radius of particles, the one utilizing Stokes' law. The risk of presuming Stokes' law to be valid was duly pointed out. In subsequent experiments Perrin checked his estimates of radius by introducing the two other methods. The fact that they agreed in effect verified the applicability of Stokes' law to Brownian particles, a fact that Perrin (1908b) stressed.
4. The density of the particles involved in this phase of Perrin's experiments was halved for increases in height of less than one tenth of a millimetre.
5. Arthur Fine (1991, pp. 91–92) has cast doubt on this. He mentions two possible explanations of the random motions of the Brownian particles that do not appeal to the kinetic theory, one attributing the motion to electrostatic forces and another assuming the motions to be spontaneously random and acausal. I describe in Section 6 how Perrin ruled out the electrostatic option. Spontaneous random motions of the Brownian particles, which rarely collide with each other, would be insufficient to explain the density distribution of particles measured by Perrin and shown to conform to straightforward predictions of the kinetic theory.
6. My analysis of Perrin's experiments has taken advantage of the instructive discussion in Mayo (1996).
7. There was a sequel to these considerations of the effect of charge. In 1914 Perrin, helped by R. Constantin, showed that concentrated emulsions obey Van der Waals equation, $(P + a/V^2)(V - b) = RT$. The surprise was that the constant 'a' was negative. The Brownian particles in contact with water were charged and repelled each other.
8. I take these examples from Mayo (1996).

Chapter 13
Experimental Contact with Electrons

Abstract Both Zeeman and Thomson conducted experiments towards the end of the nineteenth century that gave evidence for the existence of the particle now known as the electron. Their experiments were responses to specific problems in nineteenth century physics and were able to take advantage of technological advances of the latter half of that century. A variety of experiments gave similar values for the ratio of charge to mass of the particles detected that were three orders of magnitude greater than estimates of that ratio for the hydrogen molecule. Further experiments soon indicated that this was due to the minute mass of the particle rather than an excessively large value for the charge. The robust character of the arguments drawn from the experiments and the extent to which they reinforced each other made it difficult to deny the existence of the electron as a component of atoms. Whilst this achievement signals the end of the story told in this book, it marked the beginning of atomic physics and chemistry rather than their conclusion.

13.1 Introduction

Strong evidence for the existence of micro-particles preceded Perrin's experiments by a decade. It was provided by experiments that involved the detection of the negatively charged particle now known as the electron. The most notable experiments were those conducted by Pieter Zeeman and J. J. Thomson and his students in the closing years of the nineteenth century, although the compatibility of their results with those produced by others, such as Emil Wiechert and Walter Kaufmann was an important dimension of the argument.

There is a reason why experimental access to charged particles such as the electron is more readily achieved than to neutral molecules or atoms. Because of their charge, electrons and ions can be manipulated by accelerating and deflecting them in electric and magnetic fields. Also because of their charge, such particles cause ionisation and act as the locus of condensation, leading to a range of effects that are readily visible. Experiments revealed that electrons and ions have a charge, and also yielded a measure of the ratio of their charge to their mass and, eventually, of their charge and mass individually. There is an irony here. In the long tradition of atomism that we have investigated in this book, an entrenched idea was the notion of brute

A. Chalmers, *The Scientist's Atom and the Philosopher's Stone*, Boston Studies in the Philosophy of Science 279, DOI 10.1007/978-90-481-2362-9_13,

matter constituting the material of atoms, characterised by some property such as impenetrability or, after Newton, mass. From that perspective, electrical phenomena associated with charged bodies experimented on in the laboratory were treated by atomists as phenomena to be explained by reference to atomic mechanisms. This view was modified by Maxwell in the 1860's insofar as he introduced a continuous aether, in addition to, and interacting with, atoms and molecules. But his aether was a mechanical aether, governed by the fundamental laws of mechanics. For the Maxwellians, charge was a discontinuity in a strain in the aether brought about by its interaction with matter. After the experiments of Zeeman and Thomson with which this chapter is concerned, the charge of particles like the electron was treated as a primitive along with their mass. As far as access to experiment is concerned, it is the charge of micro-particles that is tangible and detectable. Mass is less so, and, as a consequence, the mass of uncharged particles is measured indirectly via experiments on the charged ones. Charge, as a primitive property of micro-particles, had not been anticipated by those seeking an account of the ultimate structure of matter. Its introduction converted atomism into an experimental science in ways that had been impossible before and in a way that had not been anticipated by philosophical atomism.

The identification of charged particles at the atomic and sub-atomic level by Zeeman and Thomson was fairly direct. However, there are identifiable reasons why it was not until late in the nineteenth century that this became possible. In part the preconditions involved the development of the necessary electrical, vacuum and spectroscopic technology. In part they involved an identification of the law governing the action of electric and magnetic fields on moving charged particles now known as the Lorentz force law. Embellishments of the program were made possible by discovery of the photoelectric effect, X-rays and radioactivity. There are historical reasons, then, why scientific versions of atomism blossomed in the late nineteenth and early twentieth centuries and not before.

13.2 Historical Background to the Experiments of 1896/7

We saw in the previous chapter that by the concluding decade of the nineteenth century the kinetic theory, although not problem-free, had considerable support. That theory, in conjunction with the experimental determination of quantities such as the diffusion rate of gases, made it possible to estimate absolute parameters of molecules, their mass and size, and also the number of them in a given mass of gas. Distinct from these arguments, evidence for atoms and molecules from a quite different direction emerged in the course of the nineteenth century. They involved the electrical properties of matter including those connected with the transmission of electricity through liquids and gases, the magnetic effects of electric currents, spectroscopy, and, after Hertz's experiments of 1888, the production of electromagnetic waves by fluctuating currents. Alongside the experimental developments were two theoretical approaches to electricity and magnetism. One in vogue on the Continent involved distance forces between elements of positive and negative electric fluids. The other, the Faraday/Maxwell approach, sought to explain electric and magnetic phenomena as the results of the action of a material aether that became identified with the aether assumed in the wave theory of light. These two approaches were eventually reconciled and their mutual strengths combined in the 'electron' theories of H. A. Lorentz and Joseph Larmor in the 1890s. I elaborate a little on the background to the experiments of Zeeman and Thomson in the remainder of this section, drawing heavily on the work of others.[1]

After Hans Christian Oersted's experiments of 1820 the magnetic effects of electric currents became an experimental fact. Ampère gave a theoretical treatment of these effects and of the forces acting between current-carrying conductors. He postulated 'molecular currents' as the cause of permanent magnetism.

A decade before Oersted detected the magnetic effect of currents, Humphry Davy had demonstrated that chemical compounds can be dissociated by passing an electric current through a solution of them. Faraday subsequently established laws of electrolysis. He observed that the weights of elements released in electrolysis by the passage of a given current for a given time are proportional to the equivalent weights of those elements. Faraday noted that his electrochemical laws combined with the atomic theory of chemical combination suggested that an equal quantity of electricity is connected with each atom, although Faraday himself was reluctant to embrace the atomic theory. Helmholtz (1881) spelt out the link between electrolytic phenomena and a fixed quantity of electricity associated with each atom in less hesitant terms half a century later. The connection between electrical and chemical phenomena had inspired Berzelius to hypothesise that the atoms and groups of them were held together in molecules by electrostatics attractions. We saw, in Chapter 9, how this idea was eventually threatened by the notion of substitution, including the substitution of electropositive by electronegative elements, that became a powerful device in organic chemistry.

Developments in spectroscopy also had links with atomism. The discovery that the emission and absorption spectra of a gas consists of light of definite frequencies characteristic of the gas in question suggested that those frequencies are associated with vibrational modes in atoms and molecules. Just as the characteristic sound frequencies emitted by a vibrating bell are determined by the size and structure of the bell, so the light frequencies emitted by a gas could be attributed to the size and structure of its component atoms and molecules. Clerk Maxwell (1965, Vol. 2, p. 463) gave clear expression to this line of thinking in the 1870s invoking the analogy with bells.

Connections between magnetism and light, such as the rotation of the plane of polarisation of light on transmission through some transparent materials subject to a magnetic field (the Faraday effect) and the rotation of the plane of polarisation of light on reflection from the pole of a magnet (the Kerr effect), were established experimentally. These results, combined with hypotheses about molecular currents as the source of permanent magnetism and atomic or molecular vibrations as the source of spectra, strongly suggested that the periodic variations that were presumed to be the source of light emitted by atoms and molecules were electrical.

Such speculations meshed with the electromagnetic theory of light developed by Maxwell from the mid 1860s. I say more about Maxwell's theory below. Here I note that it predicted that the ratio of the electromagnetic to the electrostatic units of charge be equal to the velocity of light and that the refractive index of non-magnetic materials be proportional to the square root of the constant measuring their electric polarisability, both predictions receiving experimental support. In 1888 Hertz produced electromagnetic radiation from oscillating electric currents, a possibility entailed by Maxwell's theory. As far as atomism is concerned, this reinforced the idea that spectra of gases can be traced to electrical oscillations within their atoms.

Experimental investigation of the conduction of electricity through gases, made possible by the high voltages provided by a stack of voltaic cells or an induction coil, proved to be more complex, and correspondingly less informative, than conduction through solutions. Some order emerged in the form of cathode rays. These were first produced by Julius Plücker in 1859, who took advantage of improved induction coils devised by Heinrich Rühmkorff and the possibility of producing improved vacua using the mercury diffusion pump devised by Johann Geissler, a technician in his own laboratory. Unlike the discharges at higher pressure, cathode rays are not readily visible, their presence being signalled by the fluorescence they cause when incident on a suitable target. Prior to Thomson's experiments, the nature of the rays was unclear. William Crookes and Arthur Schuster were among those who favoured the idea that they are beams of negatively charged particles whilst Eugen Goldstein and Heinrich Hertz favoured the idea that they were some kind of aether disturbance. In 1883 Hertz failed to deflect cathode rays in an electric field, thereby casting doubt on their identification as beams of charged particles.[2] Their deflection by magnetic fields was well established, however. Jean Perrin, in 1895, showed that negative charge accumulates on a collector receiving the rays.

As mentioned in the opening paragraph of this section, there were two approaches to the theoretical problem of unifying and perhaps explaining electrical and magnetic phenomena, the Continental theories that attributed them to distance forces between electrical fluids and Maxwells theory that sought to explain them by an aether and its interactions with matter. Prior to the discovery of electromagnetic radiation it was the former that had the strongest links with atomism. As well as Ampère's assumption that permanent magnetism is caused by electric fluids circulating within molecules there was the assumption that electric polarisation is due to the displacement of the fluids within molecules. By the 1870's, Wilhelm Weber, one of the most sophisticated articulators of the fluid theory, was suggesting that the electric fluids were composed of electrical particles with mass and that an atom is composed of a highly massive negative particle at its core with lighter positive particles in orbit around it. In that decade, too, Lorentz developed accounts of reflection, refraction and dispersion of light on the assumption that molecules of matter contain charge particles that execute harmonic oscillations in response to an incident light wave. We have already mentioned the incorporation of an atom of electricity associated with molecules into accounts of electrolysis. These examples show the strongest aspect of the fluid theories of electricity. They could readily be adapted to explanations of the electrical, magnetic and optical properties of materials by invoking some appropriate microstructure. However, the fields involved in electromagnetic radiation, an undeniable reality following the experiments of Hertz and which were a natural consequence of the rival theory developed by Maxwell, were alien to the Continental approach based on distance forces between current elements.

From the mid-1860s Maxwell constructed an electromagnetic theory built on Faraday's notion of lines of force and attempted to construe electric and magnetic fields as strains and vortices in an aether that he was able to identify with the medium presumed to be the seat of light waves. On this view, electric charge was the result of an interaction between the aether and matter. It was a discontinuity in a strain in the aether (the 'displacement') at the boundary between a conducting and insulating body. Maxwell's aether theory construed light as an electromagnetic wave and it received some support when the two predictions mentioned above were confirmed experimentally. The theory received a boost in 1888 when Hertz produced the radio waves predicted by it.

Maxwell construed electric charge as a discontinuity in that state of the aether that he referred to as its displacement, **D**. Electric current was equal to the rate of change of this displacement, d**D**/dt. It was this conception that made it possible for Maxwell to accommodate the idea of currents in space empty of matter (but not of aether) and to construct a theory able to predict radio waves. Displacement currents gave rise to magnetic fields, whilst changing magnetic fields gave rise to electric currents. As a consequence, changing electric and magnetic fields leapfrog each other through space, giving rise to each other and so constituting electromagnetic waves. Maxwell's theory thus readily accommodated the phenomenon that posed most problems for competitor theories based on action at a distance. However, Maxwell's theory was at a disadvantage insofar as it offered little guidance to the construction of accounts of the electrical, magnetic and optical properties of matter. For Maxwell, those properties were a result of some interaction between matter and the ether but he did not attempt to specify what that interaction might amount to. Maxwell was an atomist insofar as he accepted the kinetic theory, a theory that he did much to develop as we have seen. However, his electromagnetism involved the interaction of atoms and molecules with a continuous aether. This theme is explored in detail by Buchwald (1985).

There was a fundamental difficulty in Maxwell's theory. It involved an adequate interpretation of conduction currents, that is, of the passage of electricity through conductors. If electric currents are changing displacements of the aether, how is the current through a wire to be construed? On Maxwell's picture a conductor interacts with the aether in such a way that displacement cannot be sustained. A current through a wire consists of a constantly collapsing displacement.[3] A symptom of the problem of reconciling the notion of electric charge as a discontinuity in displacement and the idea that a current through a wire involves the transfer of electricity through it is the muddle Maxwell got into over the sign of the charge on the plates of a charged capacitor. After Faraday, a natural view is that the insulating material separating the plates of a capacitor becomes polarised, with negative electricity attracted towards the positive plate and positive charge attracted towards the negative plate. Maxwell himself frequently described the situation in this way. But this way of thinking suggests two adjacent charges opposite in sign, the charge on the conducting plates of the capacitor and the adjacent charges resulting from the polarisation of the insulator. Maxwell's identification of charge with a discontinuity in the polarisation of the aether does not leave room for this distinction between the two charges. As a consequence, he was tempted to regard the charge on the plate attached to the positive terminal of a battery as positive when considering the flow of current through the connecting wire and as negative when considering the polarisation of the aether. The resulting indecision becomes apparent in Maxwell's own writing, at one place in the form of a formal inconsistency.[4]

As Buchwald (1985, pp. 30–31) has shown, by the time he wrote his *Treatise* in 1873, Maxwell has contrived a conception of aether displacement that got around the difficulty associated with the sign of the charge on the plate of a capacitor. But other deep problems remained. As Maxwell observed on more than one occasion, a crucial feature of his theory is that all currents, including the transitory ones involved in the charging of a capacitor, flow in closed circuits. Conduction currents charging a capacitor are closed by displacement currents that involve changing displacements in the region between the plates. In Maxwell's view (1954, Vol. 1, p. 69), not only do the two currents form a closed circuit but also they are of the same kind.

> [W]hatever electricity may be, and whatever we may understand by the movement of electricity, the phenomenon which we have called electric displacement is a movement of electricity in the same sense as the transference of a definite quantity of electricity through a wire is a movement of electricity, the only difference being that in the dielectric there is a force which we have called electric elasticity which acts against the electric displacement, and forces the electricity back when the electromotive force is removed; whereas in the conducting wire the electric elasticity is continually giving way, so that a current of true conduction is set up –.

Maxwell had some picture of how changes in the elastic distortions of the aether could give rise to vortices corresponding to magnetic fields. But this picture could not apply to currents in conductors because conducting materials are presumed to negate the elasticity of the aether in some way.

In Maxwell's theory, displacement currents are readily intelligible and conduction currents are problematic. In the continental fluid theories the reverse is the case. Conduction currents involve the flow of electric fluids through conductors but displacement currents are mysterious. In the 1890's both H. A. Lorentz and Joseph Larmor came to appreciate these problems and, by slightly different routes, responded to them by separating charged bodies and the field. In their transformation of Maxwell's theory, arrangements and motions of charged bodies, which Larmor called electrons and Lorentz called ions, were the source of electromagnetic fields whilst it was those fields that exerted forces on bodies via their charge, with effects determined by their mass. The resulting theory was able to accommodate Maxwellian fields and hence account for electromagnetic radiation, the latter being an experimental fact following Hertz's experiments of 1888. It was also able to construe conduction currents and the polarisation of insulators as the flow and displacement of electrons (or ions) respectively.

One other theoretical issue needs to be mentioned. By the time Zeeman and Thomson embarked on their experiments there was general agreement on the formula for the force on a charged body moving in an electromagnetic field (now known as the Lorentz force). The force figured centrally in the theories of Lorentz and Larmor that had freed charged bodies of the field in the sense that they were no longer viewed as discontinuities in the field. The formula for the force had already been derived in the Maxwellian framework in the 1880's, through work by Heaviside, Fitzgerald and Thomson himself.[5] The force on a charge, q, moving with velocity, **v**, in a magnetic field, **H**, is q.**v**x**H**, a force that is perpendicular to **v** and **H** and proportional to each of their magnitudes. The theoretical interpretation of the force by the Maxwellians differed from that required by the theories of Lorentz and Larmor. But the important point for the present purpose is that, in 1997, Zeeman and Thomson could avail himself of an agreed-upon formula for the forces exerted by electric and magnetic fields on a moving charged body.

As the discussion of this section illustrates, nineteenth century treatments of electricity, magnetism and optics involved a range of hypotheses that attributed an atomic or molecular structure to matter. But there is a key difference between such hypotheses and those involved in philosophical theses about the ultimate structure of matter of the kind that we have discussed in detail earlier in this book. Unlike the latter, the nineteenth-century hypotheses of the physicists were not general theories of the structure of matter but specific hypotheses designed to explain specific phenomena identified in the course of experimental programmes. Ampère's proposal of molecular currents, for instance, was put forward to explain permanent magnetism taking advantage of Oersted's discovery of the magnetic effect of electric currents, not as a general matter theory. On the other hand, there was an analogy between the atomic speculations of the nineteenth-century scientists and atomic matter theories defended by natural philosophers. Both were accommodated to, rather than confirmed by the available evidence. If Amperè's molecular currents existed they could explain permanent magnetism. But did they exist? Was Ampère's explanation of permanent magnetism the right one? What was required was some detailed specification of the details concerning the molecular currents and independent evidence for them that would make it difficult to deny their existence. Similar claims could be made of the atomic hypotheses of the nineteenth-century generally. We saw in the previous chapter how Perrin was able to strongly counter objections of this kind to the kinetic theory. In the remainder of this chapter we see how Zeeman, Thomson and others were able to do likewise in the domain of electricity.

13.3 Discovery of the Zeeman Effect

Zeeman began experimental research on the interrelation between magnetism and optics in Lorentz's laboratory at the University of Leiden in 1890.[6] His focus was the interaction between magnetic and optical phenomena. His early attempts to detect the effect of a magnetic field on the sodium spectrum were not successful. He did eventually succeed to observe an effect in 1896 by taking advantage of improved spectroscopic techniques. Zeeman investigated the spectrum of the sodium in common salt situated in the flame of a Bunsen burner between the poles of an electromagnet. What he observed was that the two D-lines of the sodium spectrum, that appeared as sharply defined lines in the absence of a magnetic field, became broadened when the field was switched on. He took a range of measures to ensure that the observed broadening was indeed due a change in frequency of the emitted light rather than to some other cause such as a change in density or temperature of the sodium in the flame, which was observed to change shape under the influence of the magnetic field.[7]

Subsequent elaborations of the experiments were inspired by Lorentz's theoretical analysis of the broadening effect. After Hertz's production of radio waves it was natural to attribute the emission spectra to radiation caused by the oscillations of charged particles. Lorentz was able to spell out the effect a magnetic field would have on such vibrations by taking into effect the force, e.**Hxv** experienced by a particle with charge, e, moving with velocity, **v**, in a magnetic filed, **H**. The effect of the field depends on the direction of motion of the charged particle relative to it. Lorentz resolved the oscillations of a particle into three components, one linear oscillation parallel to the field and two oscillations circulating in opposite directions around field lines in planes perpendicular to them. The Lorentz force is zero in the first case and acts in a way that increases or decreases the frequency of oscillation in the case of the two circular oscillations. It was these changes in frequency that were held responsible for the broadening of the spectral lines that Zeeman had observed.

There were detailed consequences of Lorenz's theoretical analysis that posed an experimental challenge to Zeeman. On the assumption that the source of the radiation constituting the sodium D-lines is vibrating charged particles, Lorentz's analysis implies that when the spectrum is viewed in a direction perpendicular to the field a triplet of lines should be observed, corresponding to the three components of the vibration, one along the field and two around it in opposite directions. The light of the central line should be plane polarised and the light in the other two lines circularly polarised in opposite directions. Finally, the theoretical analysis yields a quantitative value for the line splitting. The Lorentz force, m. d^2x/dt^2, is equal to the x-component of **e**.v.**H**. The magnitude of the acceleration of the charged particle by the field is thus e/m.v.H. It depends on the ratio e/m, as well as the speed of the particle and the magnitude of the field. An acceleration of this magnitude yields a fractional change in period of vibration, T, equal to $e/m.HT/4\pi$. Since the frequency of the light and the strength of the magnetic field are known, measurement of the frequency (or period) difference between the two edges of the sodium line broadened by the magnetic field yields a value for e/m.

By the end of 1897 Zeeman had confirmed and taken advantage of these consequences of Lorentz's theoretical analysis. He improved the resolving power of his spectroscope by employing a Rowland diffraction grating and was eventually able to resolve the triplets of lines (rather than the mere broadening of a single line) using cadmium instead of sodium as a source. Lorentz's predictions about the polarisation of the light corresponding to the various lines making up the triplets were confirmed. Finally, a value for e/m for the oscillating particles was obtained.

The theoretical analysis of Zeeman's experiment rested on the assumption that the sources of light in the sodium and cadmium spectra that he observed were the vibrations of massive, charged particles subject to the Lorentz force. The fact that the spectral lines were split into a doublet when viewed in the direction of the magnetic field and into a triplet when viewed perpendicular to it were in accord with that assumption as was the experimentally-confirmed facts concerning the polarisation of the light associated with the components of the triplets. There was strong experimental support for the assumption that vibrating charged particles were the source of the spectra of sodium and cadmium.

The results of the e/m measurements were a surprise. Lorentz referred to the charged particles of his electron theory as 'ions'. While the details of the theory require only particles with charge and mass it is clear that Lorentz thought of his ions as the ions of electrolysis, that is, charged atoms or molecules. This explains why Lorentz responded to Zeeman's estimate of e/m by declaring '[t]hat looks really bad; it does not agree at all with what is to be expected'.[8] If one assumes the source of the light in atomic spectra are the movements of charged atoms corresponding to those presumed to be transported through electrolytes then 'what is to be expected' is a value for e/m derived from the mass of atoms and the value of the charge they carry, both of which can be derived using estimates of Avogadro's number readily available in 1897. The values measured by Zeeman were three orders of magnitude smaller than that! The implication, soon drawn by Zeeman and Lorentz, was that the charged particles whose vibrations are responsible for emission spectra are to be distinguished from charged atoms and molecules (ions) and are rather components of them. Zeeman was able to conclude from the direction of polarisation of the light associated with the split spectral lines that the vibrating particles were negatively charged. By the end of the century Lorentz was referring to the particles in his theory as electrons rather than ions.

13.4 Thomson's Experiments on Cathode Rays

J. J. Thomson was a follower of Maxwell. In his theoretical work he had construed charge as the opposite ends of Faraday 'tubes of force' where the tubes corresponded to vortices in the aether. This conception was the key model exploited by Thomson in his *Notes on recent researches in electricity and magnetism* that, in 1893, he presented as a sequel to Maxwell's *Treatise*. This conception repeated Maxwell's notion of charge as a discontinuity in the aether and so was not destined to remove the difficulties inherent in that conception. Thomson's experiments on cathode rays that complimented Zeeman's in leading to the introduction of the electron were not motivated by the theories of Lorentz and Larmor but the results, once available, were, like those of Zeeman, readily interpreted by and gave support to, those theories.

Opponents of the view that cathode rays were beams of charged particles countered Perrin's demonstration that interception of the rays resulted in an accumulation of charge. They argued that showing the rays to be *accompanied* by the transfer of charge did not demonstrate that they were *constituted* by a flow of charge. Thomson responded to this, in 1897, by demonstrating that when the rays were deflected by a magnetic field the flow of charge follows the deflection.[9]

The amount of deflection of a moving charged body in a magnetic field depends on the velocity, the greater the velocity the greater the deflection. The amount of deflection also depends on the ratio of the charge to the mass of the deflected body, the greater the charge the greater the deflecting force and the greater the mass the less deflection that force results in. Measuring the deflection in a magnetic field thus enabled a relationship to be deduced between two unknowns, the velocity, v, of the cathode ray particles and their charge to mass ratio, e/m.[10] One further relationship was required to enable v or e/m to be measured. Thomson provided two ways of providing the needed relationship.

In the first method Thomson built on Perrin's experiment involving the accumulation of charge. He measured both the charge accumulated and the heat generated by capturing cathode rays for a small length of time. The heat generated is equal to the kinetic energy, W, lost by the particles. If there are N of them then that energy is $1/2Nmv^2$. Thomson measured this quantity by having the rays strike a thermocouple of known mass, so that the heat gained by it, calculated from the temperature rise, gave the energy lost by the incident particles. The charge, Q, carried by the N particles is N.e, where e is the charge on a single particle. Thomson measured Q using an electrometer. Substituting Q/e for N in the expression for W yields a value for the relationship between v and e/m. The combination of this with the relationship between e/m and v deduced from the magnetic deflection experiment yielded values for e/m and for v.

Thomson provided a second way of measuring these quantities. Here he succeeded, where Hertz had failed, to deflect the cathode rays with an electric as well as a magnetic field. Thomson was able to produce lower pressures than those achievable by Hertz by taking advantage of technological advances made by manufacturers of electric light bulbs. He realised that gas is released into vacuum tubes from the solid surfaces in them, an effect that can only be countered by prolonged heating and pumping. As Thomson came to appreciate, ionisation of gas molecules remaining in the tube generates charges that swamp the effects of the electric field on cathode rays. Thomson supported this position by demonstrating that gases not adequately evacuated become conducting. In any event, Thomson showed that at sufficiently low gas pressures cathode rays are deflected by an electric field and also showed that when the pressure rose to be of the same order as Hertz was able to accomplish the deflection disappeared.

Thomson arranged for simultaneous deflections of the rays by both electric and magnetic fields. Values for these two fields that resulted in the two effects cancelling each other out leaving no net deflection enabled both v and e/m to be calculated. Various values resulting from the two separate methods differed amongst themselves and from each other by a factor of up to 3. But they were of the same order of magnitude as those obtained by Zeeman and by other researchers at the time, namely Emil Wiechart and Walter Kaufmann. As Thomson realised, they implied that the particles constituting cathode rays either have a charge that is very large compared to those they can be attributed to ions involved in electrolysis on the basis of values of Avogadro's number estimated by assuming the kinetic theory or a very small mass compared with that of atoms and molecules estimated on a similar assumption, or some combination of the two.

The conjecture that had emerged by the end of Thomson's experimental researches of the late 1890s was that the cathode particles are all alike, and are components of atoms orders of magnitude lighter than the atoms themselves.[11] Thomson had demonstrated that his values for e/m were independent of the nature of the gas in the discharge tube and of the material of the cathode from which the particles were emitted. The implication was that identical cathode particles are components of all atoms. Here Thomson's reasoning meshed with that of Zeeman. By the end of the century a range of experiments on cathode rays and other phenomena associated with what we now call the electron served to add support to the clams that what soon became known as electrons have a measurable mass and charge, are components of all atoms, and that electrolysis and conduction of electricity through gases at high pressure involve the transfer of ions, these being atoms with a dearth or excess of one or more electrons. Zeeman's experiments, in particular, supported the further conjecture that spectra are to be attributed to oscillations of electrons within atoms

Thomson himself added to the experimental support for such claims. In 1898, drawing on experimental work carried out in his laboratory at Cambridge by C. T. R. Wilson and Ernest Rutherford, Thomson (1898) devised experiments to measure the charge on ions generated in gases by the passage of X-rays. Rutherford had conducted experiments to measure the velocity of ions in conducting gases. Wilson had introduced a method, destined to have a significant future, for estimating the charge on particles. He had found that charge particles act as loci for the condensation of water droplets. By inducing condensation by rapid expansion and assuming that one water droplet was formed on each charged particle Wilson could estimate the number of charged ions by dividing the weight of water collected by the weight of each drop. He determined the later by using Stokes' law to deduce the size of a drop from its rate of fall, the method that Perrin was later to adopt in his experiments on Brownian motion that we discussed in the previous chapter. Using Rutherford's measurements of the velocity of ions and Wilson's measure of their number, Thomson estimated e/m for the ions caused by X-rays and compared the results with the value calculated for the value of the charge on the hydrogen ions presumed to carry current through electrolytes using estimates of Avogadro's number. They were of the same order of magnitude. A year later, Thomson (1899) measured e/m for particles emitted in the photoelectric effect and from incandescent filaments, achieving measurements consistent with his measurements of e/m for cathode rays. He also used the methods of his 1998 paper to measure the charge on the particles emitted in the photoelectric effect, with results of the same order of magnitude as estimates of the charge on the hydrogen ion. This latter result gave a direct indication that the large value for e/m for the particles was indeed due to a small mass rather than a large charge compared with the hydrogen ion.

Within a few years experiments by a range of experimenters in a variety of contexts gave converging evidence to support a range of claims concerning the ubiquity of electrons as small components of atoms, as the source of spectra in atoms, as comprising cathode rays and as responsible for the charge on ions through their presence or absence. Electrolysis and conduction in gases could be readily understood by appeal to ions. Already in 1898 Wilhelm Wien had employed magnetic and electric deflection to measure e/m for positive ions and this technique was to be fashioned into the mass spectrograph, so that, by 1913, Thomson's assistant, F. W. Aston, was able to distinguish the mass of isotopes. Experimental accuracy also increased to the extent that, by 1913, R. A. Millikan employed an adaptation of Wilson's techniques for measuring charge to measure the charge on the electron to four significant figures. By that time the electron was a central assumption in the Bohr theory of the atom, which met with success in a way that Thomson's own plumb-pudding model of the atom did not.

13.5 The Significance of Experiments on Charged Particles

In the introduction to this chapter I described the theoretical background to the experiments that were to vindicate atomic theory by identifying charged ions and the electron. In particular I described the theories of Lorentz and Larmor that set charged bodies free of the fields that they give rise and react to. The experimental results of Zeeman and Thomson found a ready interpretation in that theory, which has become known as the Lorentz electron theory. However, whilst experiments on electrons and ions did vindicate the Lorentz theory, formulated independently by Lorentz and Larmor, it is misleading to view the situation exclusively in that way. Theory needed to be adapted to the results of experiment in a way that was forced by the experiments and not anticipated by theory.[12]

There were two aspects of the experimental findings made by Thomson, Zeeman and others that were not anticipated. One was the large value for e/m, eventually attributable to the small value of m, for the particles now called 'electrons'. The other was the asymmetry between the role of positive and negative charges. Whilst there are positive and negative ions there appeared to be no positively charged analogue of the electron. It is significant that before 1906 Thomson did not use the word 'electron' to describe the sub-atomic particles that his experiments showed cathode rays to be. He used the word 'corpuscle'. G. Johnstone Stoney had coined the word 'electron' to describe the unit of charge, both positive and negative, presumed to accompany ions in electrolysis. Larmor had used it to describe the positive and negative vortices he considered to make up atoms in his theories of the 1890s. Lorentz adopted the term only after 1899, before that using the term 'ion' to refer to the charged particles in his theory. In the main I have tried to avoid using the term in my discussion of the history in this chapter. It was only after the experiments of Thomson and others that the referent for 'electron' in the form of a negatively charged and massive body of sub-atomic size was identified. In particular, Lorentz's 'electron theory' as developed by both Lorentz and Larmor was symmetric with respect to positive and negative charge and involved no necessary assumptions about the size of the charged particles that they assumed caused and responded to the electric and magnetic fields. However, it is clear that both authors thought of their elementary charges as corresponding to the 'ions' assumed in atomistic explanations of electrolysis. This explains their subsequent surprise at the values for e/m implied by experiment. The small mass of the electron and the asymmetries with respect to charge due to the part played by the electron in cathode rays, its contribution to the charge of ions by its presence or absence and its role as the source of spectra, were indeed experimental discoveries.

The nature of electricity and its connection with matter were seen as fundamental issues in the latter decades of the nineteenth century as we have seen. Thomson himself was inclined to the Maxwellian view that charge was some discontinuity in a state of the aether and adopted the Faraday tube as a device for expressing this. The experiments we have described put an end to the debate, but in an unanticipated way. Electromagnetic phenomena became explicable in terms of the motions of particles such as the electron with charge and mass generating and interacting via the electromagnetic field. Electrons give rise to radiation when they accelerate, constitute a conduction current when they flow through metals, and figure in electrolysis and discharges through gases either through the formation of ions or as cathode rays. The charge on the electron became a primitive along with its mass. This need not be interpreted as disallowing the question of what charge might be. Rather, the point is that a major programme had opened up and many explanations became possible in a way that did not require an answer to the question. The inability to explain the charge on the electron became a move analogous to Newton's inability to explain gravity.

Although it has not been the focus of my attention, a similar fate befell fields and their relationship with the aether. The fields in Lorenz's electron theory had been presumed to be states of a stationary aether. Increasingly, questions about the state of that aether became irrelevant, or, in the light of the Michelson-Morley and other experiments, problematic, whilst the Special Theory of Relativity was to reinforce the move to treat the electromagnetic fields as primitives. Electromagnetic waves were understood as the propagation of fields not as the propagation of states of an aether. Displacement currents in space are fluctuating electric fields that do not involve the displacement of anything. Successes of the wave theory of light in the nineteenth century notwithstanding, experiment gave no endorsement to the assumption of an aether. The same kind of experimental practice that led eventually to the confirmation of the existence of atoms and electrons led to the banishment of the aether.

The early twentieth century left many questions about atoms and electrons to be answered. Some of the answers took very unexpected forms. Wave-particle duality and the strange property of half-integral spin that it proved necessary to ascribe to the electron provide ready examples of that. The discovery of the electron was a beginning rather than the end of the story. But the experiments of Zeeman and Thomson, together with those of Perrin, mark the end of *our* story. There is no sensible reason to doubt that here we find experimental contact being made with atoms and their components. Microparticles became a part of experimental science in a way they had not been before.

The status of electrons after Thomson is no more problematic than the existence of air had been for Boyle, or, for that matter, the Ancient Greeks. Air cannot be directly seen, but its presence can be detected in a range of independent and mutually supporting ways. Air can be felt to resist compression in a syringe and can be felt in the form of a wind when forced out of the syringe. Boyle could not see air, but, by the time he had finished experimenting on it there was no doubt that there is such a thing as air, that it has a pressure, and that that pressure is responsible for the height of the mercury in a barometer. Thomson could not see electrons, but he could manipulate them in controlled and mutually reinforcing ways. By the time he had finished, and compared his findings with those of others, the existence of electrons with a specified charge and mass was as firmly established as Boyle's air pressure. What is more, the methods used to establish the two sets of results were entirely on a par. Scientific versions of atomism were the legacy of experimental science that had its serious beginnings in the seventeenth century and not of philosophical matter theories dating back to the Ancient Greeks.

Notes

1. Some key sources are Darrigol (2000), Buchwald (1985), Hunt (1991) and Arabatzis (2006).
2. For an analysis of Hertz's experiment see Hon (1987).
3. The best attempt to make consistent sense of Maxwell's view of conduction is that of Buchwald (1985, pp. 30–31) but he himself stresses the fundamental difficulties posed by this conception in Maxwell's theory as a whole. Buchwald (1985, pp. 65–70) analyses the deep difficulties the conduction current posed for Maxwell and his followers.
4. I identify the formal inconsistency in Chalmers (1973a). Siegel (1991, pp. 145–152, 180–181) has elaborated on this theme. He has identified direct evidence of Maxwell's indecision in unpublished drafts of Maxwell's 'Dynamical theory of the electromagnetic field'. The fact that there are two charges, opposite in sign, one on the conducting plate of a condenser and one on the adjacent insulator can be demonstrated experimentally by rotating either the plate or the insulator and measuring the magnetic field generated. An experiment of this kind was performed by Pender (1901). For further discussion and references see Chalmers (1973b, pp. 479–480).
5. See Hunt (1991), pp. 187–188, 236–237.
6. My account of the history of Zeeman's experiments draws heavily on Arabatzis (1992) and Arabatzis (2006).
7. These moves were analogous to those made by Perrin a decade later to establish that the motions of Brownian particles did indeed have their source in motions of the liquid in which they were suspended, as was described in the previous chapter.
8. As cited by Arabatzis (2006, p. 82).
9. Thomson, (1897). My account of Thomson's experiments draws heavily on Smith (2001c).
10. Thomson in fact gave values for m/e rather than e/m. I talk in terms of the former in line with what has become the norm.
11. A slight threat to the assumption was a small spreading of the beam resulting in a spatial extension of the fluorescent spot caused by the impact of the rays on the glass tube. The difficulty was removed within a year or two when R. J. Strutt identified the cause of the spreading as fluctuations in the voltage generated by the induction coil used by Thomson to generate the rays. See Smith (2001c, p. 42.)
12. Arabatzis (2006, p. 83) stresses this point in connection with Zeeman's experiment.

Chapter 14
Atomism Vindicated?

Abstract After the experiments of Perrin on Brownian motion and of Zeeman, Thomson and others that revealed the existence of the electron there remained little room for doubt about the existence of micro-particles way below the dimensions of direct observation. To what extent is this achievement a vindication of atomism? It was not a vindication of the philosophical tradition that had sought an account of the ultimate structure of matter. The atoms of modern science have an inner structure about which much is known and some inner structure of the electron might well be revealed by the next generation of particle accelerators. Knowledge of atoms and electrons was acquired early in the twentieth century by argumentation that was experimental rather than philosophical and was arrived at by a route that owed little to the tradition of philosophical atomism. As the twentieth century progressed properties, such as half-integral spin, needed to be ascribed to micro-particles in a way that gives rise to nothing but headaches for those inclined to base an account of the ultimate structure of matter on some general philosophical principles. To view contemporary atomic theory as a vindication of Democritus is to seriously misconstrue the nature of science and its mode of argumentation.

14.1 Introduction

As we have seen, Perrin's experiments left no room for serious doubt that there are molecules. The connections between the kinetic theory and chemistry, and the need to include rotational modes of motion for molecules to cope with specific heats of gases, further led to the recognition that molecules are made up of atoms, one kind of atom for each chemical element. The experiments of Zeeman and Thomson, and a range of related experiments that followed in their wake, established that atoms have tiny negatively charged electrons as components, that molecules can become positively or negatively charged ions by losing or gaining electrons, that electrolysis and electrical discharges through gases involve the transfer of ions, that conductivity in metals involves the transfer of electrons and that electrons are in some way implicated in chemical combination and the production of spectra.

To what extent can such dramatic developments be regarded as a vindication of atomism? The answer depends much on what is intended by the term 'atomism'. If

A. Chalmers, *The Scientist's Atom and the Philosopher's Stone*, Boston Studies in the Philosophy of Science 279, DOI 10.1007/978-90-481-2362-9_14,

atomism is interpreted suitably vaguely, as claiming that the objects and materials of our experience are composed of discrete parts with properties that serve to explain the properties of the wholes they are parts of, then it was vindicated in a significant and uncontroversial sense by the time the first decade of the twentieth century had expired. But such an assessment needs to be qualified as soon as atomism is defined in more precise ways so that, for example, distinctions can be made between the kinds of atomism involved in the theories of Democritus, Sennert, Boyle, Newton, Dalton, Maxwell and Perrin.

The atoms invoked by Ancient Greeks such as Democritus and Epicurus and by seventeenth-century mechanical philosophers such as Gassendi and Boyle were construed as the ultimate and unchanging components of material reality. Some notion of reality that was regarded as evident or as the only intelligible notion led to the conclusion that atoms possess only the properties of shape, size and motion together with a property such as solidity or tangibility or impenetrability characteristic of matter in general. Twentieth-century atoms are nothing like those envisaged in these philosophical traditions and they and their properties were discovered by experiment rather than philosophical analysis. The modern atom has an internal structure, most importantly an electron structure. Electrons have a charge as well as a mass. I am confident that according to the notion of intelligibility that Boyle employed to dismiss Aristotle's 'real qualities' the charge on the electron is unintelligible. If this is not seen as sufficient to establish the qualitative difference between electrons, on the one hand, and atoms in the tradition of Democritus and Boyle, on the other, then I remind the reader that it was not far into the twentieth century that it became necessary to attribute to electrons a half-integral spin, a quantum-mechanical notion having no classical correlate, and to understand electrons as obeying Fermi-Dirac rather than classical, Boltzmann, statistics. (There are only three ways, not four, of distributing two electrons over two boxes.) Such properties are far from anything envisaged by Democritus and Boyle and cannot be reconciled with the notions of reality and intelligibility that informed their theories. What is more, it is precisely these novel kinds of properties that are fundamental for explaining the details of such things as atomic spectra, chemical combination and the conduction of electricity through metals. The twentieth-century science of atoms is a violation of and departure from atomism in the sense of Democritus, Epicurus and the seventeenth-century mechanical philosophers.

Another tradition that attributed the properties of observable wholes to the properties of their discrete parts was the one that invoked natural minima. Those minima possessed properties sufficient to characterize them as minima of the substances they were the least parts of. Minima of blood, water and gold differed from each other in precisely this way. The reductionist character of modern atomism places it closer to the Democritean ideal than the theory of natural minima. It is true that, in the modern theory, the atoms of each element are characteristic of, and peculiar to, that element. However, the properties of atoms responsible for chemical, optical, electrical and indeed macroscopic properties generally involve electron structure. In the medieval theory and its successors key properties of macroscopic materials reappear as the properties of natural minima. By contrast, in the modern theory macroscopic properties are *explained* by appeal to a narrow range of properties such as electron charge and spin. Substances do have least parts as Sennert and his medieval precursors supposed, but the properties attributed to those least parts and the explanatory role that they play in the modern theory differ vastly from anything they envisaged.

Modern atoms differ markedly both from atoms in the tradition of Democritus and the mechanical atomists and from natural minima. What is more, they are able to perform their explanatory role because of the ways in which they so differ. The potency of atoms as characterized in the modern theory, that enables them to perform their role as fundamental building blocks, stems precisely from the difference between them and miniature stones.

14.2 Did Philosophical Atomism Play a Productive Heuristic Role?

In the previous section I was concerned to spell out the distinction between the modern scientific atom and the philosophical notions of an atom that preceded it. However great the distinction, there remains the possibility that the philosophical version played a productive role in leading to the scientific one. I suggest that reflection on the story told in this book invites a sceptical response to any such proposal.

The path to solid evidence for the electron came very much from left field (as they say in the USA where I am writing the first draft of this chapter). The rotation of the plane of polarization of light by a magnetic field and the existence of line spectra that Zeeman investigated using improved spectroscopic techniques were nineteenth-century experimental discoveries. The electrical technology that was a crucial part of what made discharge tube phenomena possible stemmed from Volta's discovery of the battery late in the eighteenth century. I am not aware that the discovery owed anything to philosophical atomism. The same can be said of the invention of the mercury diffusion pump that made possible the production of pressures sufficiently low to make the experimental production of cathode rays possible from the late 1850s. The investigation of cathode rays and related experiments on discharge tube phenomena forced Thomson and others to the conclusion that cathode rays are beams of sub-atomic particles with charge as well as mass. No philosophical theory had anticipated, nor could possibly have anticipated, the quantum mechanical behavior of electrons that was to enable them to play their full explanatory role.

The kinetic theory of gases shows more promise as an example of a productive path from philosophical to scientific atomism. The explanation of the properties of gases by appealing to the motions and mechanical collisions of its least parts seems to come close to the kinds of explanation envisaged by Democritus and the mechanical philosophers. The mass and velocity that needs to be attributed to the molecules in the kinetic theory find a ready home in mechanical atomism. The perfect elasticity of collisions is not so easily accommodated, however. As we have seen, elasticity was a notion that troubled the mechanical philosophers. They saw it as necessary to explain it away but did not succeed in doing so to their own satisfaction. The problem was less acute from the point of view of Newtonian atomism insofar as rebounds could be attributed to short-range repulsive forces rather than to the perfect elasticity of the molecules themselves.

Aside from qualms about how elastic collisions can be accommodated in philosophical atomism, there is a more significant problem with seeing the kinetic theory as one of its fruits. My first point is that the kinetic theory was first successful as a theory of *gases*. Gases, as chemically distinct substances in the vapor phase some of which are components of air, were not anticipated by philosophy. Knowledge of their existence was an experimental discovery of chemists in the second half of the eighteenth century. The second point concerns the extent to which the kinetic theory was tied to and served to explain the gas laws. These, too, were experimental discoveries made in the second half of the eighteenth century. What is more they presupposed a precise and measurable notion of temperature, another innovation of the same period.[1] The idea that the heat of a body is associated with internal motions was a speculation that goes back at least as far as Francis Bacon early in the seventeenth century. But the transformation of that speculation into a reasonably precise theory with empirical support depended on experimental discoveries that owed little to philosophical matter theories, atomistic or otherwise.

Chemistry is the other field that might well be invoked as an illustration of a productive heuristic role played by philosophical atomism. Even if I am right to endorse Thackray's view that Newtonian atomism was unproductive as far as eighteenth-century chemistry is concerned, Dalton explicitly appealed to and employed it in his formulation of an atomic theory of chemistry early in the nineteenth century. However, my detailed study in Chapter 9 was designed to downplay the productive role played by that atomism. I argued that progress in nineteenth-century chemistry was more a precondition for rather than a result of the productive introduction of atomism into chemistry. I have most difficulty defending my position when confronted by historians such as Rocke and Klein who see the nineteenth-century advances as coming about by way of a chemical atomism, rather than physical atomism in the tradition of Newton or the mechanical philosophers. Even if my opponents are right here, the chemical atomism they endorse differed from philosophical atomism. The properties of chemical atoms were meant to be filled in as a result of chemical research rather than being specified in advance. Progress in nineteenth-century chemistry cannot be claimed as an example of philosophical atomism bearing fruit.

It needs to be stressed just how different the methods were that led, on the one hand to philosophical atomism, and on the other to the scientific atomism we take for granted today. The former drew on principles that seemed plausible or self-evident given the knowledge of the day and employed them to build and rationally defend an atomistic world-view. The path to the latter involved grappling with specific problems in the scientific knowledge of the day, framing new notions to help articulate responses to them and insisting that claims made in terms of the new notions passed stringent experimental tests. The pursuit of this latter method led to a radical undermining of the philosophical principles assumed by the former. For instance, the macroscopic/microscopic analogy employed in some form or other to defend philosophical atomism turned out to be highly problematic. It is not just a question of its content, knowing in what respects the microworld resembles the macroworld and just what properties can be projected from the one to the other. There is the question of its truth. It became evident, in the light of scientific advances, that the macroscopic/microscopic analogy is radically false. From the point of view of quantum physics and relativistic mechanics that inform modern atomism, the principles extracted from our knowledge of the macroworld by classical philosophical atomists constituted obstacles to be overcome. Rather than being a source of and inspiration towards a viable scientific atomism, philosophical atomism constituted a barrier to it that needed to be transcended.

To take features of atomism in contemporary science and point out how it differs from atomism in the ancient and mechanical traditions is to adopt a modern perspective. But I resist the charge that my stance is anachronistic. I have argued that the beginnings of the distinction between philosophical and scientific matter theories were already present in the seventeenth century. The methods that Boyle employed and articulated in the context of his pneumatics were the very ones that were to eventually lead to knowledge of atoms and, in so dong, undermine the fundamental claims of his mechanical philosophy. Not all, nor even most, of the reservations held by nineteenth-century chemists concerning atoms can be properly understood as positivistic prejudices. They made sense in the context of an increased awareness of the distinction between scientific and philosophical accounts of matter. By the time experimental science had developed sufficiently to dispel the qualms of those wary of atomism most of the principles underlying atomism in the tradition of Boyle and Newton had been undermined.

14.3 Twentieth-century Atomism a Victory for Scientific Realism?

The undeniable success of the atomic theory in the twentieth century is frequently invoked as a victory for scientific realism.[2] Positivists who held it to be impossible for science to gain knowledge of the world behind the appearances, and instrumentalists who held that scientific theories should be seen as useful instruments aiding our dealings with the world rather than as adequate descriptions of it, have been shown to be wrong. Whilst there is some important truth underlying such claims they need to be qualified to take into account the extent to which modern science has undermined what I will call the 'billiard-ball realism' that was implied by philosophical matter theories that have become outmoded.

Before I expand on my qualifications about seeing the triumph of atomism as a victory for realism, let me acknowledge the element of truth in it. There are extreme positivist or instrumentalist views, occasionally voiced by nineteenth-century critics of atomism, which are difficult to reconcile with the eventual success of scientific atomism. As Nyhof (1988, pp. 87–89) reports, Mach did occasionally express the extreme view that the world postulated by scientists as lying behind the appearances ‘exists only in our understanding’ and that, ‘in our investigation of nature, we have to deal only with the connections of appearances with one another’. Duhem, (1962, p. 19) for his part, insisted that ‘a physical theory is not an explanation. It is a system of mathematical propositions, which aim to represent as simply, as completely, and as exactly as possible a set of experimental laws’. The ultimate success of atomism in the physical sciences flies in the face of such assertions.

The construal of the success of atomism as running counter to extreme anti-realist views is altogether too easy and stands in the way of our learning what there is to be learnt from the story of atomism as told in this book. The success of atomism is hardly necessary to counter extreme positivism and instrumentalism. Knowledge of the physics and chemistry of gases such as oxygen and hydrogen, for instance, is sufficient to do that. This suggests that, if there is something valid about intuitions to the effect that the confirmation of atomism was especially significant as far as the realism issue is concerned, they are not captured by the mere recognition that that confirmation runs counter to extreme positivist or instrumentalist views. There are a number of less extreme views, versions of which can be found in the writings of the likes of Mach and Duhem, although I will not document that here.[3] The mechanical view of the world as portrayed by the mechanical philosophers and Newton, what I have dubbed ‘billiard-ball realism’, is not one that is sanctioned by experimental science and should not be assumed by experimental science. The assumption that all science can be reduced to mechanics, or, more generally, to physics, is something that needs to be supported by effecting such reductions rather than assuming it in advance in experimental research. There is no good reason to expect that the world lying behind the appearances conforms to common-sense intuitions based on familiarity with the world of appearances. I suggest that all of these theses are supported by the history of atomism as I have told it.

There is a question that I like to press, an answer to which forces a realist to formulate that position in a way that is more sophisticated than a mere denial of extreme positivism and instrumentalism as I have characterized them above. That question is, if the establishment of atoms constituted a victory for realism, why didn’t the banishment of the aether constitute a defeat? A number of scientists in the latter part of the nineteenth-century sought a reality behind knowledge of chemical combination and the physics of gases and postulated atomic theories. They were ultimately successful in that endeavor. Maxwell and his followers sought a reality behind knowledge of optics and electromagnetism and postulated a mechanical aether. They were unsuccessful. If there is a deep structure underlying the electromagnetic field the current success of field theory does not depend on knowing what it is. The fluctuating displacements currents implicated in electromagnetic radiation, including visible light, are fluctuating electric fields that are not states of an aether or anything else. They are as alien to mechanistic intuitions as the half-integral spin of the electron and the non-classical statistics obeyed by those particles. If one is to get comfortable with such notions then one had better learn the relevant physics rather than engage in philosophical reflection.

There is a kind of realism I am willing to endorse because I believe it to be implicit in the practice of science. It involves two claims. The first of them is this.

The world is the way it is whether we know it or like it or not. This claim is best borne out by the failures of science rather than its successes. We cannot make the world conform to our conceptions or our wishes. My view is implicit in the acknowledgement that the claims of science need to be rigorously tested by experiment. The second claim in my version of realism is that science is indeed capable of revealing knowledge of the world. This claim is supported to the extent that science has proved to be progressive. A feature of my characterization of realism, which I suspect will be too weak to satisfy many self-proclaimed scientific realists, is that it does not include any substantive claim about what the world is like. A declaration like 'it is the job of science to discover the reality lying behind and serving to explain experimental knowledge' is too strong. Whether there is such reality is something that the world decides, not us. For all we know, there is no reality lying behind electromagnetic fields, and if it should transpire that there is, it will be advances in physics that establish this. From my point of view, versions of realism that implicitly or explicitly incorporate assumptions central to the philosophical versions of atomism with which we have been concerned with in this book assume too much, and it is such realist views that have been undermined by the success of scientific versions of atomism. I am keen to encounter any impression that the success of atomism somehow shows that philosophical atomism in its mechanical or Newtonian guises were on the right lines.

14.4 In the End is My Beginning

To all intense and purposes Perrin established that molecules exist and was able to determine their weights experimentally. After his experiments on Brownian motion chemists had no need for reservations when identifying chemical molecules figuring in chemical formulae with the molecules whose weights Perrin had measured. Further, chemists can use their chemical formulae to derive atomic weights from molecular weights. After Perrin, atoms and molecules could be counted as well as weighed. Philosophical speculations about the existence of discrete entities underlying and responsible for observable phenomena had become scientific truth. Our story has reached its conclusion.

Or has it? By 1910 at the latest, chemists could safely presuppose atoms and molecules with weight and understand the symbols in their formulae as referring to them. But the same query can be raised here as I raised in the context of Dalton's first proposal of his atomic theory. What kind of chemistry can one do armed merely with atoms and molecules with weight? The question does not lose its poignancy by virtue of the fact that, after Perrin, chemists had confident access to absolute and not merely relative ones. Nor is it of much relevance to chemistry to note that the kinetic theory yields an average value for the square of the velocity of molecules in a gas. Developing atomic chemistry to the extent that it could give an account of valency, explain chemical bonding and the stability of molecules, the periodic table and so on was, in the first decade of the twentieth century, a task for the future. Atomic physics, too, was just beginning. Explaining the details of spectra, metallic conduction, the photo-electric effect, the outstanding problems with the specific heats of gases, black body radiation, electron diffraction patterns and so on, not to mention a host of problems associated with radioactivity, all posed problems for the future. The beginning of the twentieth century in a sense marked the dawn of atomism rather than its successful conclusion. The tasks that lay ahead were tasks for the scientist rather than the philosopher. The relativistic and quantum mechanical pictures of the world that were to emerge were such as to pose nothing but headaches for the mechanistically-minded philosopher.

Because of the stringent way in which scientific knowledge is required to pass experimental tests, it is the best kind of knowledge that we have. As far as providing knowledge of the deep structure of the world is concerned, science has progressed in a dramatic way and proved itself capable of answering questions that were once supposed to be the province of philosophy. This does not render philosophy redundant. Many areas of philosophy, such as moral philosophy or philosophical logic, do not contest ground claimed by science in a way that some traditional metaphysics does. The best contemporary metaphysics takes the findings of science for granted and attempts to go beyond it, in an attempt, for example, to defend physicalism or a philosophy of perception. I presume that in this book I have been engaged in philosophy (as well as some history) but I have not been practising science. One of these days someone should write a book called *What is this thing called philosophy?*. But not me.

Notes

1. See Chang (2004) for a fascinating and instructive account of the experimental route to a workable notion of temperature.
2. See, for instance, Gardner (1979) and Nyhof (1988).
3. For a construal of Mach's opposition to atomism that differs from an extreme positivist one see Laudan (1976) and for a sympathetic account of Duhem's anti-realism see Worrall (1982).

References

Ampère, A. M. 1814. Lettre de M. Ampère à M. le Compte Berthollet, sur la determination des proportions dans lesquelles les corps se combient d'après le nombre et la disposition respectives des molecules don't leurs particles intégrantes sont composées. *Annales de chemie* 90: 43–86.

Anstey, P. 2000. *The philosophy of Robert Boyle*. London: Routledge.

Anstey, P. 2002a. Robert Boyle and the heuristic power of mechanism. *Studies in History and Philosophy of Science* 33: 157–170.

Anstey, P. 2002b. Boyle on seminal principles. *Studies in the History and Philosophy of Biological and Biomedical Sciences* 33: 597–630.

Arabatzis, Theodore. 1992. The discovery of the Zeeman effect: A case study of the interplay between theory and experiment. *Studies in History and Philosophy of Science* 23: 365–388.

Arabatzis, Theodore.2006. *Representing electrons: A biographical approach to theoretical entities*. Chicago: University of Chicago Press.

Aristotle, 1962. *Meteorologica*. H. D. P. Lee (trans.). London: William Heinemann.

Avogadro, A. 1923. Essai d'une manière de determiner les masses relatives des molecules élémentaires des corps. English translation, Edinburgh: Alembic Club Reprint No. 4.

Barnes, Jonathan. 1996. *The Presocratic philosophers*. Cambridge: Cambridge University Press.

Beguin, Jean. 1624. *Les elements de chymie*. Geneva: Jean Celerier.

Berzelius, J. 1813. Experiments on the nature of azote, of hydrogen, and of ammonia and upon the degrees of oxidation of which azote is susceptible. *Annals of Philosophy* 2: 276–284 and 357–368.

Berzelius, J. 1815. An address to those chemists who wish to examine the laws of chemical proportions, and the theory of chemistry in general. *Annals of Philosophy* 5: 122–131.

Birch, Thomas (ed.). 1744. The works of the Honourable Robert Boyle. London: A. Millar.

Bird, Alexander. 1998. *Philosophy of science*. London: UCL Press.

Boas, Marie. 1949. Hero's *Pneumatica*: A study of its transmission and influence. *Isis* 40: 38–48.

Boltzmann, L. 1877. Ueber die natur der gas molecule. *Annalen der Physik und Chemie* 160: 175–176.

Boltzmann, L. 1895. On certain questions of the theory of gases. *Nature* 51: 413–415.

Bosanquet, R. H. M. 1877. Notes on the theory of sound. *Philosophical Magazine* 3: 271–272.

Boscovich, Roger. 1966. *A theory of natural philosophy*. Cambridge, Mass.: MIT Press.

Boyle, Robert. 1990. *Collections from the Royal Society: Letters and papers of Robert Boyle*. Bethesda, Maryland: University Publications of America.

Boyle, Robert. 2000. *The works of Robert Boyle*. eds. Michael Hunter and Edward B. Davis. London: Pickering and Chatto.

Brock, W. H. 1967. *The atomic debates: Brodie and the rejection of the atomic theory*. Leicester: Leicester University Press.

Brock, W. H. 1993. *The Norton history of chemistry*. New York: Norton.

Brooke, John Hedley. 1981. Avogadro's hypothesis and its fate: A case-study in the failure of case-studies. *History of Science* 19: 236–271.

Brumbaugh, R. S. 1964. *The philosophers of Ancient Greece*. New York: Crowell.

Brush, Stephen, G. 1965. *The kinetic theory*. Volume 1. *The nature of gases and heat*. Oxford: Pergamon Press.

Brush, Stephen. 1968–69. A history of random processes I. Brownian movement from Brown to Perrin. *Archive for History of Exact Science* 5: 1–36.

Buchwald, Jed. 1985. *From Maxwell to microphysics*. Chicago: University of Chicago Press.

Cannizzaro, S. 1961. Sunto di un corso di filosofia. English translation, Edinburgh: Alembic Club Reprint No. 18.

Chalmers, Alan. 1973a. Maxwell's methodology and his application of it to electromagnetism. *Studies in History and Philosophy of Science* 4: 107–164.

Chalmers, Alan. 1973b. The limitations of Maxwell's electromagnetic theory. *Isis* 64: 469–483.

Chalmers, Alan. 1990. *Science and its fabrication*. Milton Keynes: Open University Press.

Chalmers, Alan. 1993. The lack of excellency of Boyle's mechanical philosophy. *Studies in History and Philosophy of Science* 24: 551–556.

Chalmers, Alan. 1997. Did Democritus ascribe weight to atoms? *Australasian Journal of Philosophy* 75: 279–287.

Chalmers, Alan. 1999. *What is this thing called science*? St. Lucia, Queensland: University of Queensland Press.

Chalmers, Alan. 2001. Maxwell, mechanism and the nature of electricity. *Physics in Perspective* 3. 425–438.

Chalmers, Alan. 2002a. Experiment and the growth of experimental knowledge. In *In the scope of logic methodology and philosophy of science*, eds. P. Gardenfors, Jan Wolenski and Katarzyna Kijania-Pacek, 157–193. Dordrecht: Kluwer.

Chalmers, Alan. 2002b. Experiment versus mechanical philosophy in the work of Robert Boyle. *Studies in History and Philosophy of Science* 33: 187–193.

Chalmers, Alan. 2008. Atom and aether in nineteenth-century physical science. *Foundations of Chemistry* 10: 157–166.

Chang, Hasok. 2004. *Inventing temperature: Measurement and scientific progress*. Oxford: Oxford University Press.

Christie, Maureen. 1994. Philosophers versus chemists concerning laws of nature. *Studies in History and Philosophy of Science* 25: 613–629.

Clark, Peter. 1976. Atomism versus thermodynamics. In *Method and appraisal in the physical sciences*, ed. C. Howson, 41–105. Cambridge: Cambridge University Press.

Clericuzio, Antonio. 1990. A redefinition of Boyle's chemistry and corpuscular philosophy. *Annals of Science* 47: 561–589.

Clericuzio, Antonio. 2000. *Elements, principles and corpuscles: A study of atomism and chemistry in the seventeenth century*. Dordrecht: Kluwer.

Cole, T. 1975. Early atomic speculations of Marc Antoine Gaudin: Avogadro's hypothesis and the periodic system. *Isis* 66: 334–360.

Cole, T. 1978. Dalton, mixed gases and the origins of the atomic theory. *Ambix* 25: 119–130.

Collingwood, R. G. 1960. *The idea of nature*. London: Oxford University Press.

Crossland, M. P. 1963. The development of chemistry in the eighteenth century. *Studies in Voltaire and the eighteenth century* 24: 369–441.

Dalton, John. 1805. On the absorption of gases by water and other liquids. *Memoires and Proceedings of the Manchester Literary and Philosophical Society* 1: 271–286.

Dalton, John. 1808. *A new system of chemical philosophy*. Volume 1, Part 1. London: Bickerstaff, reprinted by Citadel Press, New York, 1964.

Dalton, John. 1810. *A new system of chemical philosophy*, Volume 1, Part 2. London: Bickerstaff.

Dalton, John. 1811. Dr. Bostock's review of the atomic principles of chemistry. *Journal of Natural Philosophy, Chemistry and the Arts* 29: 143–151.

Dalton, John. 1827. *A new system of chemical philosophy*. Volume 2, Part 1. London: George Wilson.

Darrigol, Olivier. 2000. *Electrodynamics from Ampère to Einstein*. Oxford: Oxford University Press.

Descartes, René. 1983. *Principles of philosophy*. V. R. Miller and R. P. Miller (trans.). Dordrecht: Reidel.
Duhem, Pierre. 1962. *The aim and structure of physical theory*. P. Wiener (trans.). New York: Atheneum.
Duhem, Pierre. 1985. *Medieval cosmology: Theories of infinity, place, time, void and the plurality of worlds*. Roger Ariew (trans.). Chicago: University of Chicago Press.
Duhem, Pierre. 2002. Le mixt et la combinaison chimique: Essai sur l'évolution d'une idée. In *Pierre Duhem: Mixture and chemical combination and related essays*, ed. P. Needham, 2–118. Dordrecht: Kluwer.
Duncan, A. M. 1964. Some theoretical aspects of eighteenth-century tables of chemical affinity – I. *Annals of Science* 18: 177–194.
Duncan, A. M. 2001. *Laws and order in eighteenth-century chemistry*. Oxford: Clarendon Press.
Einstein, A. 1956. *Investigations on the theory of Brownian movement*, eds. R. Fürth, and A. D. Cowper. New York: Dover.
Emerton, Norma. 1984. *The scientific reinterpretation of form*. Ithaca: Cornell University Press.
Farrar, W. V. 1968. Dalton on stuctural chemistry. In *John Dalton and the progress of science*, ed. D. S. L. Cardwell, 290–299. Manchester: Manchester University Press.
Fine, Arthur. 1991. Piecemeal realism. *Philosophical Studies* 61: 79–96.
Fisher, Nicholas. 1982. Avogadro, the chemists and historians of chemistry. *History of Science* 20: 77–102 and 212–231.
Fox, John. 1968. Dalton's caloric theory. In *John Dalton and the progress of science*, ed. D. S. L. Cardwell, 187–201. Manchester: Manchester University Press.
Frankland, E. 1849. On the isolation of the organic radicals: *Journal of the Chemical Society* 2: 263–296.
Frankland, E. 1850. Researches in the organic radicals. Part II. *Journal of the Chemical Society* 3: 30–52 and 322–347.
Frankland, E. 1852. On the educational and commercial utility of chemistry. In *Introductory lectures on the opening of Owens College*. London: Longman, Brown, Green and Longmans, 112–133.
Frické, Martin. 1976. The rejection of Avogadro's hypothesis. In *Method and appraisal in the physical sciences*, ed. C. Howson, 277–308. Cambridge: Cambridge University Press.
Furley, David. 1967. *Two studies in the Greek atomists*. Princeton, New Jersey: Princeton University Press.
Furley, D. 1983. The mechanics of *Meteorologica, IV*: A prolegomena to biology. In *Zweifelhaftes Im Corpus Aristotelicum: Studies Zu Einegen Dibia*, eds. P. Morauz and J. Wiesner. 73–93. Berlin: Walter De Gruyter.
Gabbey, Alan. 1985. The mechanical philosophy and its problems: Mechanical explanations, impenetrability and perpetual motion. In *Change and progress in modern science*, ed. J. C. Pitt, 9–84. Dordrecht: Reidel.
Gardner, M. 1979. Realism and instrumentalism in 19th–century atomism. *Philosophy of Science* 46: 1–34.
Gatti, Hilary. 2001. Giordano Bruno's soul-powered atoms. In *Late medieval and early modern corpuscular matter theories*, eds. C. Luthy, J. E. Murdoch and W. R. Newman, 163–180. Leiden: Brill.
Gaudin, Marc, A. 1873. *L'Architecture du monde des atomes*. Paris: Gauthier-Villars.
Geoffroy, Etienne François. 1996. Table of the different relations observed in chemistry between different substances: 27 August, 1718. *Science in Context* 9: 313–320.
Glaser, Christopher. 1667. *The compleat chymist*. London: John Starkey.
Glymour, C. 1980. *Theory and evidence*. Princeton: Princeton University Press.
Goodman, D. C. 1969. Wollaston and the atomic theory of Dalton. *Historical Studies in the Physical Sciences* 1: 37–59.
Gouy, L. G. 1888. Note sur le movement brownien. *Journal de physique* 7: 561–564.
Hall, A. Rupert and Hall, Marie Boas. 1962. *Unpublished scientific papers of Isaac Newton*. Cambridge: Cambridge University Press.

Hall, A. and Hall, Marie Boas. 1965. *The correspondence of Henry Oldenburg*. Volume 1, 1641–1662. Madison: University of Wisconsin Press.
Helmholtz, H. von. 1881. On the modern development of Faraday's conception of electricity. *Journal of the Chemical Society* 39: 273–304.
Hett, W. S. 1936. *Aristotle: Minor works*. London: Heinemann.
Holmes, F. L. 1989. *Eighteenth-century chemistry as an investigative enterprise*. Berkeley, California: University of California Press.
Hon, Giora. 1987. 'The electrostatic and electromagnetic properties of the cathode rays are either nil or very feeble' (1883): A case study of an experimental error. *Studies in History and Philosophy of Science* 18: 367–382.
Hunt, Bruce. 1991. *The Maxwellians*. Ithaca: Cornell University Press.
Hunter, Michael. 1995. How Boyle became a scientist. *History of Science* 33: 59–103.
Inwood, B. and Gerson, L. P. 1994. *The Epicurus reader*. Indianapolis: Hackett.
Irby-Massie, Georgia, L. and Keyser, Paul,T. 2002. *Greek science of the Hellenistic era: A sourcebook*. London: Routledge.
Kekulé, August. 1867. On some points of chemical philosophy. *The Laboratory* 1: 303–306.
Kim, Sik Yung. 1991. Another look at Robert Boyle's acceptance of the mechanical philosophy: Its limits and its chemical and social contexts. *Ambix* 38: 1–10.
Kirk, G. S., J. E. Raven and Schofield, M. 1999. *The Presocratic philosophers*. Cambridge: Cambridge University Press.
Klein, Ursula. 1994. Origin of the concept of chemical compound. *Science in Context* 7: 163–204.
Klein, Ursula, 1995. E. F. Geoffroy's table of different 'rapports' between different chemical substances – A reinterpretation. *Ambix* 42: 79–100.
Klein, Ursula. 1996. The chemical workshop tradition and the experimental practice: Discontinuities within continuities. *Science in Context* 9: 251–287.
Klein, Ursula, 2001. Berzelian formulas as paper tools in early nineteenth-century chemistry. *Foundations in Chemistry* 3: 7–32.
Klein, Ursula. 2003. *Experiments, models, paper tools: Cultures of organic chemistry in the nineteenth century*. Stanford: Stanford University Press.
Klein, Ursula. 2007. Styles of experimentation and alchemical matter theory in the Scientific Revolution. *Metascience* 16: 247–256.
Klein, Ursula and Lefèvre, Wolfgang. 2007. *Materials in eighteenth-century science: A historical ontology*. Cambridge, Mass.: MIT Press.
Knight, David. 1967. *Atoms and elements: A study of theories of matter in England in the nineteenth century*. London: Hutchinson.
Knight, David. 1992. *Ideas in chemistry: A history of the science*. London: Athlone Press.
Konstan, David. 1979. Problems in Epicurean physics. *Isis* 70: 394–418.
Kuhn. Thomas. 1977. Mathematical versus experimental traditions in the development of physical science. In *The essential tension*, 31–65. Chicago: University of Chicago Press.
Kundt, A. and Warburg, E. 1876. Ueber die specifische wärme des quecksilbergases. Annalen der Physik und Chemie 157: 353–369.
Ladenburg, A. 1900. *Lectures on the history of the development of chemistry since the time of Lavoisier*, 2nd edition. L. Dobbin (trans.). Edinburgh: Alembic Club Reprint.
Langevin, P. 1908. Sur la théorie du movement brownien. *Comptes rendus* 146: 533.
Laudan, L. 1966. The clock metaphor and probabilism: The impact of Descartes on English methodological thought, 1650–65. *Annals of Philosophy* 22: 73–104.
Laudan, L. 1976. The methodological foundations of anti-atomism and their historical roots. In *Motion and time, space and matter*, eds. P. K. Machamer and R. G. Turnbull, 390–417. Columbus: Ohio State University Press.
Lavoisier, A. L. 1965. *Elements of chemistry*. Robert Kerr (trans.). New York: Dover.
Lear, Jonathan. 1979–80. Aristotelian infinity. *Proceedings of the Aristotelian Society*. 188–210.
Lear, Jonathan, 1982. Aristotle's philosophy of mathematics. *The Philosophical Review* 91: 161–192.
Le Febvre, Nicaise. 1664. *A compendious body of chemistry*. 2 vols. London: Ratcliffe.
Lemery, Nicolas. 1677. *A course of chemistry*. Walter Harris (trans.). London: W. Kettilby.

Loemker, Levoy, E. 1969. *Gottfried Leibniz: Philosophical papers and letters*. Dordrecht: Reidel.
Long, A. A. and Sedley, D. N. 1999. *The Hellenistic philosophers*, Volume 1. Cambridge: Cambridge University Press.
Lloyd, G. E. R. 1993. *Aristotle: The growth and structure of his thought*. Cambridge: Cambridge University Press.
Lucretius, 1994. *On the nature of the universe*. R. E. Latham (trans.). London: Penguin.
Maiocchi, Roberto. 1990. The case of Brownian motion. *British Journal for the History of Science* 23: 257–283.
Mandelbaum, Maurice. 1964. *Philosophy, science and sense perception: Historical and critical studies*. Baltimore: Johns Hopkins Press.
Mauskopf, S. H. 1969. The atomic structural theories of Ampère and Gaudin: Molecular speculation and Avogadro's hypothesis. *Isis* 60: 61–74.
Maxwell, J. C. 1877. The kinetic theory of gases. *Nature* 16: 242–246.
Maxwell, J. C. 1954. *A treatise on electricity and magnetism*. 2 vols. New York: Dover.
Maxwell, J. C. 1965. In *The scientific papers of James Clerk Maxwell*. 2 vols. ed. W. D. Niven. New York: Dover.
Mayo, Deborah. 1996. *Error and the growth of experimental knowledge*. Chicago: University of Chicago Press.
Mayo, Deborah. 2002. Theory testing, statistical methodology and the growth of experimental knowledge. In *In the scope of logic methodology and philosophy of science*, eds. P. Gardenfors, Jan Wolenski and Katarzyna Kijania-Pacek, 171–190. Dordrecht: Kluwer.
McGuire, J. E. and Tamny, Martin. 1983. *Certain philosophical questions: Newton's Trinity notebooks*. Cambridge: Cambridge University Press.
McKeon, Richard. 1928. *The philosophy of Spinoza*. New York: Longmans, Green and Co.
McKeon, Richard. 1968. *The basic works of Aristotle*. New York: Random House.
Meinel, C. 1988. Early seventeenth-century atomism: Epistemology, and the insufficiency of experiment. *Isis* 79: 68–103.
Meinel, C. 2004. Molecules and croquet balls. In *Models: The third dimension of science*, eds. S. De Chadarevian and N. Hopwood, 242–275. Stanford: Stanford University Press.
Melsen, A. van. 1960. *From atomos to atom*. New York: Harper and Row.
Michael, Emily. 2001. Sennert's sea change: Atoms and causes. In *Late medieval and early modern corpuscular matter theories*, eds. C. Luthy, J. E. Murdoch and W. R. Newman, 331–362. Leiden: Brill.
Murdoch, John, E. 2001. The medieval and Renaissance tradition of *minima naturalia*. In *Late medieval and early modern corpuscular matter theories*, eds. C. Luthy, J. Murdoch and W. Newman, 91–131. Leiden: Brill.
Nash, L. K. 1956. The origin of Dalton's chemical atomic theory. *Isis* 47: 101–116.
Needham, Paul. 2004. Has Daltonian atomism provided chemistry with any explanations. *Philosophy of Science* 71: 1038–1048.
Newman, William. 1985. The genesis of the *Summa Perfectionis*. *Les archives internationales d'histoire des sciences* 35: 240–302.
Newman, William. 1991. *The Summa Perfectionis of Pseudo-Geber*. Leiden: Brill.
Newman, William. 1994. *Gehennical fire: The lives of George Starkey, an American alchemist in the scientific revolution*. Cambridge, Mass.: Harvard University Press.
Newman, William. 1996. The alchemical sources of Robert Boyle's corpuscular philosophy. *Annals of Science* 53: 567–585.
Newman, William. 2001. Experimental corpuscular theory in Aristotelian alchemy: From Geber to Sennert. In *Late medieval and early modern corpuscular matter theories*, eds. C. Luthy, J. Murdoch and W. Newman, 291–329. Leiden: Brill.
Newman, William. 2004. *Promethean ambitions: Alchemy and the quest to perfect nature*. Chicago: University of Chicago Press.
Newman, William. 2006. *Atoms and alchemy*. Chicago: University of Chicago Press.
Newman, William. 2008. The chemical revolution and its chymical antecedents. Essay review of Klein and Lefèvre (2007). *Early Science and Medicine* 13: 171–191.
Newman, William R. and Principe, Lawrence, M. 2002. *Alchemy tried in the fire: Starkey, Boyle and the fate of Helmontian chymistry*. Chicago: University of Chicago Press.

Newton, Isaac. 1958. Some thoughts about the nature of acids. In *Sir Isaac Newton's papers and letters on natural philosophy*, ed. I. B. Cohen, 257–258. Cambridge: Cambridge University Press.

Newton, Isaac. 1962. *Principia*. A. Mott (trans.), revised by F. Cajori. Berkeley, California: University of California Press.

Newton, Isaac. 1979. *Opticks*. New York: Dover.

Nye, Mary-Jo. 1986. *The question of the atom*. Los Angeles: Tomash Publishers.

Nyhof, John. 1988. Philosophical objections to the kinetic theory. *British Journal for the Philosophy of Science* 39: 81–109.

Ostwald, W. 1896. Emancipation from scientific materialism. *Science Progress* 4: 430–436, reprinted in Nye (1986), 337–354.

Pais, Abraham. 1982. *Subtle is the Lord: The science and the life of Albert Einstein*. Oxford: Oxford University Press.

Partington, J. R. 1961. *A history of chemistry*. London: MacMillan.

Pender, Harold. 1901. On the magnetic effect of electrical convection. *Philosophical Magazine* 2: 179–208.

Perrin, Jean. 1908a. L'agitation moléculaire et le mouvement brownien. *Comptes rendus* 146: 967–970.

Perrin. Jean. 1908b. La loi de Stokes et le mouvement brownien, *Comptes rendus* 147: 475–476.

Perrin, Jean. 1909. Brownian movement and molecular reality. *Annalles de Chimie et de Physique* 18: 1–114. Reprinted in Nye (1986), 507–624.

Perrin, Jean. 1990. *Atoms*. D. Ll. Hammick (trans.). Woodbridge, Connecticut: Ox Bow Press.

Potter, P. 1988. *Hippocrates*. Cambridge, Mass.: Harvard University Press.

Principe, Lawrence. 1998. *The aspiring adept: Robert Boyle and his alchemical quest*. Princeton: Princeton University Press.

Pyle, Andrew. 1995. *Atomism and its critics*. Bristol: Thoemmes Press.

Pyle, Andrew. 2002. Boyle on science and the mechanical philosophy. *Studies in History and Philosophy of Science* 33: 171–186.

Rocke, A. 1984. *Chemical atomism in the nineteenth century: From Dalton to Cannizzaro*. Columbus: Ohio University Press.

Roscoe, H. E. and Harden, A. 1896. *A new view of the origins of Dalton's atomic theory*. London: MacMillan.

Russell, C. A. 1968. Berzelius and the development of the atomic theory. In *John Dalton and the progress of science*, ed. D. S. L. Cardwell, 259–273. Manchester: Manchester University Press.

Sargent, Rose-Mary, 1995. *The diffident naturalist: Robert Boyle and the philosophy of experiment*. Chicago: University of Chicago Press.

Sennert, Daniel. 1629. *De chymicotum cum Aristotelicis et Galenicus consensus ac dissensu*. Wittenburg: Schurer.

Sennert, Daniel. 1636. *Hypomnemata physica*. Frankfurt: Schleichius.

Shapin, Steven and Schaffer, Simon. 1985. *Leviathan and the air pump*. Princeton: Princeton University Press.

Shapiro, Alan, E. 1993. *Fits, passions and paroxysms: Physics, method and chemistry and Newton's theories of coloured bodies and fits of easy reflection*. Cambridge: Cambridge University Press.

Shapiro, Alan E. 2004. Newton's experimental philosophy. *Early Science and Medicine* 9: 185–217.

Siedentopf, H. F. 1903. On the rendering visible of ultramicroscopic particles and of ultramicroscopic bacteria. *Royal Microscopical Society Journal* 17: 573–578.

Siedentopf, H. F. and Zsigmondy, R. A. 1903. Siochtbatmachung und grossenbestimmung ultramikroskopscher teilchen, mit besonderer andwendung und goldrubinglaser. *Annalen der Physik* 10: 1–39.

Siegel, Daniel. 1991. *Innovation in Maxwell's electromagnetic theory*. Cambridge: Cambridge University Press.

Siegfried, Robert. 2002. *From elements to atoms: A history of chemical composition*. Philadelphia: American Philosophical Society.

Smith, A. Mark. 1981. Saving the appearances of the appearances: The foundations of classical geometrical optics. *Archive for History of Exact Science* 24: 73–99.

Smith, A. Mark. 1982. Ptolemy's search for the law of refraction: A case study in the classical methodology of 'saving the appearances' and its limitations. *Archive for History of Exact Science* 26: 221–240.

Smith, George. 2001a. The Newtonian style in Book II of the *Principia*. In *Isaac Newton's natural philosophy*, eds. Jed Z. Buchwald and I. Bernard Cohen, 249–313. Cambridge: MIT Press.

Smith, George. 2001b. Comments on Ernan McMullin's 'The impact of Newton's *Principia* on the philosophy of science'. *Philosophy of Science* 68: 327–338.

Smith, George. 2001c. J. J. Thomson and the electron. In *Histories of the Electron: The birth of Microphysics*, eds. Jed Buchwald and Andrew Warwick, 21–76. Cambridge, Mass.: MIT Press.

Smith, George. 2002a. The methodology of the *Principia*. In *The Cambridge companion to Newton*, eds. I. Bernard Cohen and George E. Smith, 138–173. Cambridge: Cambridge University Press.

Smith, George. 2002b. From the phenomenon of an ellipse to an inverse square law: Why not? In *Reading natural philosophy: Essays in the history and philosophy of science and mathematics*, ed. David B. Malament, 31–70. Chicago: Open Court.

Smoluchowski, M. 1906. Zur kinetischen theorie der Brownschen molekularbewegung und der suspensionen. *Annalen der physik* 21: 756–780.

Sorabji, Richard. 1983. *Time, creation and the continuum*. London: Dukworth.

Stein, H. 2002. Newton's metaphysics. In *The Cambridge companion to Newton*, eds. I. B. Cohen and G. E. Smith. Cambridge: Cambridge University Press.

Stewart, M. 1979. *Selected philosophical papers of Robert Boyle*. Manchester: Manchester University Press.

Sutherland, William, 1904. The measurement of large molecular masses. *Australasian Association for the Advancement of Science. Report of Meeting*. 10 (Dunedin, 1904): 117–121.

Sutherland, William, 1905. A dynamical theory of diffusion for non–electrolytes and the molecular mass of albumin. *Philosophical Magazine* 9: 781–785.

Thackray, Arnold. 1968. Matter in a nutshell. *Ambix* 15: 29–53.

Thackray, Arnold. 1970. *Atoms and powers: An essay on Newtonian matter theory and the development of chemistry*. Cambridge, Mass.: Harvard University Press.

Thackray, Arnold, 1972. *John Dalton: A critical assessment of his life and science*. Cambridge, Mass.: Harvard University Press.

Thomson, J. J. 1893. *Notes on recent researches in electricity and magnetism*. Oxford: Clarendon Press.

Thomson, J. J. 1897. Cathode rays. *Philosophical Magazine* 44: 293–316, reprinted in Nye (1986), 375–400.

Thomson, J. J. 1898. On the charge carried by the ions produced by Röntgen rays. *Philosophical Magazine* 46: 528–545.

Thomson, J. J. 1899. On the masses of the ions of gases at low pressure. *Philosophical Magazine* 48: 547–567.

Thomson, T. 1808. On oxalic acid. *Philosophical Transactions of the Royal Society* 98: 63–95.

Watts, Henry. 1879. *A dictionary of chemistry*, Volume 1. London: Longmans, Green and Company.

Wiener, Christian. 1863. Erklärung des atomischen wesens des tropbarflüssigen köorperzustandes, und bestätigung desselben durch die sogennanten molekularbewegungen. *Annalen der Physik* 118: 79–94.

Whyte, Lanclot. 1961. *Essay on atomism: From Democritus to 1960*. Middleton, Connecticut: Weslyan Univeristy Press.

Will, C. W. 1993. *Theory and experiment in gravitational physics*. Cambridge: Cambridge University Press.

Will, C. W. 1996. The confrontation between general relativity theory and experiment: A 1995 update. In *General relativity: Proceeedings of the forty sixth Scottish universities summer school in physics*, eds. G. S. Hall and J. R. Pulham, 239–281. Edinburgh: SUSSP Publications.

Wollaston, William, Hyde. 1808. On super-acid and sub-acid salts. *Philosophical Transactions of the Royal Society* 98: 96–102.

Worrall, John. 1982. Scientific realism and scientific change. *Philosophical Quarterly* 32: 201–231.

Worrall, John. 2000. The scope, limits and distinctiveness of the method of 'Deduction from the Phenomena': Some lessons from Newton's 'demonstrations' in optics. *British Journal for the Philosophy of Science* 51: 45–80.

Worrall, John. 2002. New evidence for old. In *In the scope of logic methodology and philosophy of science*, eds. P. Gardenfors, Jan Wolenski and Katarzyna Kijania-Pacek. 191–209. Dordrecht: Kluwer.

Wurtz, Charles, Adolph. 1869. *History of chemical theory from Lavoisier to the present day*. London: MacMillan.

Author Index

Subject Index

FSC
www.fsc.org
MIX
Papier | Fördert
gute Waldnutzung
FSC® C083411

Kolibri 360 GmbH
Ferdinand-Jühlke-Straße 7
99095 Erfurt, Deutschland
produktsicherheit@kolibri360.de